# Advances in Modern Sensors

Physics, design, simulation and applications

# IOP Series in Sensors and Sensor Systems

The IOP Series in Sensors and Sensor Systems includes books on all aspects of the science and technology of sensors and sensor systems. Spanning fundamentals, fabrication, applications and processing, the series aims to provide a library for instrument and measurement scientists, engineers and technologists in universities and industry.

The series seeks (but is not restricted to) publications in the following topics:

- Advanced materials for sensing
- Biosensors
- Chemical sensors
- Industrial applications
- Internet of Things
- Lab-on-a-chip
- Localization and object tracking
- Manufacturing and packaging
- Mechanisms, modelling and simulations
- MEMS/NEMS
- Micro and nanosensors
- Non-destructive testing
- Optoelectronic and photonic sensors
- Optomechanical sensors
- Physical sensors
- Remote sensors
- Sensing for health, safety and security
- Sensing principles
- Sensing systems
- Sensor arrays
- Sensor devices
- Sensor networks
- Sensor technology and applications
- Signal processing and data analysis
- Smart sensors and monitoring
- Telemetry

Authors are encouraged to take advantage of electronic publication through the use of colour, animations, video and interactive elements to enhance the reader experience.

# Advances in Modern Sensors

Physics, design, simulation and applications

**Edited by**
**G R Sinha**
*Professor, Myanmar Institute of Information Technology (MIIT) Mandalay, Myanmar*

**IOP** Publishing, Bristol, UK

ISBN 978-0-7503-2707-7 (ebook)
ISBN 978-0-7503-2705-3 (print)
ISBN 978-0-7503-2708-4 (myPrint)
ISBN 978-0-7503-2706-0 (mobi)

DOI 10.1088/978-0-7503-2707-7

Version: 20201101

IOP ebooks

British Library Cataloguing-in-Publication Data: A catalogue record for this book is available from the British Library.

Published by IOP Publishing, wholly owned by The Institute of Physics, London

IOP Publishing, Temple Circus, Temple Way, Bristol, BS1 6HG, UK

US Office: IOP Publishing, Inc., 190 North Independence Mall West, Suite 601, Philadelphia, PA 19106, USA

*Dedicated to my late grandparents, my teachers and Revered Swami Vivekananda.*

# Contents

# Preface

Sensors have become essential parts of all automated systems, modern gadgets and handy devices. Smartphones used commonly by all of us, have around 22 sensors for various purposes. There have been advances in sensor technology in the last two decades; and smart sensors have dominated in almost all applications of sensors. Wearable, smart and intelligent sensors are used in numerous applications, such as robotics, industrial robotics, space navigation, global positioning systems, smart environment monitoring systems, driverless cars and so on. Sensors are like the eyes in human vision and the capabilities of human eyes are exploited by the sensors deployed for computer vision. Computer vision is empowered due to two important components, namely machine learning methods and sensor technologies. There is literature and books on physics and characteristics of traditional sensors such as transducers, optical sensors, capacitive sensors, photo sensors etc, but most of the resources fail to highlight simulation capabilities and applications of modern sensors such as smart sensors, wearable sensors, wireless sensors, cognitive sensors etc. Therefore, this book attempts to present the basics, introduction and overview of sensors and technologies with more emphasis on modern sensors.

The book highlights: introduction and classification of sensors; overview and applications of optical sensors; chemical sensors; simulation and calibration of optical imaging sensors; smart and wearable sensors; cognitive and biosensors; and also a major application of modern sensors in self driving cars. We have also included future research directions on the emerging sensors so that more studies and research can be carried out in the area of sensor research. The purpose of this book is to bring out the fundamentals, background and theoretical concepts of sensors case studies.

The book will be very useful for a wide spectrum of target readers, such as research scholars, academia, and industry professionals. It is especially for those who are working on real-time hardware implementations employing smart sensors for merging applications like robotic arms, cognitive applications, brain–computer interface etc.

# Acknowledgments

I express my sincere thanks to my wife Shubhra, my daughter Samprati, and my great parents for their wonderful support and encouragement throughout the completion of this book on *Advances in Modern Sensors: Physics, design, simulation and applications*. This book is an outcome of focused and sincere efforts that could be given to the book only due to the great support of my family.

I am grateful to my teachers who have left no stones unturned in empowering and enlightening me, especially Shri Bhagwati Prasad Verma who is like a Godfather to me. I extend my heartfelt thanks to Ramakrishna Mission order and Revered Swami Satyaroopananda of Ramakrishna Mission Raipur India.

Sincere thanks to all the contributors for writing the relevant theoretical background and real-time applications of sensors and for entrusting me with the editorship. I also wish to thank all my friends, well-wishers and all those who keep me motivated.

My reverence with folded hands to Swami Vivekananda who has been my source of inspiration for all my work and achievements.

Last, but most important, I express my humble thanks to Dr John Navas, Senior Commissioning Manager of IOP Publishing for great support, necessary help, appreciation and quick responses. It has been a wonderful experience working with John. I would also like to thank Sarah Armstrong of IOP Publishing and the entire team of IOP Publishing for giving me this opportunity to contribute on this relevant topic with a reputed publisher.

# Editor biography

## G R Sinha

**G R SINHA (PhD, Senior Member IEEE, Fellow IETE, Fellow IE)** is Adjunct Professor at International Institute of Information Technology Bangalore (IIITB) and currently deputed as Professor at Myanmar Institute of Information Technology (MIIT) Mandalay Myanmar. He is ACM Distinguished Speaker in the field of DSP. He has more than 200 research papers, edited books and authored books to his credit. He has teaching and research experience of 21 years. He has been Dean of Faculty and Executive Council Member of CSVTU and currently a member of Senate of MIIT. He is recipient of many awards and recognitions at national and international level. He served as Distinguished IEEE Lecturer in IEEE India council for Bombay section. He is Associate Editor of 03 SCI journals including *IET-Electronics Letters*. He has delivered more than 50 Keynote/ Invited Talks and Chaired many Technical Sessions in International Conferences across the world. Dr Sinha has Supervised eight (08) PhD Scholars, 15 M.Tech. Scholars and has been supervising one more PhD Scholar. His research interest includes cognitive science, medical image processing, outcome based education (OBE) and ICT tools for developing employability skills.

# List of contributors

**Abhishek Chaubey**
Sreenidhi Institute of Science and Technology, Hyderabad, Telangana, India

**Shruti Bhargava Chaubey**
Sreenidhi Institute of Science and Technology, Hyderabad, Telangana, India

**Rahul Kumar Chaurasiya**
Maulana Azad National institute of technology Bhopal, India

**Sudhamani Chilakala**
CMR Technical Campus, India

**Suma Dawn**
Jaypee Institute of Information Technology, Noida, India

**Kusumanjali Deshmukh**
Hemchand Yadav University Durg, India

**Chih Peng Fan**
National Chung Hsing University, Taichung, Taiwan

**Naveen Goel**
Chhattisgarh Swami Vivekanand Technical University, Bhilai, India

**Preeti Goswami**
CSIR National Physical Laoboratory, India

**Govind Gupta**
CSIR National Physical Laoboratory, India

**Myat Thu Kha**
Myanmar Institute of Information Technology (MIIT) Mandalay, Myanmar

**May Thin Khaing**
Myanmar Institute of Information Technology (MIIT) Mandalay, Myanmar

**Vandana Khare**
CMR College of Engineering & Technology, India

**Phyu Phyu Khiang**
Myanmar Institute of Information Technology (MIIT) Mandalay, Myanmar

**Chan Myae Khin**
Myanmar Institute of Information Technology (MIIT) Mandalay, Myanmar

**Rangacharya Kondapaka**
CMR College of Engineering & Technology, India

**Jin Li**
University of Cambridge, Cambridge, UK

**Zilong Liu**
National Institute of Metrology, Beijing, China

**Neeraj Mehta**
Banaras Hindu university, Uttar Pradesh, India

**Samrudhi Mohdiwale**
National Institute of Technology Raipur, India

**Hema N**
Jaypee Institute of Information Technology, Noida, India

**Bhagwati Charan Patel**
Chhattisgarh Swami Vivekanand Technical University, Bhilai, India

**Valiure Rahaman**
Madhav Institute of Technology & Science Gwalior, India

**K C Raveendranathan**
APJ Abdul Kalam Technological University, Thiruvananthapuram, India

**Mridu Sahu**
National Institute of Technology Raipur, India

**Anuj Sharma**
CSIR National Physical Laboratory, India

**Sanjiv Sharma**
Madhav Institute of Technology & Science Gwalior, India

**Divya Prakash Shrivastava**
Higher Colleges of Technology UAE, United Arab Emirates

**Praveen Kumar Shukla**
National Institute of Technology Raipur, India

**G R Sinha**
Myanmar Institute of Information Technology (MIIT) Mandalay, Myanmar

**Kavita Thakur**
Pt. Ravishankar Shukla University, Raipur, India

**Silvia Liberata Ullo**
Universita Degli Studi Del Sannio, Benevento, Italy

**K P Venkat Vivek**
Sreenidhi Institute of Science and Technology, Hyderabad, Telangana, India

**Shrish Verma**
National Institute of Technology Raipur, India

**Prafulla Vyas**
Pt. Ravishankar Shukla University, Raipur, India

**Chan Myae Kyaw Win**
Myanmar Institute of Information Technology (MIIT) Mandalay, Myanmar

# Chapter 1

## Introduction to sensors

**Bhagwati Charan Patel, G R Sinha and Naveen Goel**

A sensor is a device that receives a signal or stimulus and responds to the stimulus in the form of an electrical signal. The output signals correspond to some forms of electrical signal, such as current or voltage. The sensor is a device that receives different kinds of signal i.e. physical, chemical or biological signal and converts them into an electric signal. The sensors are classified into different types based on the applications, input signal, and conversion mechanism, material used in sensor characteristics such as cost, accuracy or range. This chapter presents an overview of sensors and their classifications as thermal, magnetic optical, mechanical and chemical. The transfer functions, characteristics and specifications are also discussed with introduction to basic forms of sensors.

### 1.1 Introduction

We can find sensors everywhere, and the whole world is full of sensors and their applications. There are many types of sensors available around us, in our offices, gardens, shopping malls, homes, cars, toys etc. These sensors make our lives so easy and comfortable, starting from applications such as switching on the lights, fans, television (TV), automatic adjustment of the room temperature by air conditioning (AC), fire alarm, detecting obstacles when the car is reversing, making a thumb impression etc. A sensor is a device which receives signals as well as responding to a signal or stimulus. The stimulus signals can be defied by the measure, property, or state which is sensed. We also can say that a sensor is a translator that converts a nonelectrical value to an electrical value [1–3]. The output signal of a sensor may be in the form of voltage, current, or charge. A sensor has many forms of input properties and electrical output properties. If there is small change in the sensed quantity, it will cause a small change in the electrical output and the changes can be detected with their measuring capabilities.

All the sensors are categorized on the basis of their uses, applications, material used and some production technologies. Some sensors are classified also by their

doi:10.1088/978-0-7503-2707-7ch1  

characteristics such as cost, accuracy or range of sensor. There are two main types of sensors: passive sensor and active sensor. A passive sensor does not require any extra energy source and electric signal is produced directly in reply to stimulus of external sources. This means that the sensor converts input energy to output signal energy [1, 4, 5]. Examples of passive sensors include photographic, thermal, electric field sensing, chemical, infrared and seismic. The active sensors need external sources of energy for their response, known as excitation signal. To produce the output signals, sensors adopt necessary changes to these input signals. The active sensors are also known as parametric sensors due to their own properties which can be modified in response to an exterior effect and these properties can be afterward changed into electric signals. Active sensors have a variety of applications related to meteorology and observation of the Earth's surface and atmosphere. Table 1.1 shows differences between passive and active sensors.

The other types of sensors are based on their detection properties such as variation mechanism, analog and digital. The detection properties of sensors include electric, magnetic, physical, chemical etc, and variation mechanism includes conversion of the input signal to output signal, whose examples are photoelectric, thermoelectric, electrochemical, electromagnetic etc. Analog sensors produce an analog output, i.e. continuous output signals are produced with respect to the measured quantity, but a digital sensor is the opposite of analog sensors, with discrete characteristics and digital output in nature. Sensors are also divided by their detection properties, given in table 1.2

## 1.2 Sensor characteristics

Upon receiving the input stimuli, the sensor produces output which is obtained from several conversion steps before it produces an electric signal [1, 6–10]. The performance of sensors is described in terms of relationship between input and output

**Table 1.1.** Difference between passive and active filter.

| Passive sensor | Active sensor |
|---|---|
| Does not require external power | It requires power |
| It can only be used to detect energy when the naturally occurring energy is available | Provides its own energy source for illumination |
| No interference problem in the environment | Less interference problem |
| Can operate in the same environment condition | Can operate in different environment conditions |
| Sensitive to weather condition | Not sensitive |
| Not well suited for darkness conditions | Works well in darkness conditions |
| Difficulties in interpreting the output signals | Easy to interpret |
| Less control of noise | Better control of noise |
| Low price | High price |
| Examples: camera, Sonar | Examples: LASER, Radar etc |

**Table 1.2.** Sensors based on their detection properties.

| Types | Properties |
|---|---|
| Thermal sensor | Temperature, heat, flow of heat etc |
| Electrical sensor | Resistance, current, voltage, inductance, etc |
| Magnetic sensor | Magnetic flux density, magnetic moment, etc |
| Optical sensor | Intensity of light, wavelength, polarization, etc |
| Chemical sensor | Composition, pH, concentration, etc |
| Pressure sensor | Pressure, force etc |
| Vibration sensor | Displacement, acceleration, velocity, etc |
| Rain/moisture sensor | Water, moisture, etc |
| Tilt sensors | Angle of inclination, etc |
| Speed sensor | Velocity, distance etc |

signals. Sensors are characterized depending on the values of some of the important parameters. The characteristics of sensors are described here in this section.

### 1.2.1 Transfer function

The transfer function shows the functional relationship between physical input signal or stimulus ($s$) and electrical output signal ($S$), as $S = f(s)$, where '$S$' is response to the stimuli. This function can be linear or non-linear depending on the relation between input and output and nonlinearity may be in different forms such as logarithmic, exponential, or power function [1–3]. In most of the cases, the relationships are defined by unidimensional function which means that the relation between the output and input is associated with one stimulus. This linear relationship is described by:

$$S = a + b(s) \tag{1.1}$$

where $a$ is the intercept used by the output signal at zero input signal and $b$ is the slope, also called sensitivity $S$. It is also known as the sensor's output used by devices to acquire data and depending on the property of sensors, this can be amplitude, frequency, or phase. Other non-linear functions are given as:

$$\text{Logarithmic function: } S = a + b \ln s \tag{1.2}$$

$$\text{Exponential function: } S = ae^{ks} \tag{1.3}$$

$$S = \frac{1}{k} \ln \frac{s}{a} \tag{1.4}$$

$$\text{Power function: } S = a_0 + a_1 s^k \tag{1.5}$$

$$S = \sqrt[k]{\frac{s - a}{b}} \tag{1.6}$$

where constant numbers are given by $k$.

There are some sensors that may not fulfil the above properties and in such cases, higher-order polynomial approximation is required.

### 1.2.2 Full-scale input (FSI)

This is defined as the difference between the maximum and minimum values of input stimulus which can be represented in decibels (dB). This is also a logarithmic measurement in terms of ratios of power or force and voltage. Decibels are calculated as equal to 20 times the log of the force, current, or voltage:

$$1\ \text{dB} = 20 \log \frac{s_2}{s_1} \tag{1.7}$$

where $s_2$ and $s_1$ are the maximum and minimum values of input, respectively.

### 1.2.3 Full-scale output (FSO)

Full-scale output indicates the changes between the maximum and minimum values of electrical output signals when maximum and minimum input stimulus is applied. The FSO also includes all the deviations from the ideal transfer function.

### 1.2.4 Accuracy

Accuracy is an important characteristic in sensors which is calculated in terms of error in measurement and defined as the difference between measured value and true value. It is represented in terms of % of full scale or % of reading.

$$\text{Absoulute error} = |\ \text{Measured value} - \text{True value}\ |\ A_e = |\ M_v - T_v\ | \tag{1.8}$$

where $T_v$ is calculated by taking the mean of an infinite number of measurements and relative error can also be calculated as

$$\text{Relative error} = \frac{\text{Absolute error}}{\text{True value}} R_e = \frac{|\ M_v - T_v\ |}{T_v} \tag{1.9}$$

The accuracy rating indicates a collective effect of variation, linearity, calibration, repeatability errors, dead band etc, in measurements used in sensors.

### 1.2.5 Calibration

There are many sensors available but to get the best possible sensor with optimal value of accuracy, the sensor needs to be calibrated in the device where it will be used. It is an adjustment or set of adjustments made on a sensor or device to make that device function accurately and error free. For instance, we have to measure the pressure with an accuracy ± 5 pa, and a given sensor is rated with an accuracy of ±

10 pa. Can we use this pressure sensor? Yes, we can, but the given sensor needs to be calibrated and we have to find out its initial transfer function during calibration.

In the calibration method, we have to find out its particular variables. These variables describe the complete transfer function and should be identified before calibration. Calibration of linear devices is calculated by equation (1.1) and variable 'a' and 'b' should be determined accurately.

In order to get constant values in the equation with good accuracy, the linear transfer function is calculated as $v = a + b(p)$. To find constants '$a$' and '$b$', a sensor can be exposed with two pressure values ($p_1$ and $p_2$) with respect to their corresponding output voltages ($v_1$ and $v_2$), then we get $v_1 = a + b(p_1)$ and $v_2 = a + b(p_2)$, and the constants are calculated as

$$b = \frac{v_1 - v_2}{p_1 - p_2} \text{ and } a = v_1 - bp_1 \tag{1.10}$$

and pressure of calibration can be computed as

$$p = \frac{v - a}{b} \tag{1.11}$$

The calibration error is actually a type of inaccuracy which is accepted by manufacturers during the time when the devices or sensors are calibrated in the factory. This obtained error is not uniform and can change during the process of calibration.

### 1.2.6 Hysteresis

Hysteresis is a common phenomenon or characteristic which is caused by changing properties of a material such as frictional and structural changes. Hysteresis error is the difference between two output values that correspond to the same input depending on direction followed by the sensor, as shown in figure 1.1.

The value of hysteresis error is represented by a positive or negative percentage of the given pressure range. We can identify this hysteresis error for half of the given range of pressure reference point.

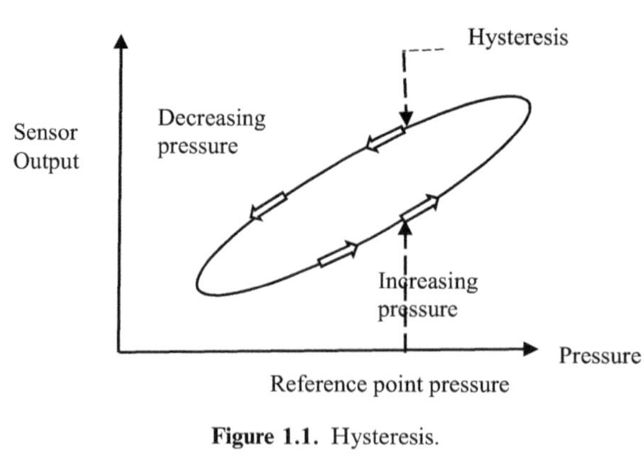


**Figure 1.1.** Hysteresis.

### 1.2.7 Non-linearity

The linearity function in a sensor is calculated by the maximum deviation by straight line given in equation (1.1), over the specified dynamic range. The non-linearity of the sensor is calculated by the measurement of the difference in $Y$-axis of two lines of equal slope, one passes through the minimum points and the other passes through the maximum points of the output curve. The total non-linearity is the difference between intercept value of $Y$-axis and a parallel line which goes through the maximum deviation point. The total non-linearity is approximated by determining the maximum deviation at the midpoint of the $X$-axis and the largest error is observed between the actual and calculated values.

The midpoint ($X_m$) is calculated as:

$$X_m = \left(\frac{X_{\max} - X_{\min}}{2}\right) + X_{\min} \tag{1.12}$$

and the line connecting end points represents the total amount of non-linearity, which is calculated by % of linearity as:

$$\%\ \text{Linearity} = \left|\frac{\left[\left(\frac{Y_{max} - Y_{\min}}{2}\right) + Y_{\min}\right] - Y_s}{\left[\left(\frac{Y_{max} - Y_{\min}}{2}\right) + Y_{\min}\right]}\right| \times 100\% \tag{1.13}$$

where $Y_s$ is the actual $Y$ value at this $X$ value and if the actual measured $Y$ values pass through zero, then it is simplified to be:

$$\%\ \text{Linearity} = \left|\frac{\left(\frac{Y_{\max}}{2}\right) - Y_s}{\left(\frac{Y_{\max}}{2}\right)}\right| \times 100\% \tag{1.14}$$

### 1.2.8 Resolution

The resolution of a sensor is defined as minimum detectable signal fluctuation while reading or measuring some quantity using a suitable sensor. This is also an ability of the measurement to obtain and notice minor changes in the characteristic of the measurement result.

### 1.2.9 Saturation

Every sensor has its certain operation limit, which is the state where the output signal of the sensor will no longer to respond at some level despite increasing the input stimuli values. This characteristic is referred as saturation of the sensor. There is a point where output of the sensor does not respond as required to an increase in the input stimuli, and that particular point is known as saturation point or threshold point.

### 1.2.10 Repeatability

Repeatability means occurrence of a value again and again when the system returns to the same position multiple times to measure the range of output signals. Direction of approach is an important issue to measure the position in repeatability, which can be represented as the maximum change among readings of output as notified by two repeated cycles unless otherwise specified. It is generally denoted as % of full scale (FS):

$$\delta_r = \frac{\Delta}{\mathrm{FS}} \times 100\% \tag{1.15}$$

### 1.2.11 Dead band

Dead band is a region where the sensitivity of a sensor does not have any effect. In this range, the output can remain nearly zero over the entire dead band ranges without any change in measurement.

### 1.2.12 Reliability

Reliability of the sensor node is the ability to perform the desired function under any given circumstances a for definite period. This can be represented in terms of statistics as a probability that the sensor will operate without failure in the stated time interval. When the performance of the sensor is exceeded under given circumstances, it causes a failure of the sensor and it can be temporary or permanent. So before manufacturing or designing the sensor, it should be properly checked, considering the various worst circumstances and conditions.

### 1.2.13 Output characteristics

Just like the input characteristics, this depends upon the type of input which the sensor is measuring, output characteristics depend upon the type of electrical output we are getting from the sensors. The output of the sensor can be voltage, current, impedance or it can be a function of any other quantities. So, the output quantity should be acceptable by the other stages of the instrumentation system, because as mentioned for measurement or in an instrumentation system, we not only have a sensor but there are also other components present in the overall system. These components should be compatible with the output of the sensor. The electrical output of the sensor should be acceptable by the other stages of the measurement system, so that the components and their evaluations can work over it.

### 1.2.14 Impedance

Impedance of a circuit is the total effective resistance or the measure of opposition to the passes of current when an AC voltage source is applied. It is represented as $Z$ and its unit is ohm ($\Omega$). This is calculated by applying a voltage source to the sensor with a resistor and measuring the changed voltage across the resistor and sensor device.

Input impedance is the impedance that is observed by the voltage source between the two terminals of the circuit. Now, we are assuming that the applied voltage source is the idle source but in an actual case this voltage source will also have some finite resistance. In the case of output impedance, in the circuit we have connected some load to the output terminal and the circuit is giving some voltage; ideally this output voltage should also appear across the two terminals of this load. From the load perspective, it will also have some finite resistance with this output voltage and this finite resistance are known as the output impedance of the circuit or device.

### 1.2.15 Excitation

Excitation is the appropriate electrical signal required to operate the active sensor. It is given by the range of voltage and/or current. The frequency of the excitation signal must also be specified in some sensors. Transfer function of the sensor may be altered by changing this excitation signal or it would cause output error.

### 1.2.16 Dynamic characteristics

Dynamic characteristics of a sensor are determined by analyzing the response of transfer function, span, calibration and input (steps, impulse, ramp, sinusoidal) of the sensor. This is when input response varies and the response does not follow it due to coupling characteristics and this is determined with a time dependent behavior, also known as dynamic characteristics. When a sensor does not respond immediately, the given value of input is not a properly received value and is different from the real time value, then dynamic error will occur.

### 1.2.17 Precision

It can be described as the nearest among a set of values and it is dissimilar from accuracy. Let $p_t$ be the true value of the variable $p$ and a random experiment measure $p_1, p_2, \ldots, p_i$ as the value of $p$. Then our measurements $p_1, p_2, \ldots, p_i$ are precise when they are very close to each other but not necessarily close to true value $p_t$. However, if we say $p_1, p_2, \ldots, p_i$ are accurate, it means that they are close to true value $p_t$ and hence they are also close to each other. Hence accurate measurements are always precise.

### 1.2.18 Environmental factors

Environmental conditions are major factors which affect input and output stimuli of the sensor. There are mainly three types of factors namely air, soil and water. Each sensor is designed to work only with certain conditions. Noise can also affect the output signal of sensor and aging can degrade the performance of the sensor. There are some other factors that reduce the performance of sensors, such as electrical, mechanical, chemical and thermal etc. Therefore, performance and long-term stability of the sensor can be improved by designing the components for extreme conditions, so that all these factors do not create an adverse effect during the operation of the sensor.

### 1.2.19 Uncertainty

Every measurement has some uncertainty. Uncertainty in the data contains some variable values that make them deviate from the correct or original values and is measured by the amount of error as mean or average value of a data set. Error is the difference between the true value and the measured value. Uncertainty is the range of values within which the true value lies in some levels of confidence.

### 1.2.20 Application characteristics

Sensors are not generally designed for general purpose and are application oriented. Sensors are required according to applications of different types of sensors such as: speed sensor for synchronizing the speed of multiple motors; temperature sensor used for controlling the temperature; ultrasonic sensor for measuring the distance, etc.

## 1.3 Types of sensors

There are many sensors commonly used in various applications [1, 11–21]. All these sensors are categorized as per their physical properties like temperature, resistance, pressure, heat flow etc. The following is a brief discussion on different types of sensors.

### 1.3.1 Temperature sensors

A temperature sensor is used to measure the amount of energy in the form of heat and cold produced by an object and system. It allows one to sense or detect any physical change to that energy and gives the output as analog or digital. Temperature sensors are used in various applications such as notification of environmental temperature, medical instruments, automobiles etc. According to application and its characteristics, many different types of temperature sensors are available. There are basically two types of temperature sensors, contact temperature sensor and non-contact temperature sensor. In contact temperature sensor, there is physical contact with the object being sensed and to monitor the change in temperature, conduction is used. It is used to sense solids, liquids or gases over a wide range of temperatures. In a non-contact temperature sensor, we use convection and radiation properties to measure the changes in temperature. It uses radiant energy in the form of heat and cold.

- *Thermostat:* The thermostat is a kind of contact temperature sensor employing an electro-mechanical component and using two thermally different kinds of metals, nickel, copper, tungsten or aluminium etc, which are stuck together to form a Bi-metallic strip. When it is cold, one of the strips is contracted and its contacts are closed and current passes through the thermostat. When it is hot, one metal strip is expanded and opens the contacts to stop the flow of current.
- *Thermistor:* The thermistor is another type of temperature sensitive device or resistance whose electrical resistance changes as the object temperature

changes. This is made up of semiconductor materials. When temperature of the object or surroundings increases or decreases, resistance will also increase or decrease. How much the resistance will increase or decrease depends on the properties of the semiconductor material. The thermistor is of two types: positive temperature coefficient, (PTC) and negative temperature coefficient, (NTC). In PTC, resistance value increases with an increase in the temperature and in NTC, its resistance value goes down with an increase in the temperature. Thermistors are used for precise temperature measurement, control and compensation. Thermistors are highly sensitive and exhibit non-linear characteristics of resistance versus temperature. Generally, these are made up of manganese, nickel, cobalt, copper and iron.

- *Resistive temperature detector:* The resistive temperature detector (RTD) is also known as resistance thermometer, and used for measurement of temperature. It is based on the temperature coefficient of sensors and generally composed of high-purity conducting metals like platinum, copper or nickel. These materials are looped into a coil whose changes of electrical resistance depend on a temperature function. The working principle of an RTD is very similar to that of the thermistor.
- *Thermocouple:* The thermocouple is a device which is used for the measurement of the temperature variation in a measurement of sensors. The thermocouples are coupled with two metals joined together forming a junction. Thus, there are two junctions in the metals, one is called hot junction and other is called cold junction, also referred as measuring junction and reference junction, respectively. These junctions are kept at different temperatures due to the change of EMF (electromotive force) induced in a thermocouple and output voltage obtained with the help of the relationship between the voltage and temperature. When the two junctions are at different temperatures, a voltage is developed across the junction which is used to measure the temperature sensor. The thermocouple is based on three main effects: Thomson effect, Seebeck and Peltier effect. It has broadest range of temperatures of all the temperature sensors, covering from –200 °C to 2000 °C.

### 1.3.2 Position sensors

The position sensor detects the position of an object either linearly or in rotation with respect to some fixed point or position. Position can be determined by the distance between two points moving away from some fixed points. We can measure the displacement of position in a straight line by linear sensor and angular displacement using rotational sensors. Position sensors are also known as potentiometers and used to measure the displacement of the object. A potentiometer can be an electrical or resistive type of sensor, because its working principle is based on change in resistance of wire with its length. This converts rotary or linear displacement to electrical voltage. The resistance of wire is directly proportional to length of wire. If the length of wire changes then the resistance of wire also changes.

Potentiometers are available as rotary and linear potentiometers in the market, and can be used to measure the angular position and linear position, respectively; through voltage division the changes in resistance can be used to create an output voltage that is directly proportional to the input displacement.

The sensors have three terminals, where the one in the middle is known as the wiper, and the other two are known as the ends. The wiper is a movable contact where resistance is measured with respect to it and either one of the end terminals. The displacement of the moving object is measured with the help of the sliding element of the potentiometer. When position of the moving body changes then its resistance between two fixed points also changes. The result is obtained in the form of differential output voltage which varies linearly with the movement of core position. The resulting output signal has both the amplitude and polarity. Amplitude is calculated as linear function of the displacement and polarity gives the direction of movement. Major advantages of the potentiometer include user friendly operation, low cost, high amplitude output and the sensors are used for measuring even large displacement, but its operating cycles are limited.

### 1.3.3 Light sensors

A light sensor is a photoelectric passive sensor which changes the light energy into an electrical signal output. It measures the ambient light which is surrounding light, room light and reflected light. The major component of a light sensor is the light dependent resister (LDR) or photoresistor. It is a resistor that depends on the light which changes its resistance depending on the amount of light incident on it. The sensors are made up of semiconductor materials and therefore when light is incident on semiconductor material it becomes low conductive and therefore has less resistance. When we increase the light intensity, its resistance decreases and vice versa which is shown in figure 1.2. Intensity of light falling on an LDR is measured in lux.

There are different kinds of light sensors such as photoresistors, photodiodes, photovoltaic cells, phototubes, photomultiplier tubes, phototransistors, charge coupled devices (CCDs) etc.

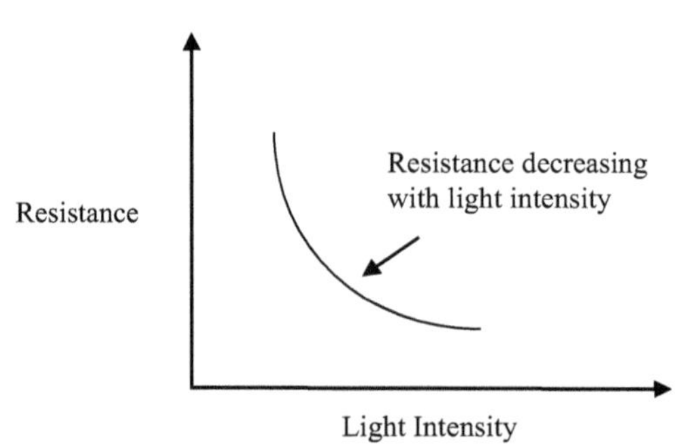


**Figure 1.2.** Resistance decreases with light intensity.

### 1.3.4 Sound sensor

A sound sensor is also known as auditory and used to detect the intensity of sound. It converts the acoustic wave into an electrical signal output. These sensors can also detect sound pressure waves which are not within the audible range, making them suitable for a wide range of tasks. Sound sensors are mostly used for security purposes.

### 1.3.5 Proximity sensor

A proximity sensor can be used for detecting the presence of a nearby object without any physical contact. It emits an electromagnetic field for a beam of electromagnetic radiation as infrared instances and changes in the field returning a signal. The object being sensed is often referred to as the proximity sensor's target. Depending on different types of proximity sensors, different targets are used. For example, an inductive proximity sensor needs a metal object, whereas a capacitive photoelectric sensor is suitable for a plastic target. A proximity sensor has high reliability due to the absence of mechanical parts and lack of physical contact between the sensor and target. It has very short range when used as a touch switch. It is commonly used in industrial applications, manufacturing of food production, mobile phones etc.

### 1.3.6 Accelerometer

This sensor is used to detect the acceleration of an object, and operates by sensing the acceleration of gravity, and the direction of the object is calculated. This sensor is a kind of microelectromechanical system (MEMS), which uses a silicon integrated circuit. These sensors convert the mechanical motion caused in an accelerometer into an electrical signal by using the piezoelectric, piezo-resistive and capacitive components.

### 1.3.7 Infrared sensor

An infrared (IR) sensor consists of two packs, one is Rx (receiver) and the other is Tx (transmitter). Transmitters are used in transmitting the rays in the infrared spectrum and the receiver receives the IR spectrum range. In the IR spectrum, the voltage is given between its terminals and then it emits rays. The main principle of working of an IR sensor is reflectivity by an object. When an object is placed in front of the transmitter it tends to reflect the rays that are coming from the IR sensor back to the IR sensor. Whenever a ray that is reflected by an object is received by the receiver it generates a voltage level across the terminal. This voltage level depends upon the intensity of light that is reflected by the object. Transmitter and receiver are placed side by side, and the IR transmitter transmits a signal within a limited range and going to a certain distance. When IR rays hit the surface, some rays are reflected depending upon the colour of the surface. The brighter the colour the more IR rays are reflected; similarly, the darker the surface the more IR rays are absorbed by the surface and fewer IR rays are reflected back.

### 1.3.8 Pressure sensor

Pressure is an external force exerted on a surface in unidirectional areas. We commonly measure the pressure of liquid, air and other gases. A pressure sensor monitors this pressure and is sometimes called a pressure transmitter as it converts pressure into an electrical signal. The most common type of pressure sensor is the strain gauge-based pressure sensor. Conversion of pressure into electrical signal is achieved through the physical deformation of strain gauge which is bound into the diaphragm of the pressure sensor. The strain will produce a change in electrical resistance which is proportional to the pressure. Change in voltage is the result of ambient pressure. A pressure sensor can also be used to measure other variables such as fluid or gas flow, speed, water level, and altitude.

### 1.3.9 Ultrasonic sensors

An ultrasonic sensor uses ultrasonic waves for the purpose of sensing and measuring the distance of a particular object. Ultrasonic waves are very high frequency waves. The sensors have two main transducers, namely transmitter and receiver. A transmitter uses 40 KHz of frequency wave transmitted in the air and when it is blocked by an object then its gets reflected and bounced back to the sensor. These reflected waves are absorbed by the receiver of the sensor. So, the total time taken by the ultrasonic waves to travel from the transmitter to the object and again from the object to the receiver of the sensor is given by the output of the sensor. Ultrasonic sensors are used in many applications such as robotics, driverless cars, for measuring distance, and also in radar systems etc.

### 1.3.10 Touch sensor

Touch sensors are sensitive to touch, pressure and force. The sensors operate as switches and when the surface of the sensor is touched the current starts to flow in the circuit just like current flowing in a closed circuit. When there is no contact, it performs like an open circuit and no flow of current is reported. There are two types of touch sensors, capacitive and resistive. The touch sensors are used popularly in modern gadgets such as smartphones, and other handy devices.

- *Capacitive sensor:* The capacitive sensor has an important element as a capacitor. Parallel capacitors are generally placed like top and bottom plates at some certain distance and between these parallel capacitor plates there is a dielectric medium. The main principle of change in capacitance is used such that it may be caused by change in overlapping area, change in distance between two plates and change in dielectric constant. Changes of these parameters can be made by the physical variables like displacement, force, pressure and flow of liquid. Capacitance and output impedance are measured with a bridge circuit. An extremely small force is needed to operate them and hence they are very useful for a small system. The sensors are highly sensitive with good frequency response and high output. As force requirement is small, thus the power requirement is also less to operate the sensors. The metallic parts of the sensor must be insulated from each other in order to reduce the effect of stray capacitance.

- *Resistive sensor:* The resistive sensor is based on the change in resistance of the material and is used to measure temperature, displacement, moisture etc. A slider is free to move between two points and at a certain point we get the zero output and at some other point we get the maximum output. The output voltage is obtained between these two points and it is directly proportional to displacement. So, the change in length of the wire causes the change in the value of the resistance. This property is utilized to measure the changes in displacement using resistivity and resistance. When the fixed voltage is applied across end terminals of the sensor, a proportional voltage is generated across the slider and this voltage can be calculated using voltage divider rule. As distance increases, the output voltage will also increase. The resistive technique used in this sensor can be used to sense or measure linear displacement.

### 1.3.11 Humidity sensor

Humidity is the amount of water present in the surrounding air and a hygrometer is the device which measures humidity directly. Humidity is a non-electrical quantity that is converted into electrical quantity by using resistance, capacitance and impedance properties. There are various parameters that change due to humidity. There are five basic types of humidity sensor: resistive hygrometer, capacitive hygrometer, microwave refractometer, aluminium oxide hygrometer and crystal hygrometer.

- *Resistive hygrometer:* In a resistive hygrometer, the main element is a material whose resistance changes with the change in humidity or relative humidity. A wire or electrode coated with hydroscopic salt (lithium chloride) can be used for measurement of the humidity. Resistance of salt changes with humidity because hydroscopic salt absorbs moisture and its resistance decreases.
- *Capacitive hygrometer:* In a capacitive hygrometer, the changes in humidity are caused by the changes in the capacitance. Dielectric medium is used in the capacitor and the capacitor consists of two electrodes or plates and a dielectric medium is there between the plates. There is also some hydroscopic material which exhibits the change in dielectric constant with the change in the humidity. Therefore, such hydroscopic material or salt can also be used for construction of a capacitive hygrometer. If the change is very small, then the capacitor includes a frequency determining element in the oscillator and another frequency is produced by the beat frequency oscillator. This frequency is heterodyned and the difference in frequency is a measure of relative humidity.
- *Microwave refractometer:* A microwave refractometer consists of two cavities, each coupled with Klystron. Klystron is a material which produces microwaves in which one cavity is filled with dry air and another cavity is filled with a mixture whose humidity is measured. In the mixture, water

vapour will be present and due to the presence of water vapour, there will be a change in dielectric constant, and frequency of one of the oscillators changes consequently. If there is no change in dielectric constant, its frequency is going to be constant, whereas in the mixture water vapours are present, and there is change in dielectric constant which results in change in its frequency. Frequency changes are measured as the measure of humidity.

- *Aluminium oxide hygrometer:* In an aluminium oxide hygrometer, aluminium oxide is coated on anodized aluminium and this aluminium oxide exhibits a change in the dielectric constant with respect to changes in humidity. There are two electrodes in which one is the inner electrode and the other is the outer electrode made from a very thin layer of material like gold. Some pores are presents in the inner layer. Due to the change in humidity, dielectric constant changes and this change can be measured to measure the humidity by bridge or electric method. The errors are much reduced and the response time is small and therefore the response is very fast.
- *Crystal Hygrometer:* In a crystal hygrometer, crystals are coated with hydroscopic materials (hydroscopic polymers). These crystals are used as frequency determination elements in the oscillator, and therefore just like with the capacitive hygrometer, if there is change in humidity then frequency also changes. Frequency changes due to the humidity as the mass of the crystal changes with amount of water absorbed by the coating. This change in frequency is measured. Humidity sensors are used in industry, agriculture, the medical field, environment monitoring etc.

### 1.3.12 Colour sensor

A colour sensor is used to detect and identify various colour patterns and convert them into desired frequency as output. It consists of four photodiodes of red, green, blue and clear (no colour). All these photodiodes are connected in parallel and work as filters. For example, if we have to detect red colour, we use red colour filter for this purpose. Colour light signals are sensed by the photodiodes and we get the square wave signals with the frequency directly proportional to light intensity and that is transferred to the microcontroller and we get the result of colour.

### 1.3.13 Chemical sensor

A chemical sensor is a device which transmits chemical information from a chemical reaction. The chemical information may be of composition, concentration and chemical activity which originates from a chemical reaction or from physical activities. It has different applications such as for home appliances and the chemical industries. The chemical sensor usually contains two basic components, which are a chemical resonance system known as the receptor and a physical chemical transducer. The receptor interacts with analytic molecules and the transducer sends the electric signal. A test sample is given to the receptor which checks composition connected with the transducer. The transducer collects the information from the receptor and sends it to the signal amplifier. This amplifies the signal from the transducer and sends it as output signals. There are two types of chemical sensors used to detect the composition: optical sensor and electro chemical sensor.

- *Optical sensor:* In the optical sensor, there are an emitter and a detector as the main elements. The emitter senses the light to the optical sensor and the light rays fall on the analyte and these rays may be reflected or refracted. These reflected or refracted lights are passed through the detector. Now the detector receives these lights and according to their intensity, the chemical compound present is analysed. Operation of an optical sensor is very simple and it uses absorption coefficient characteristics of the medium and path length travelled by the rays.
- *Electrochemical sensor:* The electrochemical sensor operates by acting on gas molecules of interest and produces an electric signal proportional to the compound present in the gas. It consists of sensing modules and electrodes, separated by a thin layer of electrolyte. There are two plates and the centre is filled by electrolytes. One plate is the cathode and the other plate is the anode. An external membrane is introduced in solution and it is absorbed by certain ions from the solution. Therefore, chemical properties of the solution change and the electromagnetic field will also change; and consequently change in the electromagnetic field ensures that the chemical composition is present in the gas.

### 1.3.14 Seismic sensor

A seismic sensor measures small movements of the ground and also amplifies and records these small movements. It is also known as a seismometer, and is mostly used in measuring the details of earthquakes, volcanic eruptions and other vibrations. There are two types of seismic sensor, inertial seismometer and strain meter or extensometer seismic sensor.

- *Inertial Seismometer:* The inertial seismometer consists of a weight suspended from a frame by a spring. The frame moves due to the vibration being measured but the mass is held stationary due to the spring. It is used to measure a large-scale vibration such as an earthquake. Now the movement of mass is converted for output as a digital electric signal. Since both types of seismic sensors most commonly output an electric signal, calibration is necessary to derive a relationship between the input and output.
- *Strain meter or extensometer seismic sensor:* In a strain meter seismic sensor, a strain gauge is used to measure the motion relative to the various points. It generally is used for smaller scale measurement and movement of mass is converted for output as a digital electric signal.

### 1.3.15 Magnetic sensor

Magnetics sensors respond to the presence or interruption of a magnetic field like flux, strength and direction by producing a proportional output. It converts magnetic information into an electrical signal for processing by the electronic circuit. A magnetic sensor is used in different types of application such as sensing position, velocity and movement of an object. There are different kinds of technology used to design a magnetic sensor. Fluxgate, Hall effect, resistive, inductive, proton processing etc, have a dissimilar approach of using magnetic sensors. A resistive magnetic sensor keeps the electrical resistance of the magnetic field and an inductive magnetic sensor uses coils surrounding its magnetic material, which have the ability to detect changes within the Earth's magnetic field. A fluxgate magnetic sensor uses the approach of changing flux parameters. Each type of technology focuses on a specific area for identifying measurements to be detected. Sensitivity of the magnet is increased by combining layers of magnetic alloys and the magnetic field is surrounded by an electric current, and variation within the field is detected. The output of a magnetic sensor increases with a strong magnetic field and decreases with a weak magnetic field.

## 1.4 Comparison of different sensors

This section presents a tabular comparison of different types of sensors [1, 10, 18–30] in terms of advantages and disadvantages, as shown in table 1.3.

**Table 1.3.** Comparison of different sensors.

| Sensors | Advantages | Disadvantages |
|---|---|---|
| Temperature sensors | • Reference temperature not required<br>• Large response time<br>• Easy display<br>• Durable | • Self-heating error from applied power<br>• Difficult to calibrate |
| Position sensor | • Accurate, reliable, and predictable measurement<br>• Higher switching rate<br>• High susceptibility to noise | • Sensing range depends on the type of metal of the target object |
| Sound sensor | • Used in speech recognition software<br>• Easy to manipulate sound in real time<br>• Does not require cabling compare to wired mic | • Sound files require more memory size<br>• Interference cancelation is required<br>• Limited coverage area |
| Light sensor | • Requires very small power and voltage<br>• Available in different shapes and sizes<br>• Easy to integrate with a lighting system<br>• Quick response time and low cost | • Nonlinear characteristics<br>• If applied voltage exceeded, it will cause irreversible damage to the photo resistor<br>• Temperature sensitive<br>• Vulnerable to surges and spikes |
| Accelerometer | • Good response at higher frequencies<br>• Withstands high temperature<br>• Small size | • Sensitive to high frequency<br>• Requires external power |

(*Continued*)

**Table 1.3.** (*Continued*)

| Sensors | Advantages | Disadvantages |
|---|---|---|
| Infrared sensor | • Operates with low power<br>• Capable of detecting presence or absence of light<br>• Does not require contact with object<br>• Not affected by corrosion or oxidation<br>• Strong noise immunity | • Requires line-of-sight deployment<br>• Gets blocked by common objects<br>• Limited range<br>• Affected by environmental conditions<br>• Transmission data rate is slow |
| Pressure sensor | • High output signal level<br>• Low cost<br>• Technological robustness | • High hysteresis<br>• Sensitive to vibrations<br>• Movable contacts |
| Ultrasonic sensor | • Sensing capability to all the materials<br>• Not affected by dust, rain, snow etc<br>• Works in any adverse conditions<br>• Higher sensing distance<br>• Not affected by colour or transparency of objects<br>• Can be used in dark environments | • Sensitive to variation in the temperature<br>• Difficulties in reading reflections from soft, curved, thin and small objects<br>• Cannot work in a vacuum<br>• Sensing accuracy is affected by soft materials |
| Smoke and gas sensor | • Simple and low-cost technology<br>• Measures flammability of gases<br>• Linear output and low power requirements<br>• Wide measurement range<br>• Higher sensitivity, resolution and reliability | • Requires air or oxygen to work<br>• Narrow or limited temperature range |
| Humidity sensor | • Does not require much maintenance<br>• Flexibility to use<br>• No ageing effects | • Sensitive to dewing and substances<br>• Limited accuracy and measurement range |
| Color sensor | • Easy to change or modify setups without even re-programing the sensor device<br>• Easy to implement<br>• Common color space, used in a wide range of devices | • Lens subject to contamination<br>• Sensing range affected by colour<br>• And reflectivity of target |
| Chemical sensor | • Linear output, low power requirements and good resolution | • Narrow or limited temperature range |

| | | |
|---|---|---|
| | • Excellent repeatability and accuracy | • Short or limited life<br>• Cross-sensitivity for other gases |
| Seismic sensor | • Detects lateral and vertical variations in velocity<br>• Produces detailed images of the subsurface<br>• Used to map stratigraphic units. | • Data processing is time consuming<br>• Equipment is expensive |

## 1.5 Modern sensors

The sensor technologies have changed a lot in the last decade in terms of compactness, smartness and sensitivity. The traditional sensors such as photosensors, optical sensors, capacitive sensors and almost all sensors have been replaced by their integrated circuit forms such as MEMS (microelectromechanical system). The sensors are embedded in all modern computing and navigation devices in compact forms and this is why an ordinary smartphone carries around 22 sensors for various purposes. The technologies of sensors have further advanced and become intelligent as smart sensors and available in wearable forms. This may be seen in smart watches, smart gadgets or a large application such as self-driving cars where hundreds of smart sensors are involved for seamless and smooth driving without assistance of a driver. The same can also be seen in robotics, medical diagnosis, brain–computer interface (BCI) and many more, where AI (artificial intelligence) has empowered the sensors with intelligence and smartness for emerging and modern applications such as industry, healthcare and sophisticated automation.

## 1.6 Conclusions

This chapter presented an overview of sensors with basics, characteristics and different types of sensors. A sensor is a device that receives a signal and converts it into an electrical signal. These sensors are classified on the basis of their applications, cost, accuracy and range. Sensors are classified also into different categories like thermal, electrical, magnetic optical, mechanical and chemical sensors. The sensor technologies have become advanced now and cognitive and smart sensors are being used in all modern applications.

## References

[1] Fraden J 1997 *Handbook of Modern Sensors: Physics, Designs and Applications* 2nd edn (College Park, MD: AIP Press)

[2] Brignell J and White N 1996 *Intelligent Sensor Systems* 2nd edn. (Bristol: IOP Publishing)

[3] Boles T T 2012 Comparing piezoelectric and piezoresistive accelerometers *J. Sensors* **7** 34–5

[4] Heerens W C 2008 Application of capacitance techniques in sensor design *J. Phys. E: Sci. Instrum.* **19** 897–906

[5] Zhao W and Chen G 2009 Comparison of active and passive damping methods for application in high power active power filter with LCL-filter *2009 Int. Conf. on Sustainable Power Generation and Supply, Nanjing* pp 1–6

[6] Tseng K-C, Huang C-C and Lian F-L 2008 Analysis of sensor characteristics and error eliminations of a wheeled mobile robot with range sensors *IEEE Int. Conf. on Robotics and Biomimetics, Bangkok, 2009* pp 228–33

[7] Oshima S, Watanabe T and Fukui T 1972 Parametric sensor having sharp beam characteristics *IEEE Trans. Magn.* **8** 590–1

[8] Thoma P, Colla J and Stewart R 1979 A capacitance humidity-sensing transducer *IEEE Trans. Compon. Hybrids Manuf. Technol.* **2** 321–3

[9] Ching-Yao C 2002 On the detection of vehicular crashes-system characteristics and architecture *IEEE Trans. Veh. Technol.* **51** 180–93

[10] Powell A and Meydan T 1996 Optimization of magnetic speed sensors *IEEE Trans. Magn.* **32** 4977–9

[11] Fraden J 2016 Sensor characteristics *Handbook of Modern Sensors* (Cham: Springer) https://doi.org/10.1007/978-3-319-19303-8_3

[12] Jagadeesh Kumar V 2017 Sensors and their characteristics ed B George, J Roy, V Kumar and S Mukhopadhyay *Advanced Interfacing Techniques for Sensors. Smart Sensors, Measurement and Instrumentation* (Berlin: Springer) pp 1–72

[13] Usamentiaga R, Garta D F, Molleda J, Bulnes F G and Perez J M 2011 Temperature measurement using the wedge method: comparison and application of emissivity estimation and compensation *IEEE Trans. Instrum. Meas.* **60** 1768–78

[14] Usamentiaga R, Molleda J, Garcia D F, Granda J C and Rendueles J L 2012 Temperature measurement of molten pig iron with slag characterization and detection using infrared computer vision *IEEE Trans. Instrum. Meas.* **61** 1149–59

[15] Tarikul Islam S C M Linearization of the sensors characteristics: a review *Int. J. Smart Sens. Intell. Syst.* **12** 1–21

[16] Abudhahir A and Baskar S 2008 An evolutionary optimized nonlinear function to improve the linearity of transducer characteristics *J. Meas. Sci. Technol.* **19** 75–86

[17] Bengtsson L E 2012 Lookup table optimization for sensor linearization in small embedded systems *J. Sensor Technol.* **2** 177–84

[18] Abdul Shaib M F B, Rahim R A, Muji S Z M, Ling L P and Abdul Jamil M M 2012 A study on optical sensors orientation for tomography system development *Sensors Transducer. J.* **140** 45–52

[19] Jinesh Mathew K J, Thomas V P N, Nampoori and Radhakrishnan P 2007 A comparative study of fiber optic humidity sensors based on chitosan and Agarose *Sensors Transducer. J.* **84** 1633–40

[20] Rajeev Jindal S, Tao J P, Singh and Gaikwad Parikshit S 2012 . High dynamic range fiber optic relative humidity sensor *Opt. Eng.* **41** 1093–6

[21] Khijwania S K, Srinivasan K L and Singh J P 2005 Performance optimized optical fiber sensor for humidity measurement *Opt. Eng.* **44** 034401

[22] Song Z, Jiang G and Huang C 2011 A survey on indoor positioning technologies *International Conference on Theoretical and Mathematical Foundations of Computer ScienceCommunications in Computer and Information Science* vol 164 (Berlin: Springer) pp 198–206

[23] Koyuncu H and Yang S 2010 A survey of indoor positioning and object locating systems *Int. J. Comput. Sci. Netw. Secur.* **10** 121–8

[24] Mishra V, Singh N, Tiwari U and Kapur P 2011 Fiber grating sensors in medicine: current and emerging applications *Sens. Actuators. A Phys.* **167** 279–90

[25] Leal-Junior A G, Marques C, Frizera A and Pontes M J 2018 Multi-interface level in oil tanks and applications of optical fiber sensors *Opt. Fiber Technol.* **40** 82–92

[26] Gulácsi A and Kovács F 2020 Sentinel-1-imagery-based high-resolution water cover detection on Wetlands, aided by Google Earth Engine *Remote Sens.* **12** 1614

[27] Patel S, Park H and Bonato P 2012 A review of wearable sensors and systems with application in rehabilitation *J. Neuro Eng. Rehabil.* 9–21

[28] Sharma A B, Golubchik L and Govindan R 2010 Sensor faults: detection methods and prevalence in realworld datasets *ACM Trans. Sensor Netw. (TOSN)* **6** 23–9

[29] Hill D J and Minsker B S 2010 Anomaly detection in streaming environmental sensor data: a data-driven modeling approach *Environ. Model. Softw.* **25** 1014–22

[30] Treiber M and Kesting A 2014 Traffic flow dynamics: data, models and simulation *Phys. Today* **67** 54

# Chapter 2

# Classification and characteristics of sensors

**P Vyas and K Thakur**

This chapter deals with various aspects of sensors. Under the following topics, we will discover the importance of sensors in our day to day life in this modern era:

- Introduction;
- Classification of sensors;
- Different types of sensors;
- Transfer function;
- Characteristics and specification of sensors;
- Comparison of sensors.

The sensor, which converts one form of energy to an electric signal for further action or processing in an automated system, plays a vital role in our day-to-day activities. Varieties of sensors are available, which have their own distinct characteristics and lots of applications. Different types of sensors are explained with their types, mechanism, implemented technique and variety of applications. Here in this chapter, the classification of the sensor is described precisely and covered in almost all possible ways. In this chapter we will discuss the transfer function, which shows the yield signal conduct concerning input factors as perfect conditions. The sensor's characteristics can be characterized by various parameters, which are explained in the characteristics and specification of sensor section. The last section explains the comparison of various sensors.

## 2.1 Introduction

In the current era, sensors are part and parcel of our daily life. They play an essential role in our day-to-day life, whether in putting our hands under a tap and water falls automatically, opening or closing a door, electricity automatically switching off/on, a mobile screen automatically activated or deactivated by touch of fingers, a gesture etc. It is a very cumbersome task to manage the power system of our homes or

doi:10.1088/978-0-7503-2707-7ch2  

offices. Almost all automated systems have sensors. They make our lives simpler. Varieties of sensors are found in our homes and offices to make our life easy.

Generally, sensors and transducers are treated as the same physical devices. However, the fundamental differences between sensor and transducer are that the transducer just converts one form of energy to another, the additional logic circuitry is not required, and also there is no need for a transducer to be calibrated with respect to some reference value [1]. The transducer is just connected to the sensor. On the other hand, the sensor, which is a combination of transducer and electronics logic circuitry, is a device which detects the physical, biological, chemical etc, changes in their quantities, and then coverts one energy format to another, primarily electrical energy, for proper amplification, analysis etc. The electrical signal, which is easy to process, display, store, transmit and quickly evaluate in terms of amplitude, frequency and phase, can be easily analyzed by microcontroller or computer. For accurate measurement, all sensors need to be calibrated concerning some reference value or standard. The sensor is a human-made device which receives a stimulus (sensed signal, whether quantity, property or condition that is sensed) and responds with an electrical signal. Sensors are impacted in almost all automation systems, chemical plants, industrial plants, our bodies, automobiles, airplanes and lots of applications which make life easier.

As shown in figure 2.1, a sensor converts non-electrical to electrical values which is compatible with an electronic circuit. A loudspeaker is treated as a transducer, and a microphone is treated as an acoustic sensor. Depending upon the amount of information provided by the sensor, there are two types of sensors—direct or simple, and complex. The most important characteristics of simple sensors are that there is a direct relationship between the physical properties (light, force, temperature etc) of stimuli and essential electric property (resistance, capacitance etc) which is demonstrated in figure 2.2. For example, a switch is a simple on/off sensor, the resistance of photoresistor directly relates to light intensity, in the case of a force sensing resistor—resistance is directly related to pressure, similarly the capacitance of a microphone is directly related to pressure etc; on the other hand, the human retina is a complex sensor consisting of more than a hundred million photosensitive elements.

A complex sensor comprises one or more transducers and a sensor, such as a chemical sensor, which has a chemical reaction to heat a sensor and then heat is converted to an electrical signal by a sensor as shown in figure 2.3. They communicate using either digital pins with pulse width modulation (PWM), serial

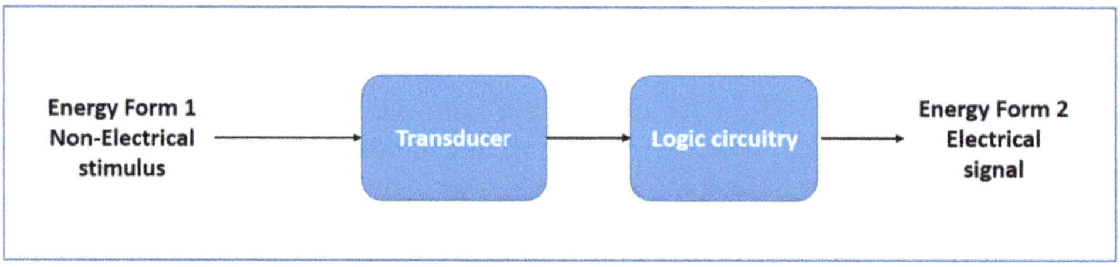


**Figure 2.1.** Block diagram of a sensor.

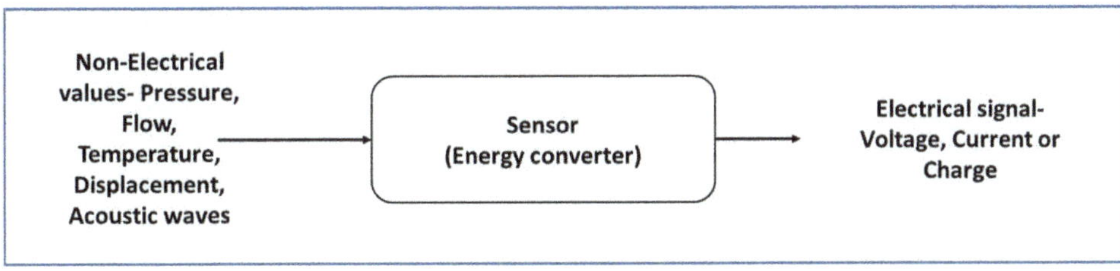


**Figure 2.2.** Direct sensor.

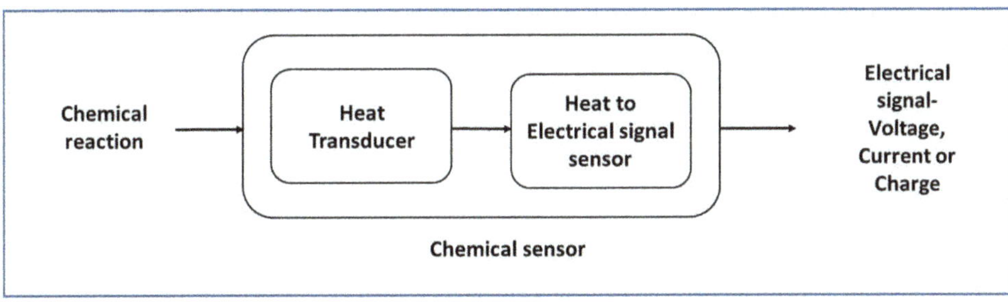


**Figure 2.3.** Complex sensor.

communication, I2C or SPI. There are many sensors within this group, namely accelerometer, global positioning system (GPS), PING ultrasonic sensor, precise temperature sensor etc.

Fundamental principles of sensor design are still the same but the sensitivity of sensor has increased, the dimensions have become smaller, the selectivity became better, and the price reduced. The sensor is still governed by the laws of nature. The microprocessor, which was invented in the early 1970s, played a dramatically important role and is an integral part of sophisticated daily used consumer appliances, namely the coffee machine, washing machine, room temperature control, music player, various automated instruments etc. In general, the microprocessor accepts, manipulates and provides a digital form of signal or data. But the real-world data/signal is analog, making it complex to analyze those data and complicated to design an electronic circuit. So, a sensor which is connected as an interface device plays an important role to convert non-electrical physical quantities into electrical signals for an electronic signal, which only understands the movement of electric charge. The sensor basically senses the physical changes, for example an accelerometer senses force/acceleration

On the other hand the transducer converts physical quantities to electric or other forms; for example, thermocouples convert variation in temperature to an electric signal. The thermistor, light detecting resistor, piezoelectric, photodiode, etc are examples of the transducer. The sensor is the core elements of a sizeable electronic system incorporated with a signal processor, memory devices, data recorders, actuators etc. Several sensors in a system may vary from a single to a thousand sensors. The sensor is the core element of the automation system. Without it, there is no automation. We can't imagine filling millions of soft drink bottles without automation; it would be somewhat difficult to test the quality of food, to control food spoilage, to monitor oxygen and trace gas content in the lungs, to measure atmospheric conditions whether measuring the wind speed and direction, radiation, humidity, etc [2].

## 2.2 Classification

Sensors may be classified into conventional and modern sensors. Conventional sensors require much additional external circuitry for signal processing and display. Modern sensors, which are also known as smart sensors, intelligently convert raw signal to a proper measuring level for further processing. Smart sensors have considerable in-built capabilities like self-testing, self-calibration, lookup table etc, to communicate suitably with other devices. These sensors are manufactured by inculcating artificial intelligence, neural network techniques to get better throughput. As wireless technology grows, modern sensors are comprised by wireless technologies. Several miniaturizing techniques namely system-on-chip (SOC), microelectromechanical system (MEMS), application specific integrated circuit (ASIC) etc, are available for wireless sensors. There are various ways to classify the different types of sensors. Some are based on applied external power, produced output signal, detection mechanism, conversion phenomenon etc. Broadly, sensors may be classified as direct and indirect. A direct sensor can convert a non-electrical stimulus into electrical signal with intermediate stages, e.g. thermocouple directly converts temperature to voltage.

On the other hand, indirect sensors have multiple transformation steps to transform the measured signal into an electrical signal, for instance, a fiber-optic displacement sensor that converts light into photons and then current. Sensors may also be classified according to their significant use, viz domestic appliances, scientific measurement, information and telecommunications, energy, meteorology, security, food processing, military, commerce, automobiles, civil engineering, etc. Classification of sensors may also be based on various specifications or characteristics of the sensor, namely sensitivity, accuracy, response time, frequency response, hysteresis, resolution, temperature, weight, conduction material etc. Transducer type is another approach to scan for a specific sensor. In the present era, semiconductor-based ICs are extensively used with the integration of microprocessor, microcontroller, logic circuit and other digital components to develop sensors. IC sensors are available in the form of microsystem, intelligent sensors, nanosensors, and others. Based on micrometer range dimensions, these sensors are used to measure temperature, pressure, radiation etc, as well as mechanical, chemical, environmental, biomedical and biological variables. Figure 2.4. shows the classification chart of sensors.

(A) *Based on an external supply*—this is the major basis on which to classify the sensor, whether a primarily electric external signal is required or not.

- *Active sensor*, which is shown in figure 2.5, requires an external excitation signal or power supply to generate an amplified output signal. It adds energy to the measurement environment as a component of the estimation procedure. The active sensor is sometimes called

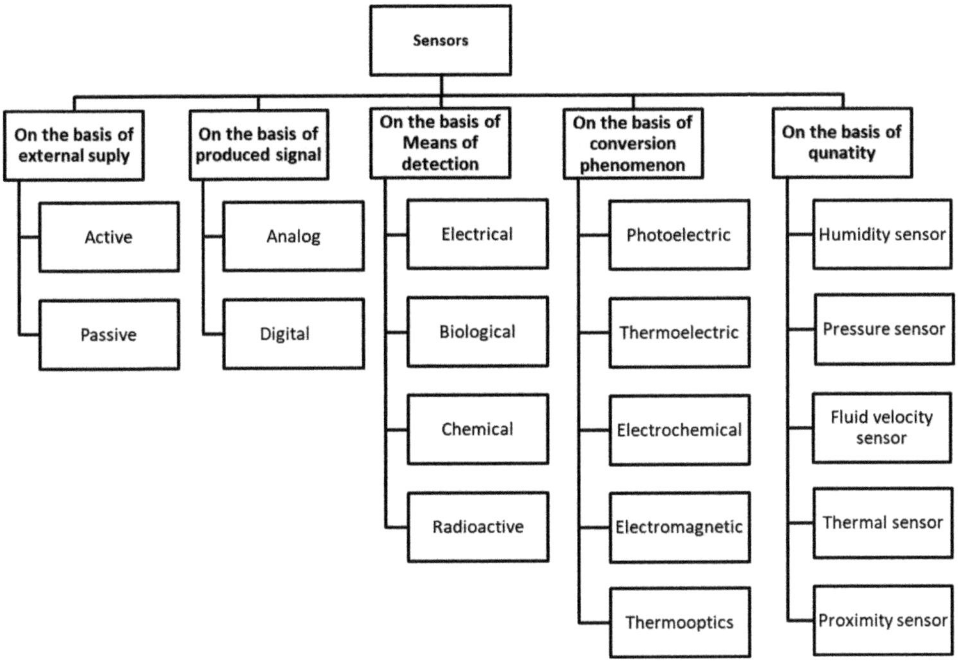


**Figure 2.4.** Classification of sensors.

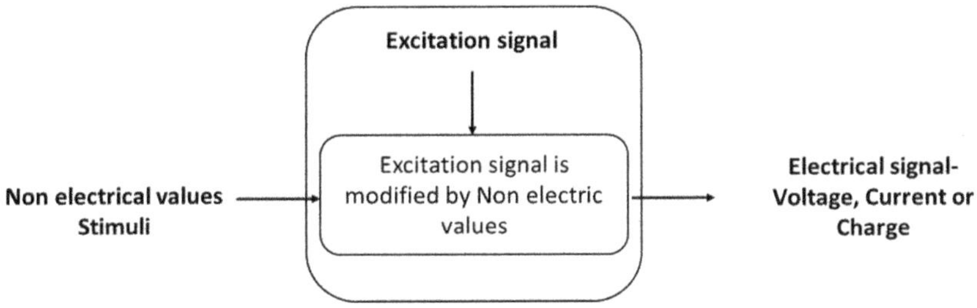


**Figure 2.5.** Active sensor.

parametric because its own properties change in response to an external effect and these properties can be subsequently converted into an electrical signal, e.g., the thermistor is a temperature-sensitive resistor. It doesn't create any electric signal, yet by passing an electric flow through it (excitation signal), its resistance can be measured by detecting variation in current and /or voltage across the thermistor. Variation in resistance is straightforwardly identified with temperature through a known function. Another example of an active sensor is a resistive strain gauge in which electric flow must be applied to it from an external power source. Examples of active sensors are proximity, accelerometer, gyroscope, barometer, tactile sensor, active IR sensor etc [3].

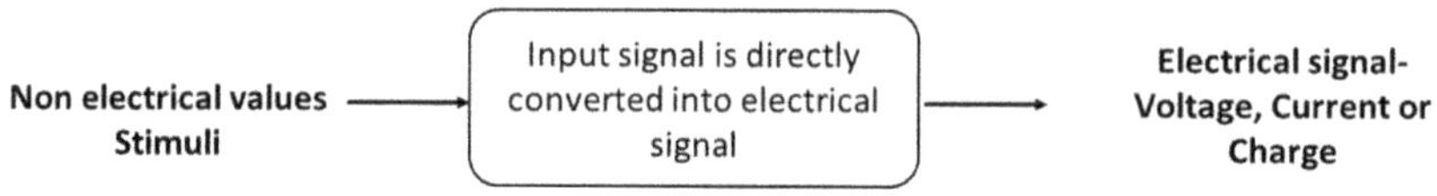


**Figure 2.6.** Passive sensor.

- *Passive sensors* do not require any external excitation signal or power supply, i.e. they are not have an external energy added as part of the measurement process. They directly generate output response and do not control produced electrical signal that is represented by figure 2.6. These have no controlling mechanism to control the flow of current in a circuit—for example, thermocouple, piezoelectric, photodiode, passive IR detector etc. The piezoelectric effect occurs in materials where an externally applied elastic strain causes a change in electric polarization which generates a charge and a voltage across the material. Passive sensor technologies can't be detected by observed parties as they only sense what is in the environment rather than relying on a transmitter whose activity might be detected with equipment. A small DC motor can be used as a passive sensor which senses the speed of the rotation. It generates voltage when rotated, i.e. behaves as a dynamo. In the case of a piezo sensor, this is a passive sensor that works in two ways. Firstly, it converts pressure into an electric signal, and in another way, when the electrical signal is applied, it generates an environmental signal like a beep [3].

(B) *Based on the selected reference*—Another classification of sensors, which is based on their selected reference condition, are absolute and reference sensors.

- In the case of **absolute sensor**, an absolute physical scale that is independent of the measurement condition is considered as a reference value to detect a stimulus. For instance, a thermistor—a temperature sensitive sensor, is an absolute sensor where electric resistance straightforwardly identifies with the absolute temperature scale of Kelvin. Barometer and altimeter are also examples of absolute sensors. An absolute pressure sensor produces a signal about vacuum—an absolute zero on the pressure scale.
- On the other hand, a **relative sensor** measures with reference to the ambient environmental conditions. The output signal of relative sensors may be influenced directly because of the variation of atmospheric parameters. A gauge pressure sensor measures gauge relative to the ambient atmospheric pressure.

(C) *Based on a produced signal*—The sensor generates an electrical signal which may be analog or digital. Dependent upon the generated signal, sensors are categorized into analog and digital sensors.

- *Analog sensors* sense the external parameter and produce a corresponding analog output signal in the form of the voltage of stimuli quantity which is being measured. An example of this type of analog sensor is demonstrated in figure 2.7. The output voltage may be in the range of 0–5 V. They return multiple values and convert a variable physical quantity into a signal that the control system can understand —voltage or current. Three basic components that are standard for analog processing are resistor, capacitor and inductor. The operational amplifier is a basic building block of the signal processing system. Analog signals are mostly affected by noise and created errors. These are very sensitive and face parallax and shadowing errors. Thermocouple, humidity, analog proximity, pressure, speed. The analog sensor namely resistance temperature device (RTD) has higher resolution. That is why these are suitable for wide temperature range and stability [4].

- *Digital sensors* produce a signal that is digital in nature. These are also called binary sensors, and return an on/off signal. They can be implemented for anything that can be divided into two states—ON/OFF, True/False, Is/Is not. An example of a digital sensor is shown in figure 2.8. Digital sensors are useful in performing mathematical analysis, storing, numeric display, transmitting etc. These are most the commonly used sensors in an industrial application. These are suitable for application where multiple measurements are necessary like process industries. A digital signal can be easily stored, communicated and analyzed. Digital sensors are rugged and immune to dynamic error. For example, in position, full, empty, power on and running. The major drawback of digital sensors is that they have low calculation range. For example, the SHT digital relative humidity and temperature sensor has lower temperature range. But these are better if you want better accuracy, throughput and stability. Some examples of digital sensors are Hall effect, limit switch, rotary encoder,

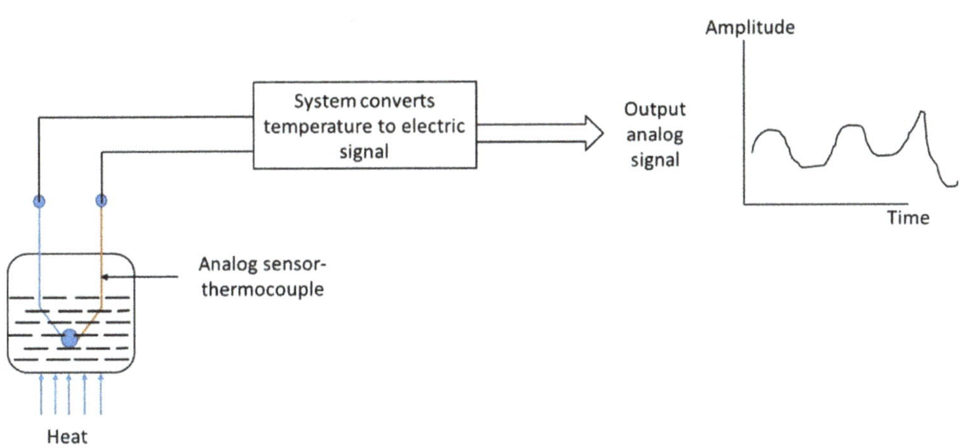


**Figure 2.7.** Analog sensor.

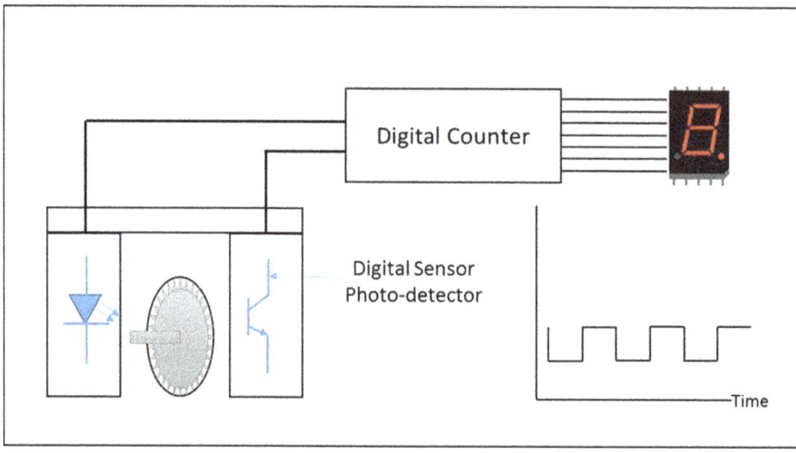


**Figure 2.8.** Digital sensor.

proximity sensor, ultrasonic sensor, capacitive sensor, auxiliary sensor, push button etc. Digital pressure sensors have inbuilt an analog-to-digital converter that converts the analog input to digital pressure output [4].

(D) *Based on means of detection*—This type of classification depends upon the internal interaction type, whether chemical, biological, electrical, mechanical, etc.

- In the case of **biosensors**, the analyte, i.e., the substance which is to be detected, identified or separated, is internally interacted with a biological material namely enzyme, antibody, nucleic acid, artificial binding protein, bacteriophage, antibody etc. This interaction produces a physical or chemical change which is detected by the transducer and converted into an electric signal. The basic block diagram of this type of sensor is as shown in figure 2.9.

  Here, the receptor interacts, catalyzes, and participates together with the analyte. The receptor layer can respond selectively to particular substances or to a group of substances known as molecular recognition. Biosensors may be categorized on the basis of sensor devices as well as the biological material into electrochemical, amperometric, potentiometric, impedimetric, voltametric biosensor, physical sensor, piezoelectric, thermometric, optical biosensors etc. Biosensors are the most popular and extensively usable sensors today. They are applicable in a broadly diversified field, namely water quality management, toxins of defense interest, food quality monitoring, soil quality monitoring, environmental monitoring, disease detection, prosthetic devices, drug discovery, screening for sickness, insulin treatment, clinical psychotherapy and diagnosis of disease etc [5].

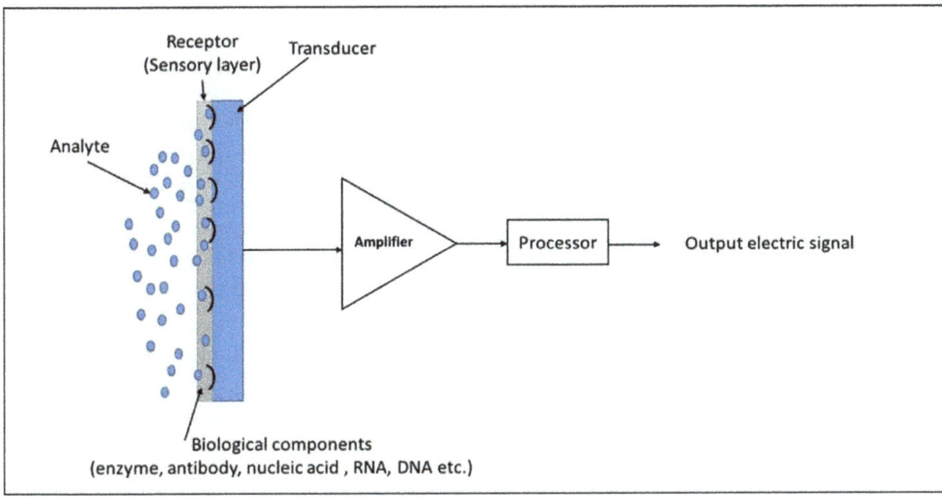


**Figure 2.9.** Block diagram of a biosensor.

- *Chemical sensors* are also categorized as electrochemical (amperometric, conductimetric, ootentiometric etc), electromagnetic (optical, mass etc). At the interface (receptor layer), chemical reactions, namely oxidation/reduction, chemisorptions, ion exchange, adsorption, acid–base, physiosorption, precipitation etc, occur after interaction with the analyte, namely reducing gases, oxidizing gases, humidity, alcohol, organic vapor etc. Chemical sensors are widely used to test the quality of food substances, presence of harmful gases in the atmosphere, to help test the quality of substances, to identify the composite and the concentration of reaction in a solution [6].
- *Electric sensors* such as a photo electric sensors, capacitance proximity sensors, humidity sensors, pressure sensors, RTD high quality sensors and custom pressure control sensors are available on the market.
- Pyrometer which is a radiation thermometer is one kind of **radioactive sensor**.

(E) *Based on conversion phenomenon*—The sensor basically converts a non-electric signal to an electrical signal by using various transformation techniques. These techniques are used to convert one form of energy to another form of energy and finally an electric signal. Direct sensors directly convert the non-electric signal to electric signal, on the other hand a complex sensor involves multiple steps to convert it into an electric signal. There are several transformation techniques which may be broadly categorized into physical, chemical and biological.

- *Physical sensors* identify the variations in physical properties, namely temperature, moisture, density, thermal, stiffness, thickness, presence of irregularities, heat, pressure, magnetic field and parameters related to mass or energy etc of the sample. There are various physical transformations which are used to convert stimuli into electric signal directly. Some common physical conversion effects are thermos-electric, thermos-elastic, photo-magnetic, electro-elastic, electro-magnetic, magneto-electric etc. Physical sensors are used in various fields like in biomedical applications, such as blood pressure, muscles displacement, blood flow, bone growth [7].

(F) *Based on quantity*—This type of classification depends upon the type of the quantities which have to be identified, measured and analyzed. There are various quantities namely magnetic, electrical, pressure, temperature, fluid, proximity, humidity, radiation, position, thermal, chemical, radio waves etc [8]. Some typical quantities and their related sensors are as follows:

- temperature quantity measures by resistance temperature detector (RTD), thermistor, thermocouple;
- pressure—Bourdon tube, manometer, diaphragms, pressure gauge;
- Force/torque—strain gauge, load cell;
- Speed/position—tachometer, encoder, LVDT;
- Light—photo-diode, light dependent resistor.

## 2.3 Commonly used sensors and their features

Sensors are widely used devices or components. They are an indispensable component of an auto-operated system. Their basic purposes are to gather information and to control the system. A sensor provides data for display purposes. This gives an understanding of the current status of the system parameters e.g. speedometer of an automotive system records and displays the speed of a vehicle with respect to time. The output signal of a sensor is also applied for controlling other systems. Intelligent sensor innovation is progressively revolutionizing our regular day-to-day existence. Ultrasound, non-contact radar and optoelectronic arrangements and laser advances are the present patterns in industry that drive development. They are present everywhere in our everyday lives. They give security, save lives and improve our personal satisfaction. The modern automobile is an example which provides critical safety features like anti-breaking, collision avoidance, traction control etc. Sensors in the home and garden make our life comfortable and secure with the connection of the Internet. In a smart home solution, devices can be interlinked together and increase security and convenience by using the Internet of Things. technique. There are a wide range of applications in the home, from sensor controlled burglary and fire protection as well as personal protection against intruders, through heat control and lighting to modern household control. Modern sensors provide sustainability in the case of energy and water monitoring, automated appliance and lighting control, smart grid energy load balancing etc. The sensor plays an important role in sustainable smart cities [17, 18].

As an ever increasing number of everyday issues are mechanized and arranged, the significance of creative sensor innovations will likewise increment later on. It will be fascinating to perceive what our regular day-to-day existence will resemble in five to ten years' time. Recent development in the field of the wireless smart sensor provides accurate, inexpensive and greatly simplified methods of instrumenting structures for structural health monitoring. A list of applications of mentioned particular sensors helps to find out suitable circumstances where a particular sensor can be chosen. There is some overlap among the sub-categories. For example, the chemical sensor and biosensor can both be used to test quality of food substances. In this section, mostly usable sensors are described with their description, types, implementation technique or working mechanism and prominent fields. Images of various day-to-day applicable sensors are shown in figure 2.10. Table 2.1. lists the different kinds of sensors which are commonly used in a variety of applications.

## 2.4 Transfer function

Transfer function, which is also known as network function or system function, describes the theoretical mathematical function of a sensor's output with respect to input in an ideal condition including ideal material, ideal tools, ideal workers etc. In general, ideal function may be explained in the form of a table of value, graph, mathematical equation etc. Transfer function demonstrates the ideal relationship between output and stimuli signal. The output electric signal $Y$ that is produced by a sensor, which depends upon the stimulus signal (input signal) $x$, can be mathematically represented by transfer function as represented below [28]:

$$Y = f(x)$$

Function $f$ may be a simple linear function or nonlinear function namely logarithmic, exponential, or power function. Often, this relationship is unidimensional: one output versus one input stimulus. The following equation represents the unidimensional linear relationship where $a$ represents the value at zero input signal,

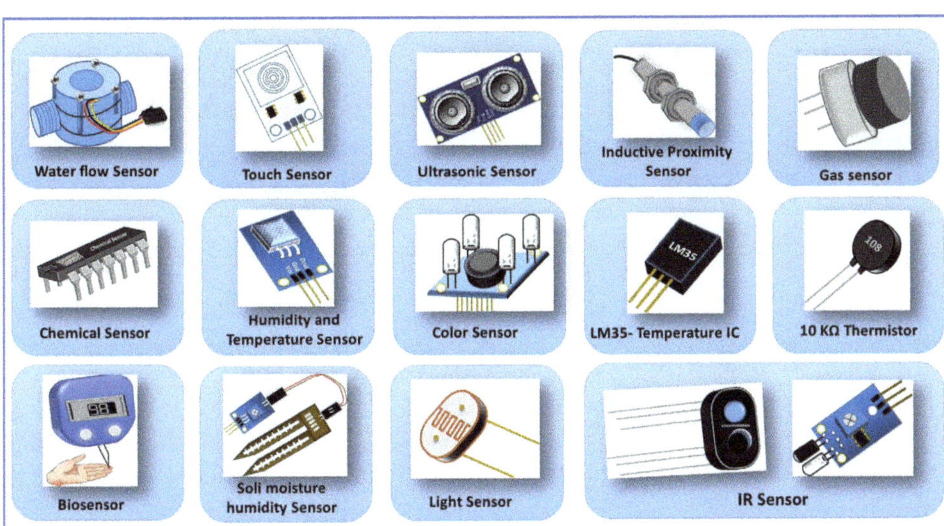


**Figure 2.10.** Images of various sensors.

**Table 2.1.** Comparison among variety of sensors—description, types, implementation techniques and exclusive applications.

| Name of sensor | Description | Types | Implemented technique | Applications |
|---|---|---|---|---|
| Temperature sensor | Most common and popular physical contact and non-contact sensor, senses the temperature which is the most commonly measured variable. | • Thermocouple<br>• Resistor temperature detector<br>• Thermometer<br>• Thermistor<br>• Thermowells<br>• Infrared thermograph<br>• Optical pyrometer<br>• Radiation thermometers | Temperature can be sensed directly i.e. direct contact or remotely (non-contact). Working with a temperature sensor depends upon the type of temperature sensor. In case of thermocouple, temperature differences between two joints (hot and cold points) of two dissimilar metal rods generate a small voltage. Similarly, in RTD, resistance of device changes with temperature [19]. | • Computers, mobile phones, automobiles, air conditioning systems, industries |
| Proximity sensor | It is a non-contact, physical sensor. It detects the presence of an object. | • Proximity sensors<br>• Capacitive proximity sensors<br>• Ultrasonic proximity sensors<br>• Photoelectric sensors<br>• Hall-effect sensors | A proximity sensor produces an electromagnetic or electrostatic field or a light emission radiation (for example, infrared), and waits for the return signal or changes in the field. The object which is being sensed is known as the proximity sensor's target [9]. | • Detection of object<br>• Parking lots<br>• Sheet break sensing<br>• Conveyer systems<br>• Smart cell phone<br>• Counting of pieces<br>• Detection of liquid level<br>• Distance measuring<br>• Detection of metal object |

| | | | | |
|---|---|---|---|---|
| Accelerometer sensor | Detects changes in position, velocity, orientation, shock, vibration, and tilt by sensing motion. | • Piezoelectric<br>• Piezoresistive<br>• Capacitive | The accelerometer sensor is the electromechanical device used to measure the proper acceleration of the object. It resembles dumped mass on a spring. At the point when acceleration is experienced by this gadget, the mass gets uprooted till the spring can rapidly move the mass, with the same rate equivalent to the acceleration it sensed [10]. | • In smartphone as shake motion, screen orientation changes, and so on<br>• Biometric gait recognition<br>• For inertial navigation system<br>• For detecting and monitoring vibrations in rotating machinery<br>• It is used to detect the position of the device laptop and mobile<br>• To display images in an upright position on screens of digital camera<br>• To monitor active volcanoes |
| Humidity sensor | It is one of the most extensively usable sensors in consumer, industrial, biomedical and environmental applications for measuring and monitoring humidity [20]. | • Absolute humidity sensor<br>• Relative humidity sensor<br>• Capacitive humidity sensor<br>• Electric conductivity or resistive humidity sensor<br>• Thermal conductivity Humidity sensor<br>• Optical hygrometer<br>• Oscillating hygrometer<br>• Gravimetric hygrometer | Humidity sensors are made up of material which shows the variation with respect to resistance, capacitance with change in humidity. | • Printer and fax machines<br>• Automobiles<br>• Food processing<br>• Refrigerator, oven and dryers<br>• Pharmaceutical plants<br>• Clothes dryers and drying machine<br>• Food dehydration |

*(Continued)*

**Table 2.1.** (*Continued*)

| Name of sensor | Description | Types | Implemented technique | Applications |
|---|---|---|---|---|
| Distance sensor | Noncontact sensor | • Optical distance sensor<br>• Acoustic distance sensor<br>• Inductive distance sensor<br>• Capacitive<br>• Photoelectric distance<br>• Ultrasonic distance sensor | Emit ultrasonic chirp or light, radio frequency field or irradiating light and then detect the decrease or change in transmission of light or return echo or alterations of radio frequency field or changes in electric field or reflected light. | • Highly accurate distance measurement<br>• Measuring wind speed and direction<br>• Fullness of tank<br>• Detecting a small object at long distance |
| Chemical sensor | Measures and characterizes chemical compounds. In some applications it is treated as an electronic nose. | • Photo chemical<br>• Photometric<br>• Metal oxide gas sensor<br>• Electrochemical sensor —Potentiometric, amperometric, conductometric<br>• Thermal sensor<br>• Optical chemical sensor<br>• Pellister catalytic | Converts targeted analyte chemical quantity into analytical electrical signal by interaction with the chemically sensitive receptor layer of sensor [6]. | • To test the quality of food substances<br>• To test the presence of harmful gases<br>• Identify the composite and the concentration of reaction in a solution<br>• Testing and controlling of food spoilage, distribution of pesticides in agriculture and to grade beverages<br>• It is used in variant fields from atmospheric monitoring of pollutant emission to the detection of explosives.<br>• In plastics manufacturing, for processing and quality controlling |

| | | | | |
|---|---|---|---|---|
| | | | | • In medicine, these are used to determine patient health by monitoring oxygen and trace gas content in the lungs and in blood samples<br>• In the military, these are used to detect fuel dumps and airborne chemical warfare agents |
| Biosensor | It is a special class of chemical sensor. Measures and characterizes organic material. | • Enzyme sensor<br>• DNA sensor<br>• Biochemical<br>• Tissue-based<br>• Immunosensors<br>• Thermal and piezoelectric biosensor<br>• Fluorescent biosensors | Deploy biologically active materials in conjunction with several physical sensing elements.<br>Recognition element is biological based typically protein, DNA, RNA, whole cell or living cell [7]. | • Drug detection<br>• Monitoring galactose content<br>• Methyldopa detection<br>• To check food quality and safety<br>• To distinguish between the natural and artificial<br>• In the fermentation industry<br>• In the saccharification process (to make into sugar)<br>• Drug discovery and in cancer<br>• Plant biology sector to find out missing link required in metabolic process [21] |
| Optical sensor | It is contactless, converts light rays into electric signal, measures physical quantity | • Photoconductive<br>• Photovoltaic cell<br>• Photodiode | Light wave may change its properties, i.e. intensity, phase, polarization, | • Ambient light sensor in smart phone |

(*Continued*)

**Table 2.1.** (*Continued*)

| Name of sensor | Description | Types | Implemented technique | Applications |
|---|---|---|---|---|
| | of light. These are compact, lightweight, electrical isolation, wide dynamic range [11]. | • Through beam sensor<br>• Retro reflective sensor<br>• Diffuse reflection sensor<br>• Strain sensor<br>• Biomedical sensor<br>• Chemical sensor<br>• Rotation sensor<br>• Electrical and magnetic sensor<br>• Temperature sensor<br>• Pressure sensor<br>• Displacement and position sensor [22] | wavelength and special distribution. It is based upon luminescence. These sensors have the ability to convert light into an electrical signal. | • Optical sensors are an integral part of computers, photo copiers, light fixtures<br>• Alarm system, synchros for photographic flashes<br>• Detection of the presence of an object<br>• Extensive use in the biomedical field. Breath analysis, optical heart-rate monitors, liquid level indicator<br>• Military and law enforcement<br>• Night vision camera<br>• Detecting partial discharges in electrical transformers |
| Gas sensor | It is just like a chemical sensor but specifically designed for monitoring the variation in air quality and presence of various gases. | • Carbon monoxide detector<br>• Breathalyzer<br>• Carbon dioxide sensor<br>• Catalytic bead sensor<br>• Ozone monitor<br>• Hydrogen sensor<br>• Air pollution sensor<br>• Oxygen sensor<br>• Hygrometer | Gas sensors are generally constructed by gas sensing layer, heater coil, electrode line, tubular ceramic and electrode [23]. | • It is utilized in different ventures, to be specific assembling, horticulture, health, and so on for air quality observing, recognition of the harmful or burnable gas.<br>• Risky gas checking in coal mines, oil and gas industries<br>• Chemical research<br>• Manufacturing—paints, plastics, rubber, pharmaceutical and petrochemical etc |

| | | | | |
|---|---|---|---|---|
| Gyroscope sensor | Necessarily for route and estimation of angular and rotational velocity in 3-axis direction. | • Micro-electro-mechanical systems gyroscopes<br>• Rotary gyroscopes<br>• Vibrating structure gyroscope<br>• Optical gyroscopes [24] | It is used to measure angular rate or angular velocity which is defined as an estimation of speed of rotation around an axis. | • Car navigation<br>• Vehicle control<br>• Consumer electronics<br>• Robotics control<br>• Helicopter or UAV control<br>• Drone and RC control |
| Tactile (touch) sensor | It acts as a variable resistor; it is cost effective and durable. It measures the coming information in response to the physical interaction with the environment [25]. | • Force/torque sensor<br>• Dynamic sensor<br>• Thermal sensor | It is made of fully conductive substances such as copper, insulated spacing material such as foam or plastic and partially conductive material. Partially conductive material restricts the progression of current. Accordingly, the resistance of the material is shifted by changing the situation at which it reaches the completely conductive material. | • It provides new texture and level of control in several consumer products<br>• Controlling of human touch in automation applications<br>• Cockpit fabrication and streaming control<br>• Fluid level sensor<br>• Position and liquid level sensing<br>• In various appliances, namely oven, washing machine, dishwashers, refrigerators<br>• In medical, fitness and gaming<br>• Robotics<br>• Computer hardware<br>• Security system<br>• Touch screen devices on mobile phone<br>• Computing |

*(Continued)*

**Table 2.1.** (*Continued*)

| Name of sensor | Description | Types | Implemented technique | Applications |
|---|---|---|---|---|
| Image sensor | It is used to observe and transmit information to make an image. It is used to determine image size, resolution, low-light performance, depth of field, lenses, camera's physical size [12]. | • Charge coupled device<br>• Complementary metal oxide semiconductor | It is a solid state device that captures light or photons that is focused on using a lens and converts into electron, image and voltage or digital value using ADC. Understanding the plan of this sensor is necessary to choose a lens and optics of a camera. Every lens is intended for a specific sensor format and resolution [26]. | • Essential component in vision camera<br>• In mobile communication<br>• Image-sensor-based visible light communication for automotive applications<br>• In night vision tools like radar, thermal imaging devices, sonar etc |
| Acoustic sensor | It not only detects an acoustic wave but also mechanical vibration of a solid. Audible range sensors are commonly known as microphones [27]. | • Resistive microphones<br>• Condenser microphones<br>• Fiber-optic microphone<br>• Piezoelectric micro-phone [16]<br>• Electret microphone<br>• Solid-state acoustic detectors | It is one kind of pressure transducer. The acoustic wave goes through or on the surface of the senor's material; any variation in the attributes of the propagation path influences the velocity and/or amplitude of the wave. Changes in the velocity can be checked by estimating the frequency or phase characteristics of the sensor and would then be able to be correlated to the relating physical quantity being estimated [13]. | • In oceanographic data collection, pollution monitor-ing, offshore exploration, disaster prevention, assisted navigation and tactical sur-veillance application<br>• Mostly communications underground and underwater where radio frequency is suitable [14]<br>• Biosecurity application |

$b$ is the slope known as sensitivity, $Y$ is one of the output characteristics of the sensor. The value of $Y$ may be amplitude, frequency, phase, depending upon the type of sensor. Transfer function may also be logarithmic, exponential or power, which are shown below [1]:

Logarithmic function:

$Y = a + b \ln x$

Power function: $Y = a + e^{kx}$

Power function: $Y = a_0 + a_1 x^k$

Where $k$ is the constant number

In some cases, higher order polynomial approximation is considered when none of the above approximations fits sufficiently well. Slope value $b$ is not fixed in a nonlinear relationship equation. At any particular input value, $x_0$, it can be stated as:

$$b = dY(x_0)/dx$$

In the case of multi-dimensional where output of sensor is affected by more than one stimulus signal, transfer function has multi-dimensions. For example, a thermal radiation sensor is represented by multi-dimensional transfer function. The function interfaces two temperatures ($T_b$ the absolute temperature of an object of measurement, and $T_s$, the absolute temperature of the sensor's surface) and the output voltage $V$: $V = G\ (T_b^4—T_s^4)$ where $G$ is a constant. The connection between the object's temperature and the output voltage (transfer function) is not just nonlinear (the fourth-order parabola) but additionally relies upon the **sensor**'s surface temperature. To decide the sensitivity of the **sensor** with respect to the object's temperature, a partial derivative will be determined as: $b = \partial V/\partial T_b = 4GT_b^3$. The following figure 2.11, demonstrates the fitting linear equation of the effect of temperature on an Si photovoltaic cell ($Rs = 1K$). Similarly, figure 2.12. shows the output power of a CdSe photovoltaic cell dependent upon the two input variables temperature and humidity.

Some common function polynomials are $Y = a_2x^2 + b_2x + c_2$, $Y = a_3x^3 + b_3x^2 + c_3 + d_3$. RTD response is represented by third degree polynomial. Nonlinear transfer function has different sensitivity at different levels.

## 2.5 Characteristics of sensors

Sensors are characterized by some of the following important parameters, which may be broadly categorized into input characteristics, transfer characteristic and output characteristics [15, 29].

1. *Range*—It important characteristic of a sensor indicates the lower and higher values of stimulus signal which can be measured by the sensor. It determines the values between which a sensor works well. A sensor requires additional or special calibration to keep it in the mentioned range, otherwise it will not work well or will give erratic results or may be damaged or have significantly modified sensitivity etc.

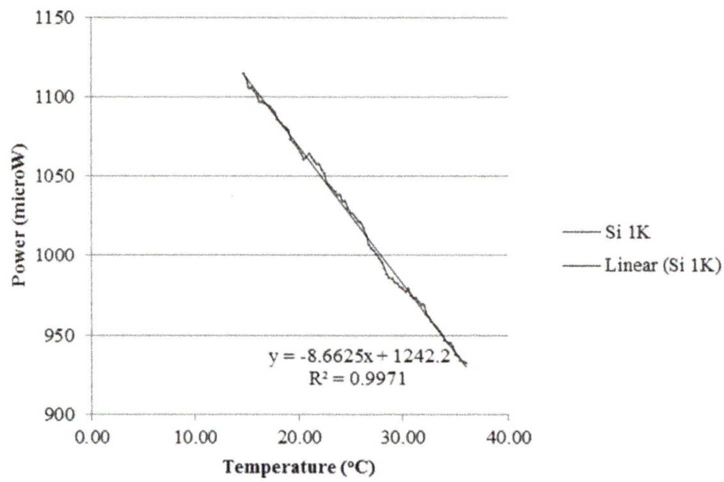


**Figure 2.11.** Simple one-dimensional transfer function of a temperature sensor.

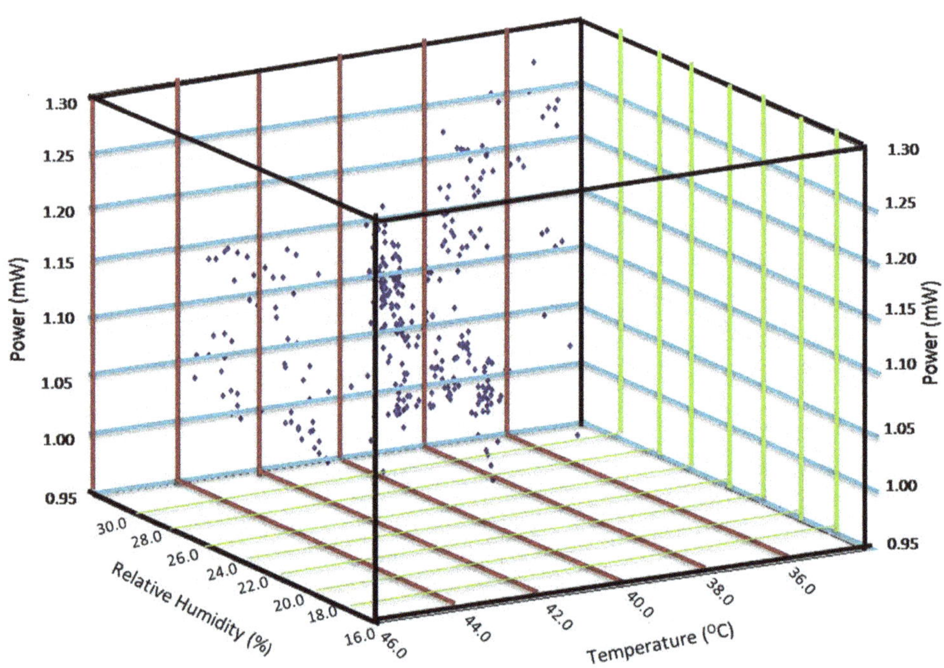


**Figure 2.12.** Two-dimensional transfer function of a humidity–temperature sensor.

2. *Span*—It is referred to as the difference between the maximum and minimum values of input.
3. *Accuracy*—It is an important characteristic of a sensor. In general, we say the term accuracy but actually it is an error. It is represented numerically as inaccuracy or the difference between measured and actual standard values,

known as error. It represents the highest deviation from the ideal values at the input. It determines the closeness of the measured value to the actual or true value. It is expressed in percentage (%) of a full-scale reading. Accuracy defines maximum error in measurement. Accuracy can be expressed directly in terms of the error tolerance, or as the percentage of the difference between measured and actual input values of the sensor's full scale. For example, if a 100 psi full-scale pressure sensor measures 76 psi but the actual pressure is 75 psi, then the error is 1 psi or 1% when this error is divided by the full scale. Mostly industrial sensors have ∓0.25% or ∓0.1% of full scale. Two error modes—non-linearity and hysteresis are involved to express accuracy of sensor. Non-linearity is basically the maximum deviation from the value predicated by the best-fit line.

On the other hand, hysteresis is the differences in the values of the reading at the same conditions up-scale and down-scale, i.e. clockwise or anticlockwise. So, basically the accuracy of the sensor is expressed to be the sum of the non-linearity error and the hysteresis error.

ISO 525–2:2019 standard provides the guidelines to measure accuracy by estimating precision error and trueness. Precision error is calculated as the average distance from the mean value. In the field of statistics, it is observed by calculating root mean square (RMS) or standard deviation (SD). Trueness is basically showing the closeness of agreement of the average of many replicated measurements relative to a reference standard, expressed numerically as an 'intrueness', that average error is known as bias. Figure 2.13. shows the trend of increasing accuracy, in general we say that: Increasing precision + increasing trueness = increasing accuracy.

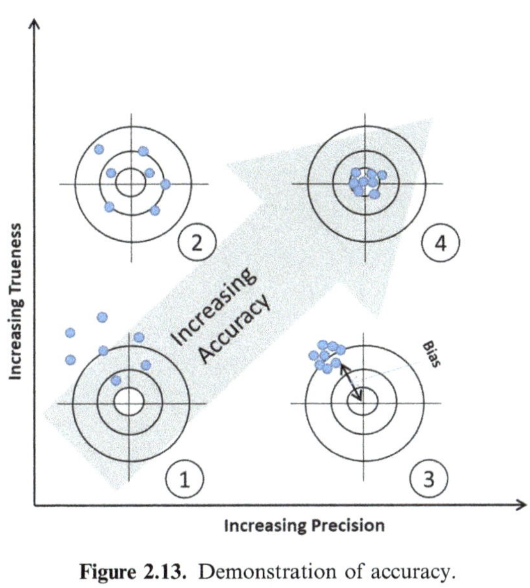


**Figure 2.13.** Demonstration of accuracy.

4. *Precision*—It shows closeness among a set of values. It is not the same as accuracy. Let $X_t$ be the true value of the variable $X$ and an irregular test measures $X_1$, $X_2$, ..., $X_i$ as the estimation of $X$. We will say our estimations $X_1$, $X_2$, ..., $X_i$ are precise when they are exceptionally close to one another yet not really near true value $X_t$. However, if we say $X_1$, $X_2$, ..., $X_i$ are accurate, it means that they are close to true value $X_t$ and hence they are also close to each other. Hence accurate measurements are always precise. Figure 2.14. demonstrates the accuracy and precision parameters.
5. *Sensitivity*—It is characterized as the proportion of rate of change in output with respect to input. Sensitivity may vary as per external factors like temperature. This is called sensitivity drift. Sensitivity is expressed in terms of percentage. If $Y$ is the output quantity in response to input $X$, then sensitivity $S$ can be expressed as:

$$S = \frac{dy}{dx} = \frac{\Delta y}{\Delta x}$$

6. *Non-linearity*—Non-linearity determines the maximum deviation between the measured values of a sensor from the ideal curve, which is shown in figure 2.15. It actually measures the nonlinearity of the sensor. Non-linearity is determined by the five point calibration method, e.g. pressure standard is implemented to apply pressure at 0%, 50%, 100%, 50% and 0% of the sensor full scale. A best fit straight line is fitted to these points, and the maximum deviation of any of the five points from the value predicated by the best-fit line, is defined as the non-linearity error.
7. *Hysteresis*—This property shows the difference in output when input is applied in two directions—increasing and decreasing. It is simply the difference in the value of the reading at the same conditions, up-scale and down-scale, which is shown in figure 2.16.

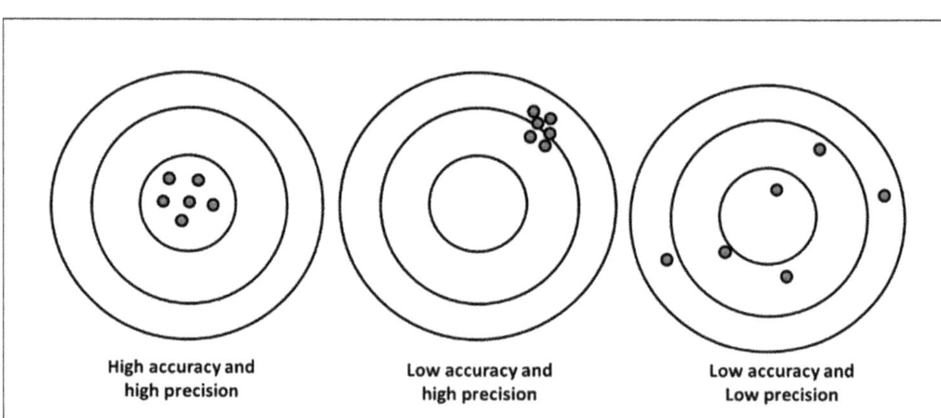


**Figure 2.14.** Demonstration of the accuracy and precision parameters.

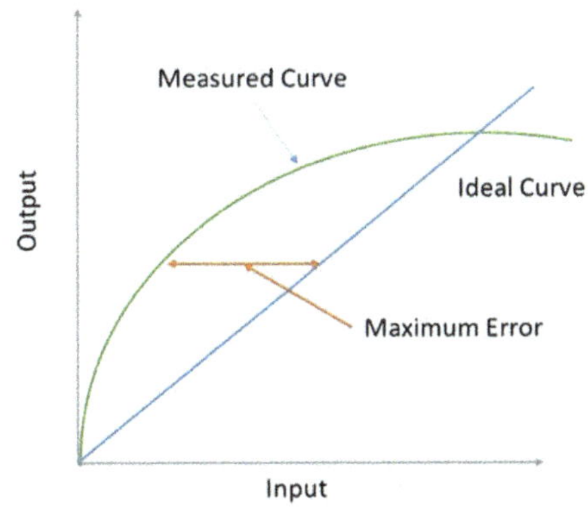


**Figure 2.15.** Demonstration of the non-linearity.

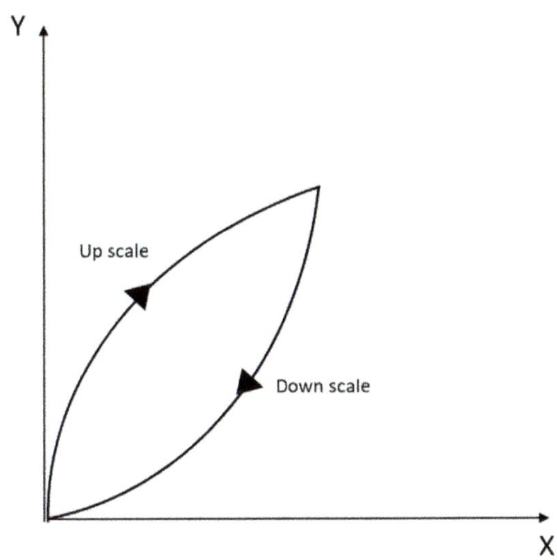


**Figure 2.16.** Hysteresis graph.

8. *Resolution*—In general, resolution discriminates the smallest change in input. In other words, it is the smallest change in input that can be sensed by the sensor. For example, relative humidity sensor SHT7x has resolution 0.4 minimum and 0.05 maximum in terms of %RH or 8 and 12 bit minimum and maximum, respectively.
9. *Reproducibility*—It shows the ability of a sensor to generate the same output value when the same input value is applied under the same operating conditions in either direction.
10. *Repeatability*—It is characterized as the capacity of the sensor to create the same output every time when similar information is applied, and all the physical and estimation conditions kept equivalent, including the operator,

instrument, ambient conditions etc [29]. In the case of relative humidity sensor SHT7x, it is $\mp 0.1\%$RH [30].

11. *Response time*—It shows the speed of the sensor, how fast it responses. It is determined by the time at which some certain percentage, like 95% of the final output value, is achieved after applied input.

A suitable sensor should have the following characteristics:

- *High sensitivity*—It determines the number of changes in output concerning a unit change in input.
- *Linearity*—It determines how much the output gets changed linearly with respect to input.
- *Resolution*—Resolution is measured in terms of the smallest change in the input that can be detected by the device. Good sensors have high resolution.
- *Power consumption*—Sensors must have the property to consume little power. Good sensors consume little power.
- *Complexity*—Smart sensors have less complex systems architecture.
- *Interconnection*—Good sensors have small interconnectivity of multiple sensors and control system.
- *Adaptability*—Good sensors have better adaptability to calibrate themselves and compensations.
- *Fault tolerance system*—Good sensors have a better fault tolerance system.
- *Reliability*—Good sensors are most reliable. They are less affected by external interference and failure rate is negligible.
- *Maintenance*—Good sensors have low down time and lower maintenance.

## 2.6 Sensors should meet the following basic requirements

Sensors are designed to fulfil the requirement of measuring stimuli under the specified criteria. The following basic requirements should be considered:

- *Ruggedness*—Capable of withstanding overload; some safety arrangements should be provided for overload protection.
- *Linearity*—Its input–output characteristics must be linear.
- *Repeatability*—It should reproduce the same output signal when the same input is applied again and again.
- High output signal quality.
- High reliability and stability.
- Good dynamic response.

Today, the top three trends seen in sensors are miniaturization, digitization and sensor fusion. Sensors are growing rapidly across countless applications in our lives. Among the applications of sensors, many require multiple sensors in a small area without degradation of performance and often with low power requirements, e.g. smartphones have multiple sensors, namely gyro sensor, proximity sensor, accelerometer, ambient light sensor etc. The second trend is digitization which is required

**Table 2.2.** Various major and sub factors for selecting proper sensors.

| Major factors | Sub factors |
|---|---|
| *Environmental factors*<br>These show the kind of environmental conditions that a selected sensor will be exposed to. For example, submersible sensors which are operated in high pressure and moisture are manufactured from those materials that can tolerate these harsh conditions. | • Temperature range<br>• Humidity effect<br>• Corrosion<br>• Size<br>• Over range protection<br>• Susceptibility to electromagnetic interference<br>• Ruggedness<br>• Power consumption<br>• Self-test capability |
| *Economic factors*<br>These are also countable factors to choose a suitable sensor. In some applications, it is not required to select very sensitive, highly precise and expensive sensors so economically we have to decide on the right one. | • Cost<br>• Availability<br>• Lifetime |
| *Sensor characteristics*<br>Sensors are implemented by some principles like inductive, capacitive etc. Each implementing technique has its merits and demerits. For example, sensors which are based on the principle of inductive mechanism are robust and suitable for harsh conditions but these are less sensitive than others. | • Sensitivity<br>• Range<br>• Stability<br>• Repeatability<br>• Linearity<br>• Error<br>• Response time<br>• Frequency response |

to fulfill the requirements of Internet of Things space in which smart sensors not only capture sensing data, but also interpret that data for a variety of applications. The third major trend is sensor fusion, which integrates multi-sensing elements in extremely tiny packages. Pressure sensing, temperature sensing, force sensing and position sensing are four major growing fields in the sensor industry today. These are generally used in wearable and medical devices, drones, home appliances, industrial applications, transportation, automotive machinery, high-power batteries, aerospace, medical instruments, elevators, high power electric motors, money handling systems etc.

## 2.7 Factors for choosing sensors

As per our electronic system or applications for automation, it is necessary to adopt the best suitable sensors. There are various factors mentioned in table 2.2. for selecting a particular sensor.

## 2.8 Conclusion

Here, we have seen various kinds of sensors, their characteristics and effective uses in a wide spectrum of applications. The major objective of showing a comparison among various kinds of sensors is to adopt a suitable sensor in an effective way, identification of potential uses, and assessment of the degree of suitability. This chapter summarizes the various bases of classification, namely requirement of external signal, types of produced signal, means of detection, conversion phenomenon etc. This chapter also describes the features, implemented techniques, and applications etc of modern sensors. Uses of sensor technology are limitless. Realistic implementation of the smart cities concept cannot be imagined without sensors. In general, the picture of smart cities includes smart infrastructure, smart electricity and water distribution, smart surveillance, smart services, smart healthcare, smart transportation etc. Sensors are used to make optimal use of available resources to fulfil major aspects of smart cities. This chapter also explains the fundamental characteristics of sensors like transfer function, sensitivity, accuracy, precision, non-linearity, hysteresis etc. It shows the selection criteria of selecting a good sensor—various factors viz environmental, economical, characteristics etc.

## References

[1] Fraden J 2004 *Handbook of Modern Sensors: Physics, Designs and Applications* (Woodbury, NY: Springer)

[2] Zhang P 2010 Sensor and actuators *Advanced Industrial Control Technology 1st edn* (Amsterdam: Elsevier)

[3] AbAbhishek S 2015 The study of active and passive transducer *Int. J. Innov. Res. Elect. Electron. Instrum. Contr. Eng.* **3** 70–2

[4] Codrey Electronics 2020 Different types of sensors-analog and digital (Retrieved Feb 2020) https://codrey.com/electronics/different-types-of-sensors/

[5] Mehrotra P 2016 Biosensors and their applications – A review *J. Oral Biol. and Craniofacial Res.* **6** 153–9

[6] Kassal P, Steinberg M D and Steinberg I M 2018 Wireless chemical sensors and biosensors: A review *Sens. Actuators* B **266** 228–45

[7] Ahmad R and Salama K N 2018 *Physical Sensors for Biomedical Applications* (New Delhi: IEEE Sensor) pp 1–3

[8] FINOIT 2020 Top 15 sensor types being used most by IoT application development companies (Retrieved Feb 2020) https://finoit.com/blog/top-15-sensor-types-used-iot/

[9] Electrical 4U 2020 Sensor-type of sensor (Retrieved Feb 2020) https://electrical4u.com/sensor-types-of-sensor/

[10] Electronics Projects Focus 2020 Accelerometer sensor working and application https://elprocus.com/accelerometer-sensor-working-and-applications/

[11] Elprocus 2013–2020 Optical sensor basics and applications (Retrieved Feb 2020) https://elprocus.com/optical-sensors-types-basics-and-applications/

[12] Yamazato T 2014 Image-sensor-based visible light communication for automotive applications *IEEE Commun. Magaz.* **52** 88–97

[13] Akyildiz I F, Pompili D and Melodia T 2005 Underwater acoustic sensor network: research challenges *Ad Hoc Netw.* **3** 257–79

[14] Hoummady M, Campitelli A and Wlodarski W 1997 Acoustic wave sensor: design, sensing mechanisms and application *Smart Mater. Struct.* **6** 647

[15] Morris A S 2001 Instrument types and performance characteristics *Measurement and Instrumentation* (3rd edn) (Oxford: Butterworth Heinemann)

[16] Regtien P and Dertien E 2018 Piezoelectric sensors *Sensors for Mechatronics* ed P Regtien and E Dertien (2nd edn) (Amsterdam: Elsevier) pp 245–65

[17] Electronicsforu.com 2020 Everything you wanted to know about modern sensors https://electronicsforu.com/electronics-projects/electronics-design-guides/everything-wanted-know-modern-sensors

[18] Irish J 2005 Sensor/instrument specification definitions, ocean instrumentation course 13.998, lecture on instrumentation specification (WHOI) (https://ocw.mit.edu/courses/mechanical-engineering/2-693-principles-of-oceanographic-instrument-systems-sensors-and-measurements-13-998-spring-2004/readings/lec2_irish.pdf)

[19] Bentley J P 1984 Temperature sensor characteristics and measurement system design *J. Phys. E: Sci. Instrum.* **17** 430

[20] Electronics Hub 2020 Humidity sensor-types and working principle https://electronicshub.org/humidity-sensor-types-working-principle/

[21] Vigneshvar S and Sudhankumari C C 2016 Recent advances in biosensor technology for potential applications-an overview *Front. Bioeng. Biotechnol.* **4** 11

[22] Regtien P and Dertien E 2018 Optical sensors *Sensors for Mechatronics* ed P Regtien and E Dertien 2nd edn (Amsterdam: Elsevier) pp 183–243

[23] Components 101 2020 Introduction to gas sensors: construction types and working https://components101.com/articles/introduction-to-gas-sensors-types-working-and-applications

[24] Passaro V M N, Cuccovillo A, Vaiani L, Carlo M D and Campanella C E 2017 Gyroscope technology and applications: a review in the industrial perspective *Sensors* **17** 2284

[25] Park M, Bok B G, Ahn J H and Kim M S 2018 Recent advances in tactile sensing technology *Micromachines* **9** 321

[26] Electronics Projects 2020 Image sensor working and its applications https://elprocus.com/image-sensor-working-applications/

[27] Sim J M, Lee Y and Kwon O 2015 Acoustic sensor based recognition of human activity in everyday life for smart home services *Int. J. Distrib. Sens. Netw.* **2015** 679123

[28] Yumpu 2020 Exercise 3, transfer functions https://yumpu.com/en/document/read/28780971/exercise-3-transfer-functions-solutions

[29] Electrical 4U 2020 Characteristics of sensors https://electrical4u.com/characteristics-of-sensors/).

[30] Sensirion-the sensor company 2020 Digital humidity sensor https://sensirion.com/en/environmental-sensors/humidity-sensors/pintype-digital-humidity-sensors/

# Chapter 3

## Optical sensors: overview, characteristics and applications

**Kusumanjali Deshmukh, Naveen Goel and Bhagwati Charan Patel**

Contactless detection and quantification of varied properties of light, namely intensity, wavelength, frequency and polarization is the prime role of optical sensors, and relies on materials and compounds that have the capability to convert photons into electric signals directly or through an intermediate device. Optical sensors are increasingly in demand in the biomedical field, biotechnology, environmental monitoring, pharmaceuticals, and clinical analysis due to their ease of use, precision of measurements, miniature size and many such advantages. Researches in this field are exponentially rising to prepare such materials and then fabricate and design devices. Sensitivity of these sensors may be range specific or may have a broad spectrum from IR to UV wavelengths. As alternatives for other investigative devices, they are becoming quite accepted due to their less cost, low power operation, and longer lives. The processes of optical sensors are fundamentally based on the selective interaction of optical signals with an organic or inorganic matrix, which results in production of an electrical signal, indicating the quality of analyte. Optical sensors offer several other advantages, such as independence from electrical noise, opportunity of remote sensing, and fairly low cost of production. This chapter addresses optical sensor types in their workings, and takes account of all features of optical sensors from the mechanism employed, their detection schemes and algorithms, and application of sensors.

### 3.1 Introduction

Optical systems involve physically less space for articulating information as needed by other electronic and magnetic based devices. This feature bestows optical detectors with an edge over traditional devices. Optical systems with better and larger bandwidth enable more data transmission, which is positively valuable for transmission of high-speed data and high-resolution videos.

doi:10.1088/978-0-7503-2707-7ch3  

'Optical sensors are not only replacing conventional sensors in many areas in science, engineering, and medicine but researchers are also creating new kinds of sensors that have unique properties,' comments Technical Insights Analyst Joe Constance. 'These properties relate to the ability of the sensors to measure physical, chemical, and biological phenomena.'

Data relayed by a thermocouple by conventional techniques have a high chance of getting distorted due electromagnetic interference. Optical sensors, particularly fiber-optic sensors, demonstrate superior resistance compared to thermocouples towards impervious surroundings and electromagnetic interference. Such distinctiveness makes them the best option as thermal sensors for numerous device applications. Fiber-optic sensors put forward a broad range of rewards over conventional sensing systems like miniaturization and longer effective functional period, resistance to electromagnetic intrusion, ability of multiplexing, and elevated sensitivity. Numerous fields, especially for medico care devices and aerospace sectors prefer fiber optics sensor technology.

Scientists have been working to successfully prepare fiber-optic reflector sensors for measuring very high temperatures in boilers of power plants, aircraft, ships and nuclear plants etc. This is intended to further improve the connection between the fiber and the reflector, and to reduce the essential electronics for data acquisition and analysis.

Numerous research and present day commercial applications such as medico technologies, quality and process control, imaging, metrology and remote sensing, use optical sensors. In addition, newer sensor techniques that enable added advanced sensing are constantly being developed by utilizing new frequency bands such as, for example, THz radiation, or by using novel materials like micro-, nano-structured materials and meta materials [1, 2]. Improvement in photonic technologies could drastically lower the price as has happened in the case of telecommunications industries.

## 3.2 Optical sensors: fundamentals

Optical signals or light rays are deciphered into electronic signals by an optical sensor. The working principle of an optical sensor involves a few compulsory steps like evaluating a physical parameter of light such as its intensity, before and after interaction with an object to be detected and then finally decoding it into some form readable by an integrated circuit to measure and display it [2]. For contactless detection of objects and temperature, counting and positioning of parts, optical sensors are adopted as a better option. They can be both internal and external as per the requirements of the devices. When measurands are placed externally and sensors only collect and send out the light required for detection they are known as external sensors. Internal sensors that are embedded or are part of the fiber to measure the bends or other small changes like stress or pressure are termed as internal or intrinsic sensors. Some devices which we observe in our daily life are:

- automatic lamps that turn on in dim light, also used in headlights of bikes nowadays;

- sensors to detect position by interruption of light beam;
- barcode readers in provision shops and others;
- body temperature sensors, very often seen nowadays in the wake of COVID-19.

In the present era, an optical detector is quite a common component of many complex systems that incorporate a light source, a sensor and a circuit for measuring and display. An electrically coupled trigger responds to a modification in the light signal. Variation in one or many light beams can be assessed and detected by an optical sensor. The light sensor activates on any variation and creates a photoelectric trigger that accordingly modifies the output electrical signal. For instance, a photoresistor acts in response to the intensity of light and switches the circuit. Electrical signals can be selectively chosen to be switched from one optically integrated circuit to another with the assistance of an optical switch. Different methods and effects like electro-optic effects, magneto-optic effects or mechanical means can activate an optical switch. Integrated and discrete microelectronic circuits can be equipped with optical switches.

Measurements and assessment made by optical sensors are generally based on changes in the following optical properties: absorbance, reflectance/transmittance, refractive index, phase shift, polarization, wavelength (fluorescence, Raman etc). The most general equation of plane wave, electromagnetic wave is:

$$\vec{E} = E_0 e^{i(\omega t - kx + \varnothing)}\hat{z} \tag{3.1}$$

Any of the parameters like amplitude $E_0$, or $\omega$ or $k$ or $\varphi$ or $z$ may vary and can be monitored and measured. Although the electromagnetic spectrum is very broad, the optical sensing mostly relies on visible, UV and FIR regions only. Different parameters like chemical species, temperature, radiation, velocity, pH-value, liquid level, strain, pressure, acoustic fields, displacement (position), electric field and vibrations can be measured by different optical sensors.

Thus, the operating principle of an optical sensor is the transmitting and receiving of light, reflection from an interruption or hindrance by an object, modification of a light beam emitted and transmitted by a light source, mostly a light emitting diode (LED). This disruption of light beam is detected and measured based on the type of sensor used. It hence becomes feasible to identify or spot objects irrespective of their composition and construction (be they plastic or wood or metal or something else). Detection and recognition of transparent objects or coloured ones with the slightest of contrast can be easily detected by specific devices.

### 3.2.1 Modes of operation

Single point or distributed points of operation may be adopted by optical sensors [3]. In the single point operation, a complete change in phase is desirable to set the sensing element into motion. In the distributed point operation, the sensor is responsive along an extensive chain of points or an array of single optical fibers. An added characteristic of optical sensors takes account of the feature that they are placed internally or externally in a device. Different modes of operation of optical sensors are broadly classified and explained as below (figure 3.1).

#### *3.2.1.1 Through-beam sensors*

This arrangement consists of working components separated by some distance, wherein, a transmitter and a receiver are placed exactly in line facing each other. A light beam is sent by the emitter or transmitter towards the receiver. Any disturbance or modification of the incident light beam is recorded for a response as a switching signal at the receiver. The point or position of interruption is hence irrelevant here (figure 3.2).

***Advantages***

Surface structure, color and reflectivity of an object's surface is irrelevant in this detection technique and it can operate for appreciably large distances.

However, to promise an elevated operational dependability, a sufficiently large surface of the item is necessary to modify the incident beam of light.

#### *3.2.1.2 Retro-reflective sensors*

This arrangement is made concise by placing transmitter and receiver together at the same place. An incident beam of light from the emitter is reflected towards the receiver back in the incident direction. The switching operation is initiated by any interruption in the light beam. In this arrangement too, position of the interruption is of no significance (figure 3.3).

***Advantages***

Large operating distances are facilitated by retro-reflective sensors by the use of several switching points that are accurately identical with very little space between

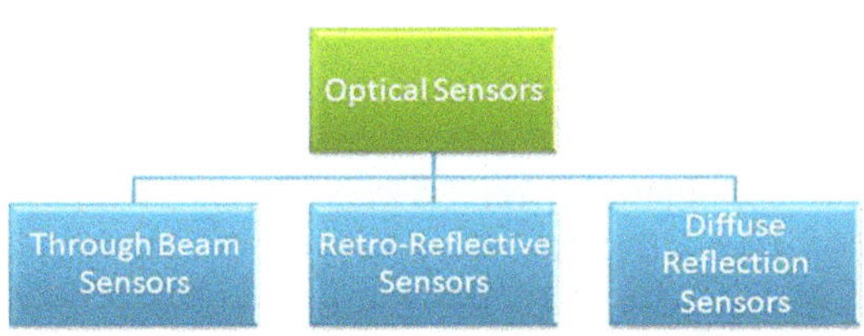


**Figure 3.1.** Different modes of operation.

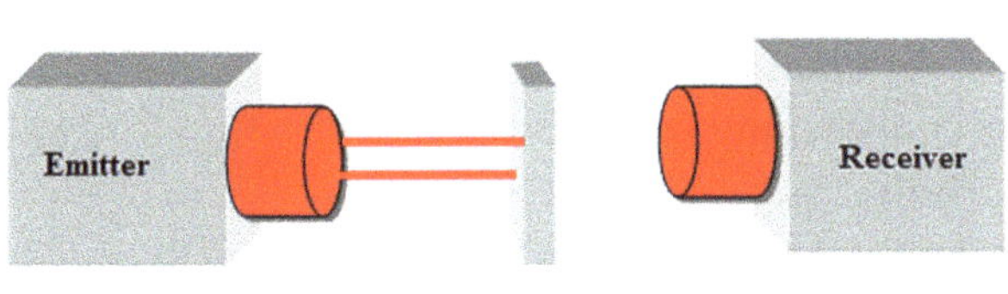


**Figure 3.2.** Through-beam sensor: emitter and detector are placed in line and interruption in the path is recorded.

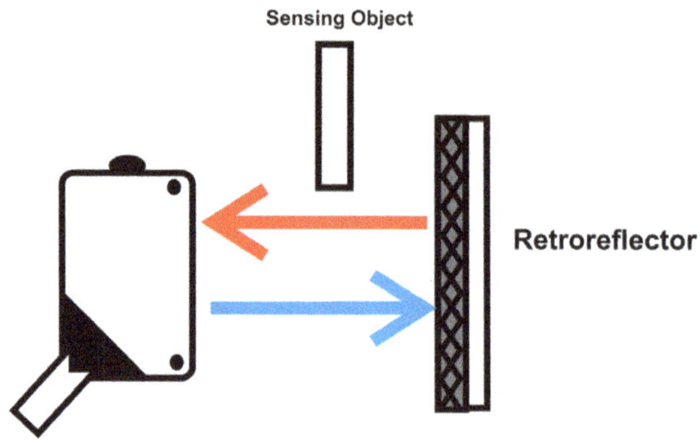


**Figure 3.3.** Retro-reflective sensors: emitter and reflector are positioned at the same point and an external reflector is required and interruption in these light signals is recorded.

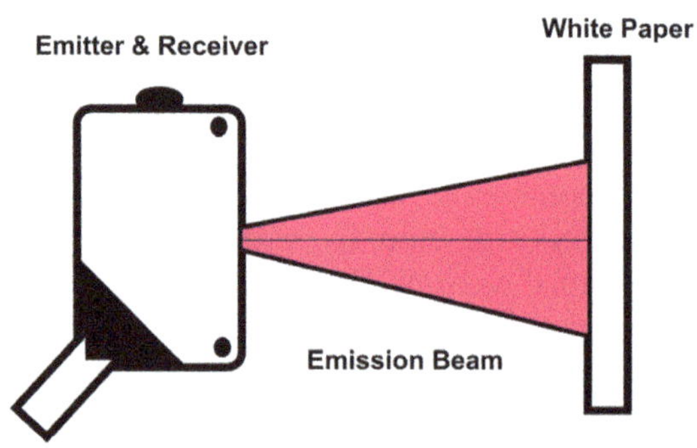


**Figure 3.4.** Diffuse reflector sensor: emitter and detector are placed in the same cabin; reflection may occur at different points.

them. Any interruption in light beam due to any object is precisely detected and its surface structure or shape and color do not affect the process.

#### *3.2.1.3 Diffuse-reflection sensors*

This arrangement too has transmitter and receiver in one cover but placed at different positions. The light incident upon an object is reflected by it and is then detected by the detector (figure 3.4).

***Advantages***

The switching operation is initiated by receiver as per diffused light intensity received at its end. In spite of the sensitivity settings some erroneous switching operations need to be taken into account as the reflection is always better from the rear part than the front.

### 3.2.2 Light sources for optical sensors

A variety of light sources as per their varied uses can be observed in our homes, offices and surroundings [3]. Each have their specific and unmatched features which make them useful for us. Have you ever thought of it? The foremost light sources available to investigate and learn optics were the Sun, the ultimate source of light, and burning candles' flames. Light emitted or produced from excited matter like sodium and other halide ions provides the reference lines in the optical visible electromagnetic spectrum and they are mostly the sources of light used to date. A vital requirement in optical sensing is the monochromatic light source and it must be compact, and long lasting. The most commonly used light source for all possible ranges of spectrum in present day applications are LEDs and light amplification by stimulated emission of radiation (LASERs).

*3.2.2.1 Light emitting diode*

In semiconductor diodes, the process of recombination of majority charge carrier, electrons of the n-doped region with majority carriers, holes of p-doped semiconductors at the junction, releases energy in the form of light, as shown in figure 3.5. Applied external voltage provides these charge carriers with the necessary excitation energy to cross the barrier potential and recombine with each other. The photons released may stimulate another charge carrier. This supports pairing the LED with an optical device.

*3.2.2.2 Light amplification by stimulated emission of radiation*

LASER is actually a series of phenomena that produces an intense narrow beam of light as a result of amplification using stimulated emission of radiation. Laser beams are extremely coherent, directional and monochromatic. These properties of a laser beam result in an intense and highly focused point. Almost all modern technology from communication to sensing, cutting and welding, exploit this property of the laser. Precise and minimal invasive eye surgery can be performed due to its properties and achievability of controlled production [4].

In most of the conventional lighting solutions like flash lights or LEDs, conversion of electrical energy to optical or light energy takes place either directly or via some intermediate process. Light from lasers, on the other hand is produced by initially exciting electrons to higher energy states and then by stimulating them to emit during transition to lower energy states. A flashlight emits by spontaneous

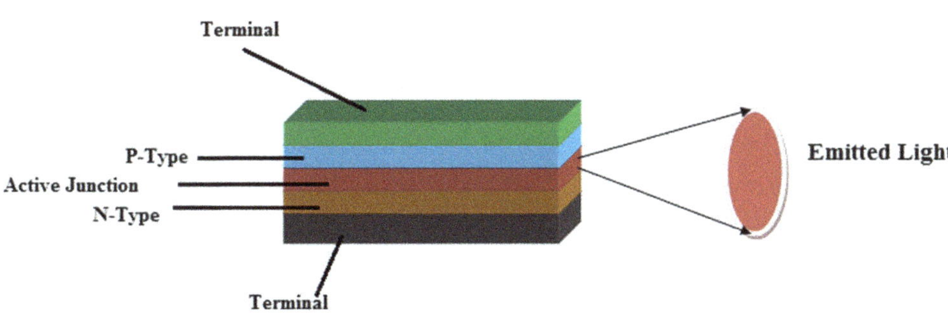


**Figure 3.5.** Schematic diagram for the workings of an LED.

emission, whereby electrons emit photons during random transition from a state of higher energy to lower energy. Lasing action of light is initiated via stimulated emission which is then processed in a resonator cavity to obtain coherent photons, all in phase with each other, as can be seen in figure 3.6(a) and (b). A flashlight or any other source of light, including sunlight, produces non-coherent photons. Also, lasers do not spread out while propagating as the light from any conventional source or flashlight would, which makes it extremely focused and

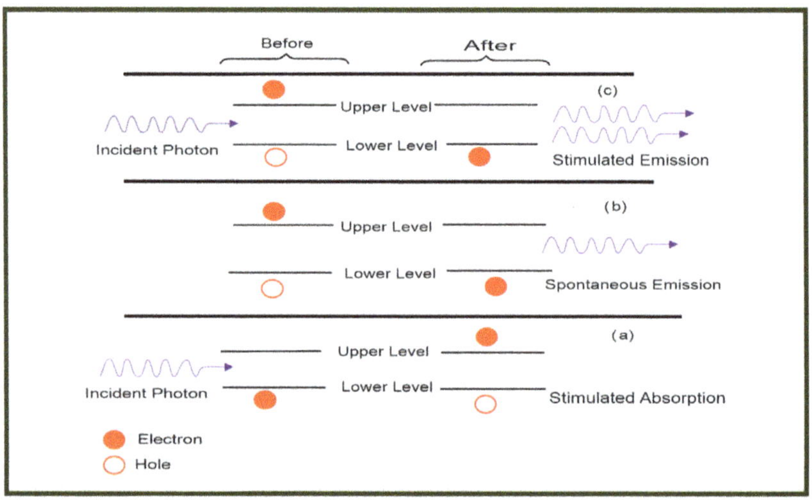


(a)

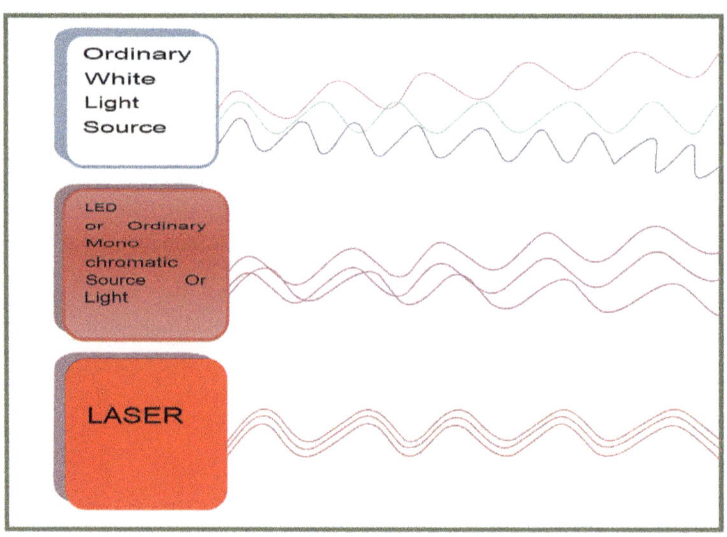


(b)

**Figure 3.6.** (a) Lasing action due to stimulated emission. (b) Difference between ordinary light and LASER. A photo tube or photoemissive cell.

intense. This property allows the light to travel long distances and thus they are used in navigation and sensing systems.

### 3.2.3 Advantages of optical sensors

Optical sensors offer considerable advantages over conventional sensor types due to their extraordinary performance. Illustrative advantages of optical sensors over non-optical sensors are [5]:

- better sensitivity;
- electrical passiveness;
- broad dynamic range;
- flexibility of configuration;
- capability to multiplex;
- negligible electromagnetic interference.

An array of sensors and network of sensors could be created due to the capability of optical fibers to transmit optical signals over large distance [6]. This facilitates bypassing complexity in translating electronics into photonics individually at every sensing site, thus reducing expense and improving flexibility. An extensive use of optical fibers has minimized the difficulty by reducing interference from various effects.

## 3.3 Optical sensing devices (detectors)

The working component of an optical sensor is the material or a device which receives light energy from a source or emitter or is reflected from the material (to be sensed) and converts it into electrical energy. Conversion of electrical signal and any change in it due to perturbation caused by an object in the path are monitored and measured as a function of distance, temperature, pressure or other parameters as required. Photodetectors may be classified by their mechanism for detection:

- photoresitors (light dependent resistors): photons generate electron–hole pairs in a pure or appropriately doped semiconducting layer.
- photodiodes: absorbed photons generate electron–hole pairs in the depletion region. Works in near UV and VIS region of EM spectrum.
- phototransistors: amplifying photodiodes.
- Thermal detectors: photons cause electrons to transition to mid-gap states then decay back to lower bands, causing phonon generation and thus heat. Works in IR region of EM spectrum.
- photoemission (photocells): photons cause electron transition from the conduction band of a material to free electrons in a vacuum or gas. Works in UV range of EM spectrum.
- fiber optics: a type of photoelectric sensor that enables detection of objects in narrow locations by transmitting light from a fiber amplifier unit with a fiber unit.

Working principle, construction, types and applications of these components are discussed briefly.

### 3.3.1 Photoemissive cells (photoemitters)

A device which works solely on photoelectric effect is a photoemissive cell, normally famous as a phototube. Light radiation of the appropriate wavelength when striking a metal surface ejects out electrons, ready to carry current in the tube and outer circuit. Photo cells or electric eyes are better nomenclature for photoemissive cells for these names appropriately elucidate their working. Actually, photo cells are gas-filled or vacuum tubes that are sensitive to light. In the past, i.e. before semi-conductor diodes were developed, phototubes were widely used, however, they are now substituted in various devices by semiconductor photodetectors. Nevertheless, one of the extremely sensitive light detectors extensively employed in a variety of research purposes is the photomultiplier tube.

The phototube is essentially a vacuum tube, enclosing a curved cathode and a cylindrical anode or plate (figure 3.7). Actually, the photoelectric emission occurs at the cathode surface.

A schematic of phototube operation is shown in figure 3.8, where incident photons of appropriate frequency hit the photocathode, knocking out electrons from its surface, which then accelerate and are attracted towards the positive anode.

Thus photocurrent is reliant on the frequency of incident photons and their intensity. The current through the device is generally in the microamperes range unlike photomultiplier tubes, as there is no arrangement for amplification to occur.

Sensitivity of the device for a given wavelength range is determined by the material employed as the photoemissive cathode. For example, a cathode made of cesium–antimony offers a device that is extremely sensitive between the violet to UV region, while sensitivity falls off to almost zero for red light. While, an infrared to red

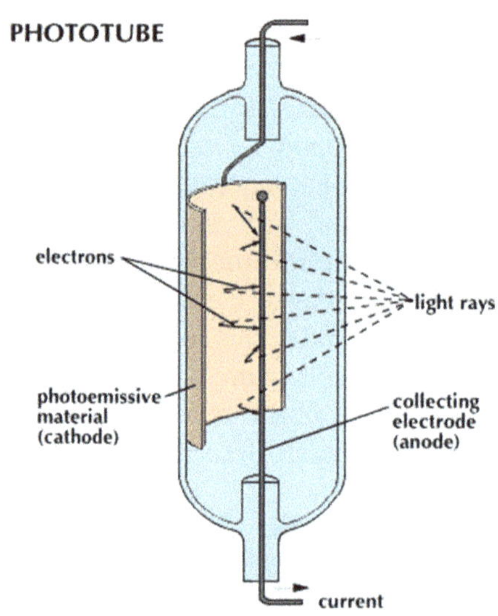


**Figure 3.7.** A phototube or photoemissive cell.

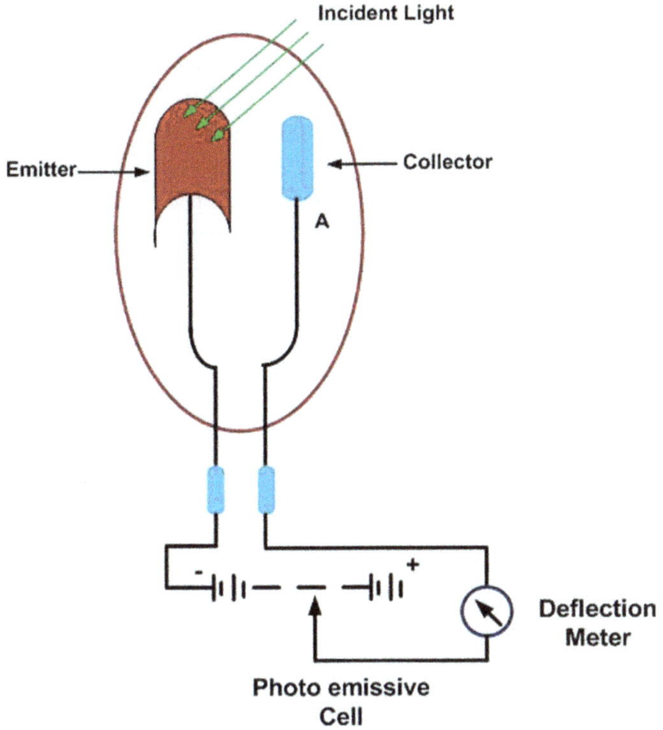


**Figure 3.8.** Schematic of the workings of a photoemissive cell.

sensitive cathode is obtained from a cesium cathode prepared on oxidized silver. This cathode is quite insensitive to the blue region but not zero.

Nearly constant anode current relative to plate voltage for a given level of illumination is a characteristic feature for vacuum devices. Further, as compared to vacuum devices, sensitivity of gas-filled devices is better; however, the response to frequency modulated illumination reduces at lower frequencies. The time required for these photo generated electrons from the cathode to reach the plate, transit time, restricts the frequency response of the photocell in general to some extent.

An important use of the photocell, in earlier days, was to interpret an optical audio track for film projection. Numerous light-sensing functions employed phototubes until they were replaced by photo-assisted semiconducting diodes and resistors. In the earlier days, phototubes were used as light detectors too, but they were reasonably cumbersome, elaborate, and expensive. Smaller and inexpensive electronic components viz. LDRs are now widely utilized as light detectors. Photomultipliers are used today in scientific applications, such as for detecting radiation of all varieties, and in gadgets such as night vision goggles, where they intensify the faint light of a dark place so that it can be viewed with added clarity.

### 3.3.2 Photoresistor or light dependent resistors

A photoresistor also known as LDR is a photoconductive device. The most used materials for fabrication of LDR are thin films of cadmium sulfide, cadmium selenide and other II-VI semiconducting materials [7], which exhibit change in resistance on exposure to light radiation. A power source is required for the operation of an LDR and hence to generate photocurrent, thus it is an active sensor; this photo effect is evident as the change in the circuits' electrical resistance.

The phenomenon of dark and photo current is well understood by band structure of semiconducting materials. Difference in energies between valence band, fully occupied and the conduction band, fully unoccupied is termed as energy band gap $E_g$. No allowed energy states lie in the gap for pure semiconductors. However, a few new levels within this gap appear just below the conduction band and above the valence band corresponding to donor ions and acceptor ions, respectively, for the doped materials. In dark, the valence band is fully occupied by the electrons and the conduction band is fully vacant, thus the crystal has only thermally generated charge carriers whose concentration is a function of temperature alone. Hence, at absolute zero, a high resistance is offered to the current in the semiconductor circuit.

When light rays of an appropriate wavelength are made incident on the photoconductive crystal, electrons in the valence band absorb photons and jump to higher energy states of the conduction band. This movement of electrons increases the number of free electrons in the conduction band and hence the number of holes in the valence band, enhancing the electrical conductivity. For instance, cadmium sulphide crystal has an $E_g$ of 2.41 eV and the absorption spectral edge wavelength lies in the visible range at 515 nm [7]. Hence, cadmium sulphide identifies and absorbs light rays of wavelength shorter than 515 nm. Other photoconducting materials have their own specific absorption-edge wavelengths. As discussed, CdS is most perceptive at the visible and UV region, whereas Si and Ge respond effectively in only the near IR region.

Semiconductors' conductance is hence governed by charge carrier density, their nobilities and life times; as given by

$$k\sigma = fe\left(\mu_n\tau_n + \mu_p\tau_p\right) \tag{3.2}$$

where, $\mu_n$ is the mobility of free-electrons and $\mu_p$ the mobility for holes (cm $V^{-1}$ $s^{-1}$), $\tau_n$, $\tau_p$ are the free-electron and hole life times, charge of an electron $e$, and number of photo generated carriers per unit volume per second $f$. Typically a CdS cell has $\mu_n\tau_n \gg \mu_p\tau_p$; thus, contribution to total conductance by free holes is negligibly small as compared to free electrons. Thus, for an $n$-type semiconducting cell:

$$k\sigma = fe\mu_n\tau_n \tag{3.3}$$

In a semiconductor, sensitivity $b$ is the number of electrons generated by one photon in its life time $\tau$:

$$b = \frac{\tau_n}{t_t} \tag{3.4}$$

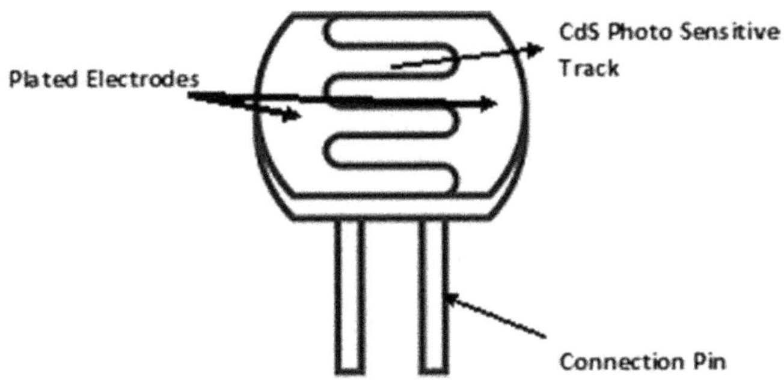


**Figure 3.9.** A typical commercially available photoresistor (LDR).

where time spent by the electron between the electrodes, the transit $t_t = l^2/V\,\mu_n$ and $l$ *and* $V$ are, respectively, the electrodes separation and the voltage applied. Then, sensitivity can be rewritten as:

$$b = \frac{\mu_n \tau_n V}{l^2} \tag{3.5}$$

For example, if voltage between electrodes is $V = 2$ V and distance between them is kept $l = 0.2$ mm for a CdS film with $\mu_n = 300$ cm$^2$ V$^{-1}$ s$^{-1}$ and $\tau_n = 10^{-3}$ s, the sensitivity will be 1500, which would suggest just one photon let free 1500 conducting electrons, converting a photoresistor into a photomultiplier device. Thus, quite sensitively and efficiently a photoresistor detects photons and any change in their intensity.

Consequently, separation $l$ between electrodes ought to be reduced and effective width $d$ must be large for decreasing resistance and increasing cell sensitivity. This is done by designing a sensor in a zigzag pattern (figure 3.9) with the electrodes connected to the ends.

*3.3.2.1 Types of photoresistors*

The photoresistive cells are classified into the following types on the basis of the techniques used for their manufacturing:

- sintered;
- single-crystal;
- evaporated.

Sintered photoresistors offer high sensitivity and large area production of sensitive layers, which finally put forward lower-cost devices.

Spectral response of a photoresistor is specific to the material, as discussed in the above paragraph, but it can be very well tailored by bringing some variation in the composition of the pure material. An illustration for this is the addition of a selenide in CdS which replaces some of the S ions and forms CdSe and this causes red shift of the spectral response i.e. towards the longer side of the wavelength band (orange and red).

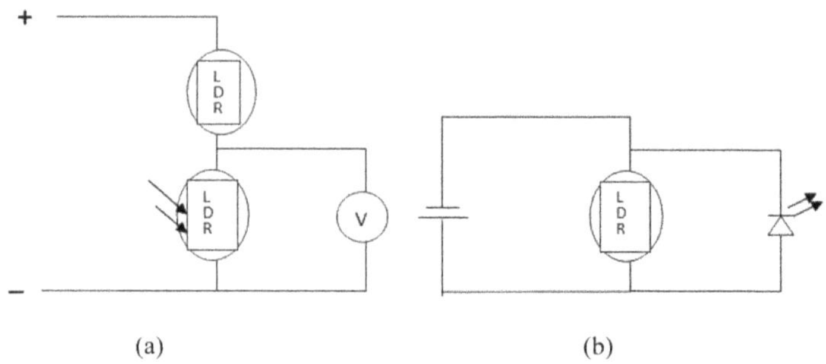


**Figure 3.10.** (a) An LDR as a potential divider (b) An LDR as a switch.

*Application of photoresistor (LDR):* figure 3.10(a) shows a typical circuit for determining the alteration in intensity of light by means of an LDR. Two resistors are needed to be placed in series connection in the circuit. One needs to measure potential drop across the LDR that is subjected to light. The entire system functions like a potential divider, as the current and voltage in the circuit remain proportional across the LDR.

Circuit (b) in figure 3.10 illustrates an automatic light switch; lights turn-on as soon as incident light intensity falls off in the surroundings. The LED turns on as light intensity falls below a minimum as the LDR starts offering high resistance across it. This is already in commercial use for automatic street lights and automobiles headlights.

### 3.3.3 Photodiodes

Photodiodes are an important class of semiconductive optical sensors, since in photodiodes the method of optical sensing involves direct conversion of optical signal (photons) into an electrical signal (electrons and holes). Solar batteries are also a type of photodiode and are best suited for power conversion but we are here only concerned about detection and measurement of incident lights.

Photodiodes connected in various circuits operate mainly in two modes. Forward biased p-n junction mode, wherein, even with illumination by light of suitable frequency, the photocurrent remains small as compared to the dark current. In other words, current generated by light (photocurrent) does not contribute much to the forward current. But when the junction is reverse biased (figure 3.11), the current increases reasonably on illumination, as shown in figure 3.12. Incident photons increase the total number of charge carriers by releasing electrons from covalent bonds on both the sides of junction. These electrons get accelerated towards the positive terminal of the battery due to the applied field and cross the junction. Similarly, the photo generated holes stream towards the opposite terminal, such that the photocurrent $i_p$ flows through the junction and in the circuit. Thus, the photodiode works under reverse-biased condition and its electrical equivalent circuit needs to be represented by dual current sources and an RC network.

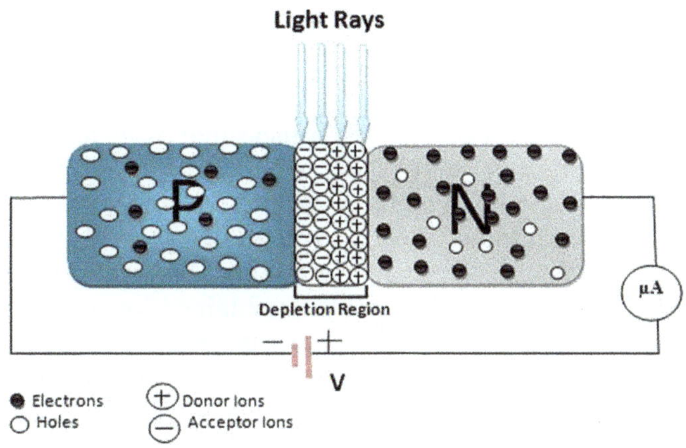


**Figure 3.11.** Structure and biasing arrangement of photodiode.

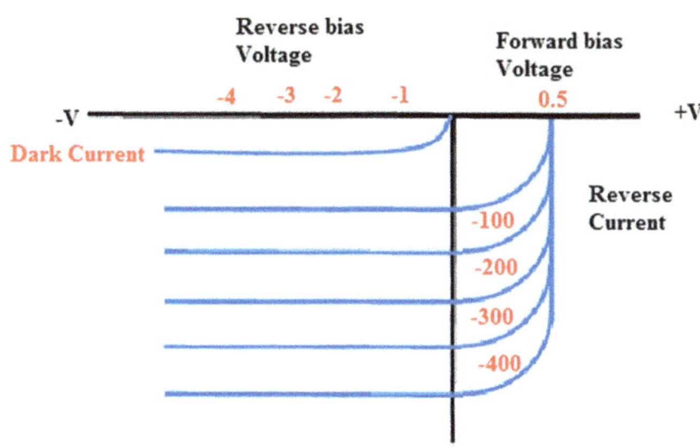


**Figure 3.12.** Characteristic curve of photodiode.

The current- to-voltage response of the photodiode is given by

$$i = i_0(e^{eV/KT} - 1) - i_s \tag{3.6}$$

where, the reverse 'dark current' due to the thermal generation of electron–hole pairs is $i_0$, photo current through the diode $i_s$, $k_b$ is Boltzmann constant, and $T$ is the absolute temperature.

Optical power and electrical current in the detector circuit are linearly dependent:

$$i = \gamma(e) = \frac{neP}{hv} \tag{3.7}$$

Change in plate current is a result of change in input power. Since electrical power is proportional to square of current, the detector's power varies quadrtically with incident input power, hence it is also called a 'square-law' detector.

Combining equations (3.7) and (3.6) yields the final equation for current:

$$i = i_0(e^{eV/k}b\ T - 1) - \eta eP, \tag{3.8}$$

The input–output gain of a detector system is measured by its responsivity, i.e. it is a measure of electrical output per unit of optical energy [8]. It is typically expressed in ampere units or volts per watt of incident radiant power. A unique responsivity is obtained for systems that respond linearly to their input [9]. Value of slope at a given point gives an estimation of responsivity for nonlinear optical cells. Mostly a linear response as a function of the incident power is obtained for optical detectors.

An optical detector's responsivity is entirely dependent on band gap of sensor material and incident wavelength of photons [10, 11]. The straightforward final expression elucidating the functional dependence of responsivity and wavelength of incident photons of a photo detector is

$$R = \frac{\eta e}{h\nu} = \eta \frac{\lambda(\mu m)}{1.23985(\mu m X \frac{W}{A})} \tag{3.9}$$

where, $\eta$ is the detector's quantum efficiency at a given wavelength (i.e., the conversion efficiency of photons to electrons) for a given wavelength, with symbols having usual meaning.

Increase in width of depletion region increases the probability of pair production and hence the efficiency of a photodiode. On the other hand, this would shoot up the response time for the photodetector since the carrier transit time would increase with the distance travelled and this in turn affects the frequency response.

Conversion efficiency of optical power into electrical power was quite small, typically, in the range 5%–10% for very early devices; however, efficiencies as high as 25% in some experimental photocells were reported in the late 1990s.

Though the term photodiode is extensively used, there are in fact a number of photodiode technologies that are in use which offer different properties and are hence used in different areas [12].

- *PN photodiode*: the earliest form of photodiode was the first form of photodiode developed. Its performance is not as advanced as some of the other types and therefore it is used less in sensitive devices.
- *PIN photodiode*: this is a type of photodetector in which the intrinsic layer between the P and N-type regions increases the distance between them. This allows efficient collection of light photons and width of the region is inversely proportional to their capacitance. Thus for small frequency signals, it obeys the characteristic diode equation, however, at higher frequencies a PIN diode behaves something like a perfect resistor. In the intrinsic or undoped region there is a deposit of stored charges. At low frequencies, the charges can be isolated and the diode is switched OFF. The PIN photodiode mostly used may have band widths as large as 100 GHz.
- *Avalanche photodiode*: applicability and effectiveness of APD is determined by numerous parameters. Quantum efficiency and total leakage current are two such factors. Number of primary charge carriers produced per photon absorbed is defined as quantum efficiency. Dark current, photocurrent and

noise add up to give total leakage current of the photodiode. The APD produces a much higher level of noise than a PN photodiode; series noise and parallel noise are components of this system. These are basically proportional to APD capacitance and fluctuations of the APD bulk and surface dark currents, respectively, for series and parallel noise.

- *Schottky photodiode*: a Schottky diode is formed by combination of an n-type doped semiconductor region and a metal such as gold. A metal–semiconductor junction is hence created rather than an actual p–n junction diode. The forward voltage drop is normally around 0.3 V because there is almost no depletion region as present in a usual p–n junction diode. Schottky diodes offer quite high current to the detector circuit for high frequency and fast switching application for numerous devices. As a result, this form of photodiode is regularly used in high bandwidth optical communication systems, e.g. fiber optic links.

*Modes of Operation*: photodiodes usually operate in *photoconductive* (PC) and *photovoltaic* (PV) modes.

- *Photovoltaic mode*: in PV, mode to eliminate the possibility of any dark current, the voltage across the photodiode is kept at zero volts. Thus the photovoltaic mode of operation requires zero bias voltage. Thus in the absence of any dark current, only thermal signals are recorded and detected. Even at low light intensity levels, this arrangement facilitates much better sensitivity but the response speed is compromised. Figure 3.13(a) illustrates an arrangement for PV mode of a photodiode. The circuit configuration consists of a diode working as a component to generate current, represented by a current source $i_p$ in the equivalent circuit. Voltage developed at the input of the amplifier is decided by the load resistor $R_b$ and the slope of load characteristic curve is proportional to the resistance, figure 3.13(b).

- *Photoconductive mode*: the photodiode is connected to a voltage source in a reverse bias to operate it in photoconductive (PC) operating (figure 3.14(a)). This causes broadening of depletion region and hence smaller junction capacitance $C_j$, series resistance too reduces. Subsequently, photocurrent

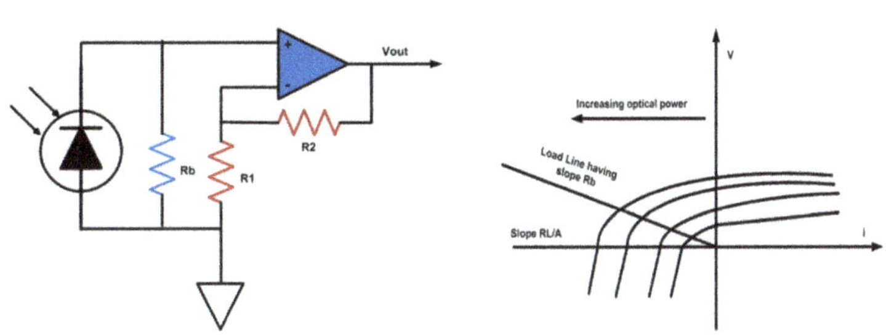


(a) (b)

**Figure 3.13.** (a) Photodiode connections in PV mode; (b) characteristic curve of photodiode and representative load line in PV mode.

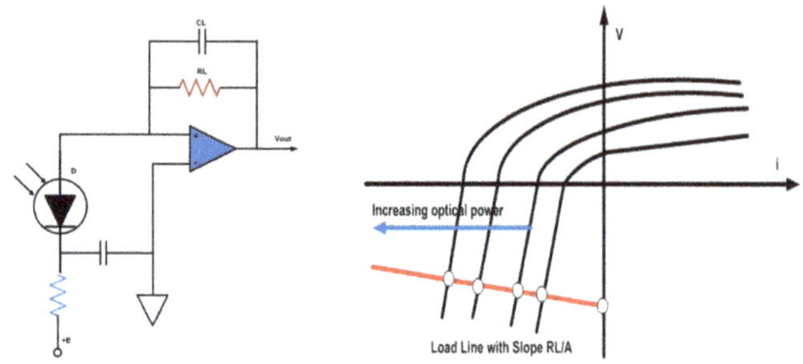


(a) (b)

**Figure 3.14.** (a) Photodiode connections in PC mode; (b) characteristic curve of photodiode and representative load line in PC mode.

exhibits a shorter rise time and a linear response is achieved over a wide range of intensities of light, figure 3.14(b). On the other hand, dark current and noise level in the circuit increases with increased reverse voltage. Hundreds of MHz bandwidths are offered under the PC mode, given that it is accompanied by increased signal-to-noise ratio. The circuit below illustrates modeled series detectors. As stated, a linear response to the incident input intensity of light is obtained as the detector is employed in reverse biased mode. Thus, the intensity of incident light and its wavelength determines the level of photocurrent generation. Output photocurrent across a load resistance is then measured and studied by an oscilloscope. All high frequency noise signals in input are filtered by an RC filter fitted in the input circuit to avoid any contribution due to it.

### 3.3.4 Phototransistor

Phototransistors are similar to photoresistors but generate both current and voltage, whereas photoresistors produce current alone due to change in internal resistance. A phototransistor is said to exhibit the combined operation of the photodiode as well as a normal transistor. It operates on the principle of photoconductivity effect. As it changes light signal incident on its surface into its electrical equivalent form.

Transistors in which the base terminal receives illumination such that the photons activate the base instead of current are phototransistors. Thus, a phototransistor is a bipolar semiconductor and converts the optical energy that is incident on it (figure 3.15).

*3.3.4.1 Construction of the phototransistor*

Construction of phototransistors is the same as for the usual transistor only with the difference that the base collector junction for phototransistors is moderately large in contrast to the regular transistor. Larger photocurrent generation is expected with this modification as this facilitates large area exposure to photons. Ion-implantation and diffusion techniques are used for formation of collector base regions. Germanium and silicon were the semiconducting materials used in former years of evolution of the

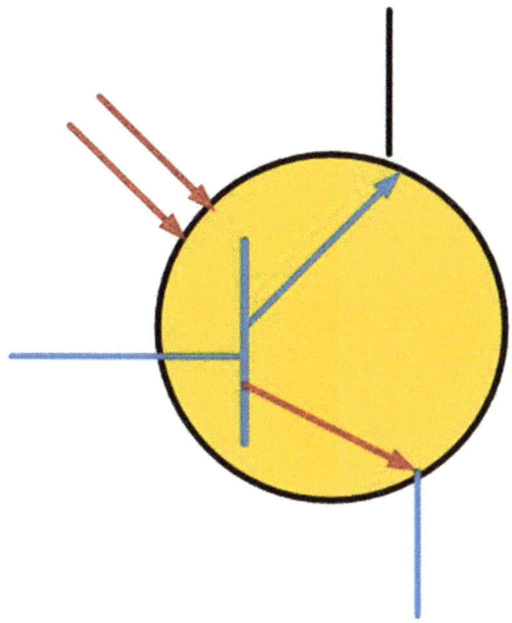

**Figure 3.15.** Circuit symbol of a phototransistor.

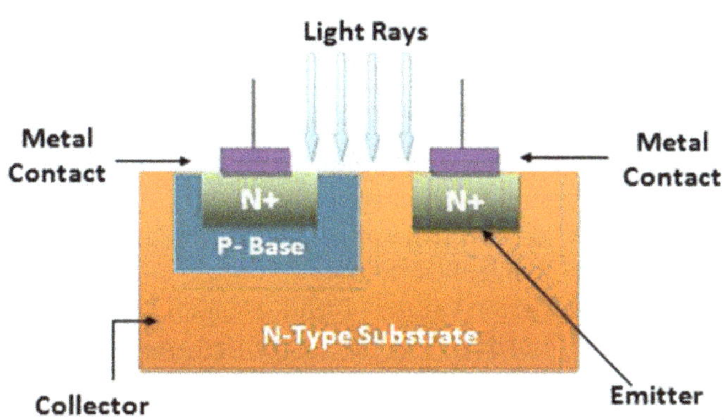


**Figure 3.16.** Homo junction structure of a phototransistor.

technology and the final structures mostly were a homogeneous material made from either of them (figure 3.16).

*3.3.4.2 Workings of the phototransistor*

Light radiation is allowed to be incident onto the base region and the output current is drawn from the emitter terminal. Thus the output photocurrent produced by the phototransistor is determined by the incident photon intensities upon the transistor. All three terminals or just two terminals (excluding base) can be connected to the circuit as per the circuit configuration, as shown in figure 3.17. The three regions of operation, namely the cut-off region, active region, and the saturation region are accessible from

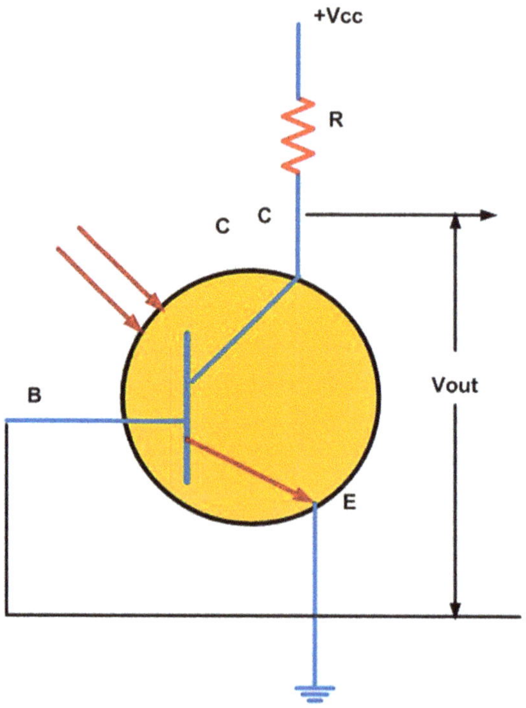


**Figure 3.17.** Phototransistor as a photodiode.

the characteristic curve of phototransistors. Operation of the transistor as a switch is achievable only in the cut-off region and saturation regions. Generation of current can be achieved in the active region of the curve only.

Current generated from phototransistor is determined by numerous factors other than luminous intensity as mentioned below:

1. *DC current gain:* superior values of photo generated current are attainable with the higher DC current gain of the transistor;
2. *Time constant:* efficiency of phototransistor is dependent on response time of the transistor;
3. *Luminous sensitivity:* ratio between the photocurrent and incident luminous flux determines the luminous sensitivity;
4. *Collector-base junction:* area of base collector junction is yet another crucial parameter, the larger the area of the collector-base junction the better the magnitude of photocurrent produced;
5. *Wavelength of the incident photons:* photocurrent generated is also governed by the incident wavelength of the light on the phototransistor, as the energy of photons is dependent on wavelength or frequency of the photons.

### *3.3.4.3 Output characteristics*

Figure 3.18 presents the characteristic output curve of a phototransistor which implicitly states all output features. Variation in output collector–emitter voltage is

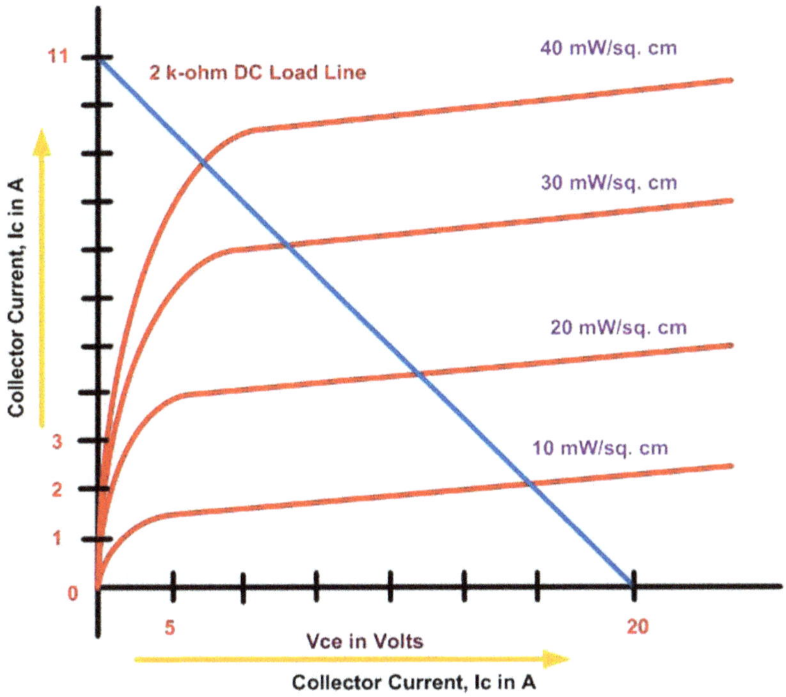


**Figure 3.18.** $V_{CE}$-$I_C$ characteristic curves of phototransistors.

recorded with change in collector current. Shape of the curve then presents all possibilities of its use in various devices as per requirement.

*3.3.4.4 Advantages of a phototransistor*

1. *Highly efficient:* as compared to photodiodes, phototransistors exhibit higher efficiency owing to better current gain. Thus, larger photocurrent is obtained from phototransistors for equal amount of light photons incident on similar semiconducting devices.
2. *Faster response:* shorter or faster response time of phototransistors is advantageous over photodiodes.
3. *Less noise interference:* the major shortfall due to noise interference in photodiodes is absent in phototransistors as they are unaffected by noise interference.
4. *Economical:* the phototransistor is cost effective and is hence economical to use in light-sensitive devices for varied applications.

*3.3.4.5 Shortcomings of phototransistors*

1. *Electromagnetic signal noise:* with the interference of electromagnetic field in the limits of operation, the phototransistor efficiency gets affected and consequently results in poor conversion efficiency.
2. *Frequency dependence performance:* at higher frequencies conversion of light into photocurrent minimizes due to greater junction capacitance.

3. *Electric spikes:* phototransistors are more prone to electric spikes as compared to other photodetectors.

*3.3.4.6 Applications of phototransistors*

Light detection or photo sensing is one of the most fundamental applications of phototransistors. Even the slightest amount of light of an appropriate wavelength is effectively detected owing to high efficiency. Another associated common application is in item counters in the manufacturing industries or sale counters etc. Level indicators and relays, encoder sensing and object detection, printers and optical remote controls are also some widely used applications. Moreover, they are a crucial component of all optoelectronics devices and are also used in optical fibers.

### 3.3.5 Infrared sensors

An IR sensor can sense its ambience using IR radiation usually emitted by the body itself to be detected. Heat or IR radiation are continuously radiated by objects at all temperatures, hence their position and motion can be detected and measured by IR sensors [13]. Thermal detectors are much less sensitive in the mid-IR range, whereas quantum detectors work efficiently in this region. The operating principle of IR sensors is carried out by a step-by-step conversion of thermal radiation into heat, heat level or heat flow into an electrical signal.

An IR temperature sensor in noncontact mode has:

1. A sensing component, which is sensitive and receptive to the IR wavelength range of the electromagnetic spectrum. Thus the component must have rapid strong response to thermal radiation, and a high-quality long-standing stability.
2. An arrangement to support and carry the sensing component and provision to illuminate it to the radiation and small value of thermal conductivity to reduce the heat loss.
3. A suitable IR LED or source of light of appropriate wavelength which can be made incident.

*3.3.5.1 Types of infrared sensors*

IR sensors can be categorized into active and passive sensors [14]:

*Thermal infrared sensors*—IR radiation emitted by objects, also called heat, is detected and photosensitivity is not dependent on the wavelength being detected. Thermal detectors can be operated at room temperature or the object temperature, however, they have sluggish response times and lower detection abilities. The most widely used thermal IR sensors are:

1. Golay cell.
2. Thermopile.
3. Pyroelectric detectors.
4. Bolometers.
5. Active FIR sensors.

*Quantum infrared sensors (cooled detectors)*—these offer rapid response detection speed and superior performance. Photosensitivity of such detectors is wavelength and temperature reliant. Quantum IR detectors need to be brought to some very low (near absolute zero) temperatures to arrive at precise assessments. The cooling is essential for the function of the semiconductor materials used. Depending on detector material and technology, the usual operating temperatures range from 4 K to just below room temperature. Thermal signatures of the handprint can be recorded by such cooled cameras only, as the uncooled cameras could record only partially. Evidently, even the smallest of temperature differences can be recorded for longer time as compared to uncooled cameras.

#### *3.3.5.2 Working principle of infrared sensors*

The laws of physics that govern the working of infrared sensors are [15]:

1. *Planck's radiation law:* every physical body spontaneously and continuously emits electromagnetic radiation which is specific to its temperature.
2. *Boltzmann's law:* describes the behavior of blackbody radiation across all wavelengths. It states that the power irradiated is directly proportional to the fourth power of temperature ($T^4$) of the blackbody thermal temperature ($T$).
3. *Wein's displacement law:* the wavelength for which a blackbody has maximum irradiation intensity is different at different temperatures and there is an inverse dependence between them, i.e. the peak position shifts towards a shorter wavelength with increase in temperature.

In a nutshell, thermal energies possessed by objects radiate as IR radiation and behave as an IR source above absolute zero. Blackbody radiators, tungsten lamps etc, are sources of IR radiation, however, in most applications IR lasers and LEDs of specific wavelengths are used as IR sources in IR sensors.

Similar to all optical sensors, an emitter circuit and a receiver circuit are an integral part of an IR sensor [16]. Thus this combination is recognized as a photo-coupler or an opto-coupler. An IR LED is the photo emitter and an IR photodiode is the detector in this arrangement. The IR photodiode must have an appreciable responsivity for all IR radiations be it from an LED or any object. Net resistance of the circuit and thus the resultant output voltage of a photodiode responds linearly to the IR radiation reflection from a surface [17].

IR sensors too work on either direct or indirect incidence. The source of IR, an LED, is positioned in line to an LDR with no hindrance in the path, or is placed alongside with a reflecting surface opposite to it. The radiation transmitted strikes the opaque surface and reflects back to the photodiode, as shown in figure 3.19.

#### *3.3.5.3 Applications of infrared technology*

##### ***3.3.5.3.1 Proximity sensors***

Intensity of the incident radiation on the photodiode would be dependent on the closeness or proximity of the object, which is then converted into voltage. A sensor arrangement which utilizes the reflective or indirect detection principle for estimation of

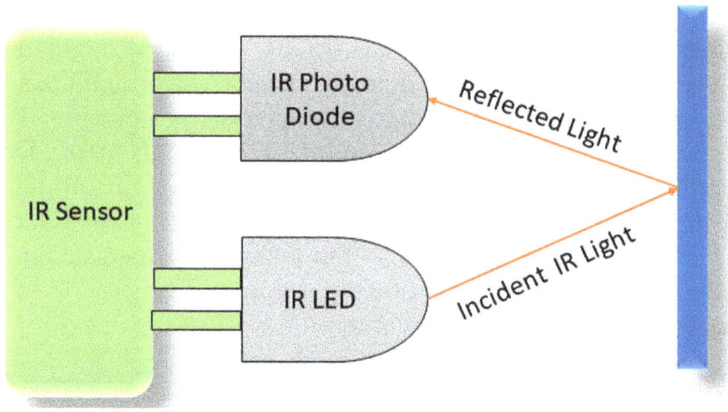


**Figure 3.19.** Schematic of a retro-reflective IR sensor.

position of an object is termed a proximity sensor [18]. The object is detected at the photodiode by the reflected signal. Proximity sensors are also used in touchscreen phones and assure that the display gets disabled for the duration of calls, when the person speaking is almost in touch with screen.

#### *3.3.5.3.2 Line follower robots*

A self-assisted robot which recognizes either a black line among white or vice versa and keeps following it needs an IR sensor component. The arrangement makes use of reflective as well as non-reflective indirect incidence. IR sensors detect its own light reflected by a surface and distinguish the color of the surface onto which it moves to transmit signals to an integrated microcontroller circuit which process the information and makes decisions as per a set of algorithms fed by the programmer.

#### *3.3.5.3.3 Item counter*

The principle of direct incidence is used to operate an item counter; as soon as an item hinders the path of radiation or obstructs the invisible signal of IR radiation, the circuit associated is triggered and the value of a stored parameter in a computer/microcontroller is increased by one unit. This is then displayed by LEDs, LCDs and seven segment displays. These counters are used for counting items on conveyor belts in manufacturing and packing industries.

#### *3.3.5.3.4 Burglar alarm*

The burglar alarm circuit requires a straight incidence of radiation on the photodiode. One wall of the door frame is fitted with an IR LED and a photodiode is placed on the opposite wall. Under normal circumstances IR radiation from the LED falls upon the photodiode directly but when a person or an object blocks the IR path, the alarm in the circuit turns on. This machinery is adopted broadly in security systems and is imitated on a minor scale for smaller objects too.

#### *3.3.5.3.5 IR music transmitter and receiver*

Audio musical notes can be produced by using an IR transmitter/receiver and generator that can be heard up to a distance of several meters. A 9 V battery is required for an IR music transmitter to work and the IR music receiver operates at the regulated power of 9 V–12 V.

#### *3.3.5.3.6 Night vision devices*

Insufficient light, especially at night may result in poor vision; hence night vision equipment is fitted with IR technology. This assists in conversion of ambient photons into electrons which undergo amplification under some chemical and electrical process, consequently resulting in visible light and thus enhancing the vision.

#### *3.3.5.3.7 Infrared astronomy*

IR astronomy is a field of astronomy with which astronomers are able to examine matter in the Universe that is impracticable to identify using detectors that utilize the visible range of the electromagnetic spectrum. By using telescopes and solid-state detectors, IR observatories use telescopes fitted with solid state detectors capable of detecting in the IR region. The Spitzer Space Telescope and Herschel Space Observatory are two of the many space stations working with this technology.

Apart from the above, many more devices in various application use IR sensors, be it for climatology or meteorology, photo biomodulation, gas detectors, water analysis, anesthesiology testing, petroleum exploration, rail safety etc. IR tracking, also known as IR homing, is a missile guidance system, often known as 'heat-seekers' because hot bodies are an ideal emitter of IR. IR reflectography reveals concealed layers in paintings and to decide on paintings' originality or reproduction or alteration status as a result of restoration work.

### 3.3.6 Fiber optic sensor

Invention of optical fibers has revolutionized not only communication technology but also sensing and detection. Physical quantities like mechanical strain, displacements, pressure, temperature, displacements, concentration of chemical species, are quite sensitively and accurately measured by optical fiber based sensors [19]. The arrangement shown in figure 3.21 consists of laser source, an optical fiber through which the laser travels to propagate any slight change to light within the fiber and then arrive at the detector which translates the change and displays it [20].

When the fiber itself acts as a sensor it is termed an intrinsic sensor, while extrinsic sensors use the fiber for transmission of laser light only. Furthermore, based on types of optical fibers, fiber optic sensors are mainly of two types (figure 3.20). Single fiber sensors and bundle fiber sensors. For instance, in an extrinsic sensor, a laser is sent through some of the fibers of a bundle while others collect the reflected light and transmit it back to detector.

Advantages of fiber optic sensors as compared to other types of sensors are [21]:

- This type of sensor is used in high voltage conditions, because they consist of electrically insulating materials;
- as there is no risk of electrical sparks these sensors are used in explosive environments;
- they do not disturb other devices because they are immune to electromagnetic interference;

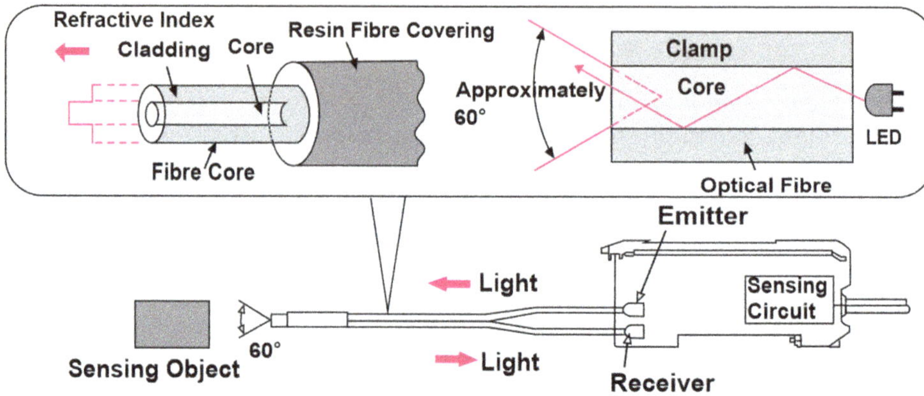


**Figure 3.20.** Construction and working of a typical fiber optics sensor.

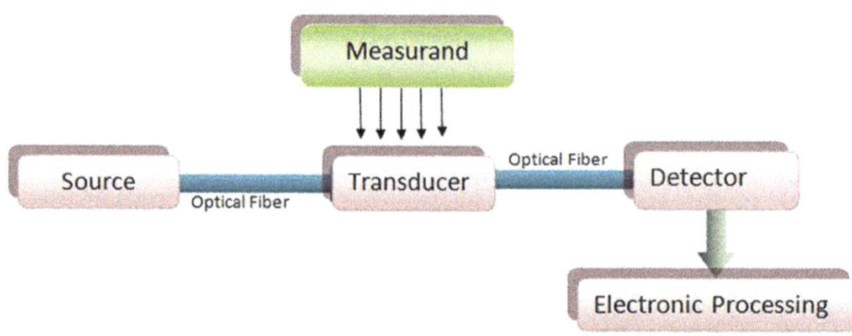


**Figure 3.21.** Working arrangement of an extrinsic fiber optic sensor.

- materials of optical fibers are chemically passive, i.e., do not infect their surroundings and are not subject to corrosion;
- the range of operating temperature in these optical sensors is very large;
- they have multiplexing capabilities: as discussed below.

The cores are divided into the following types.

*Plastic type*

Plastic optical fiber uses polymethylmethacrylate as core material through which light will transmit. Plastic fibers are light, cost-effective, and flexible, which is why they are the most common type of fiber sensor.

*Glass type*

The glass-fiber consists of a bundle of very thin strands to measure diameter of 0.51 mm upwards. A flexible stainless steel sheath or PVC is usually added to protect the bundle of cladded fibers. Theses fibers can withstand an operating temperature up to 450 °C.

*3.3.6.1 Types of optic sensor*

(a) Parallel: used for plastic fibers.

(b) Coaxial: coaxial optical sensors emit and receive the sensing light beam along a single narrow axis.

(c) Separate: these sensors have different areas for the transmitter and receiver; they contain several 10 μm glass fibers in diameter.

*3.3.6.2 Applications*

As stated in the beginning of the section, use of the optical fiber has given an edge to optical sensor technology and it is very efficiently employed for various device applications. It is used for testing of mechanical properties: to measure mechanical strain; acceleration; velocity; pressure; temperature; and displacement etc. It is used to evaluate post-seismic damage, analyze cracks, and monitor restoration and displacement, and commonly in dams to detect and monitor leakage, foundations defects and measure spatial displacement [22–30].

## References

[1] Sabri N, Naseer S A, Aljunid M S, Salim R B, Ahmad and Kamaruddin R 2013 Toward optical sensors: review and applications *J. Phys. Conf. Ser. Int. Conf. Sci. Eng. Math. Chem. Phys. (ScieTech 2013)(24–25 January 2013) (Jakarta, Indonesia) vol 423*

[2] Elham S and Mohammad H 2109 Recent advances on the biosensing and bioimaging based on polymer dots as advanced nanomaterial: analytical approaches *TrAC* **118** 840–52

[3] Suhartomo A, Simatupang J W, Widjaja B and Gapsari F 2019 Feasibility study on structural health monitoring systems using fiber-optic sensors (FOS) technology for transportation infrastructures in Indonesia *IOP Conf. Ser. Mater. Sci. Eng.*

[4] Cowen A R 2018 *Textbook of Catheter-Based Cardiovascular Interventions* (Berlin: Springer)

[5] Usman F, Dennis J O, Meriaudeau F and Ayodele O B 2018 Modelling and simulation of surface plasmon resonance breathe acetone sensor based on doped polyaniline–graphene composite *Int. Conf. Fundamental Appl. Sci. A Conf. World Eng. Sci. Technol. Cong. Kuala Lumpur Malaysia J. Phys. Conf. Ser.* 1123 012 020

[6] Huizhi K, Lin W, Meghan O and Cao Y C 2008 Nanoparticles for biosensors *Optical Biosensors Today and Tomorrow* ed F S Ligler and C Rowe Taitt 2nd edn (Amsterdam: Elsevier) pp 583–621

[7] Jacob F 2016 Light detectors *Handbook of Modern Sensors* 3rd edn (New York: Springer)

[8] Zeng Y, Ye Z, Lu B, Dai W and Pan X 2016 Enhanced photoelectric performance in self-powered UV detectors based on ZnO nanowires with plasmonic Au nanoparticles scattered electrolyte *Appl. Phys.* A **122** 4

[9] Sciammarella C A and Sciammarella F M 2012 Optical methods—computer vision *Experimental Mechanics of Solids* (Hoboken, NJ: Wiley) ch 9

[10] Stotts L B 2017 *Free Space Optical Systems Engineering: Design and Analysis Optical Receivers* (New York: Wiley) p 528

[11] Dandan P, Qingmei S, Guiqing Z, Ming W and Hongdong X 2016 Impact localization based on FBG sensing system with broadband light source *IEEE Info. Technol. Netw. Electron Auto. Contr. Conf. Chongqing* pp 986–9

[12] Malik A K 2013 Photodiode *Himalayan Phys.* **4** 95–8

[13] Singh O and Sisodia T S 2017 Solar LED street light system with automatic scheme *Int. Conf. Energ. Commun. Data Anal. Soft Comput. (ICECDS)* pp 3415–9

[14] Dasthagiraiah N V, Subbareddy N, Sonali S K, Thasleem P and Muntaz V 2018 Smart toll plaza system using IoT *Int. J. Adv. Sci. Technol. Eng. Manag. Sci.* **4** 3

[15] Patade A S, Gandhi H P and Sharma N 2019 IOT solutions for hospitals. *11th Int. Conf. Commun. Syst. Netw. (COMSNETS) (Bengaluru, India)* pp 813–6

[16] Chaulya S K and Prasad G M 2016 *Mine Transport Surveillance and Production Management System* (Amsterdam: Elsevier)

[17] Shinde S H, Munot M V, Kumar P and Boob S 2019 Intelligent companion for blind: smart stick *Int. J. Innov. Technol. Explor. Eng.* **8** 2278–3075

[18] Lahari P S, Mohammed M F, Lingaraju K and Amulya K 2018 Density based traffic control with emergency override *2018 3rd IEEE Int. Conf. Recent Trends Electron. Informat. Commun. Technol. (RTEICT)*

[19] Ogundare J O 2015 Deformation monitoring and analysis: geotechnical and structural techniques *The Precision Surveying: Principles and Geomatics Practice* (Hoboken, NJ: Wiley) 329–76 pp

[20] Firoozbakht M, Vosoughifar H and Ghoran A G 2019 Coverage intensity of optimal sensors for common, isolated, and integrated steel structures using novel approach of FEM-MAC-TTFD *Int. J. Distrib. Sens. Netw.* **15** 8

[21] Laborde C C 2019 *Application of Optical Fiber for Sensing* (Rijeka: InTech)

[22] Normandin F, Sawan M and Faubert J 2005 A new integrated front-end for a noninvasive brain imaging system based on near-infrared spectroreflectometry *IEEE Trans. Circ. Syst. I: Regular Papers 52* **12** 2663–71

[23] Potdar A S, Shinde S, Nikam P H and Kurumkar M 2017 Wireless sensor network for real time monitoring and controlling of railway accidents *Int. Conf. Trends Electron. Info. (ICEI) (Tirunelveli)* pp 190–7

[24] Gophane K C and Bhaskar P C 2018 FPGA based adaptive IoT framework for distinct applications *Fourth Int. Conf. Comput. Commun. Contr. Auto. (ICCUBEA)* pp 1–6

[25] Coca O and Fodor A 2019 Energy harvesting wireless sensor nodes with augmented reality application support *IEEE 25th Int. Symp. Design Technol. Electron. Packag. (SIITME)* pp 202–7

[26] Baghayalaksmi R and Valarmathi K 2019 A survey on sensor based detection technologies for wireless applications *Third Int. Conf. I-SMAC (IoT in Social Mobile Analytics and Cloud) (I-SMAC)* pp 129–33

[27] Ramesh S 2014 Rail parameters monitoring for the fire safety system in the compartments using automation technology global *J. Res. Eng. F Electr. Electron. Eng.* **14** 3

[28] Dorman M G, Prisbe M A and Meindl J D 1985 A monolithic signal processor for a neurophysiological telemetry system *IEEE J. Solid-State Circ.* **20** 1185–93

[29] Pancrazio J J *et al* 1998 Description and demonstration of a CMOS amplifier-based-system with measurement and stimulation capability for bioelectrical signal transduction *Biosens. Bioelectron.* **13** 971–9

[30] Mohseni P and Najafi K 2002 A low power fully integrated bandpass operational amplifier for biomedical neural recording applications *Proc. IEEE EMBS/BMES Conf* pp 2111–2

**IOP** Publishing

Advances in Modern Sensors
Physics, design, simulation and applications
**G R Sinha**

# Chapter 4

# Recent applications of chalcogenide glasses (ChGs) based sensors

**Neeraj Mehta**

Chalcogenide glasses (ChGs) belong to a matchless category of glassy alloys, especially when mid-infrared (IR) applications are concerned. Although this class of glasses is still rising from its infancy as compared to oxide glasses, noticeable attention has been paid to ChGs as a result of their unique physical/chemical properties and promising applications. In recent years, ChGs have drawn the attention of various research groups over the globe due to their fascinating, novel and vastly significant sensing applications in different areas like physics, chemistry, biology, environmental science, space science, medical science etc.

The optical fibers, solid-state thin films and lenses of ChGs along with implementing fiber evanescent wave spectroscopic techniques have been found useful for developing different kinds of sensors. These sensors ensure simple, cost-effective sensing in various areas of daily life like monitoring of pollutants in groundwater, chemical sensing of ions in solutions, biological sensing of live cells for early cancer diagnostics, and thermal-imaging cameras for night-vision in cars and for security purposes in the military. The present chapter covers a brief introduction to ChGs based sensors along with their fundamental aspects. An attempt has been made to cover the entire novel and significant applications of ChGs based sensors. Finally, the future scope and the on-going research activities have been described in the conclusion section.

## 4.1 ChGs based sensors: a brief introduction

ChGs mainly consist of the chalcogen elements Te, Se and S with the inclusion of other elements such as Ge, As, Sb, Ga, etc, as foreign atoms or chemical modifiers [1]. They are usually translucent from the visible range to the infrared range of the electromagnetic spectrum. They can be doped by native metal elements and several active and passive applications of optical devices have been proposed for

doi:10.1088/978-0-7503-2707-7ch4

rare-earth-doped ChGs [1–6]. Further, ChGs are responsive to the exposure of electromagnetic energy and particle irradiation. Consequently, they show photo-darkening, photo-bleaching, photo-crystallization, photo-plastic effects, etc [1], and most recently, the photo-dielectric effect [7].

In recent years, ChGs have drawn huge attention from glass technologists and scientists over the entire globe due to their exceptional, novel and significant applications [8, 9]. ChGs show matchless light-induced structural changes or a variety of photo-induced effects when normal or laser light of suitable wavelength is used for exposing their samples. They possess resistance-type switching behavior between the crystalline phase and the non-crystalline phase (i.e., glass or amorphous phase) [10–12]. A huge alteration in the index of refraction or absorbance followed by the structural modifications is generally observed in these glasses. Further, they can exist in both the crystalline phase and the non-crystalline phase and remain stable for a long period under the working ecological conditions.

These glasses are the potential members of the family of multi-functional materials. They behave like electronic materials (e.g., resistive switching) as well as displaying a variety of photo-induced properties (like optical materials). ChGs are also well-known as phase-change materials [2–6]. They have excellent stability against nuclear radiation (suitable for nuclear waste applications) [13]. ChGs based optical instruments, elements or devices (e.g., IR optical fibers, IR lenses, membranes, thin-films, etc) are prospective materials for diverse types of sensing applications. Table 4.1 shows the various kinds of sensing applications that can be performed by using suitable ChGs.

The IR optical fibers of ChGs have been found more useful as compared to the IR waveguides of other materials because ChGs based IR optical fibers offer the benefit of flexibility in shaping the geometry of the fiber-based sensors [14–18]. This becomes possible due to the favorable temperature-dependent characteristics of viscosity that permits a huge alteration in the fiber geometry in the temperature region close to the drawing temperature [19–21]. IR optical fibers of ChGs are ideal for remote chemical sensing [22–24]. Sulfur or tellurium containing ChGs are used as IR waveguide or fiber. Halogen doped glasses (such as Ge–S–I and Ge–Se–I) are also possible materials in IR fiber applications. Selected halogen-doped ChGs display high optical nonlinearity and so they are found suitable for all kind of optical switching devices.

Solid-state chemical sensors of ChGs can be synthesized in the form of membranes or thin-films by employing the pulsed laser deposition (PLD) technique. This technique preserves the composition of the glassy alloys even in the samples of multi-component ChGs. Thus, it is now feasible to prepare 2D layered based sensors of ChGs by using desirable composition. This was not possible in the case of earlier known techniques (e.g., thermal evaporation, chemical vapor deposition, irradiation-induced deposition, etc).

**Table 4.1.** List of some potential sensing applications of ChGs.

| Nature of sensing application | Examples |
|---|---|
| Industrial | • Analytical repair (hot spot detection)<br>• Search of heat losses in constructions<br>• Night vision arrangement in vehicle<br>• Touchless pyrometry<br>• Development and manufacturing management |
| Safety and security | • Night vision systems for surveillance assignments<br>• Targeting systems<br>• IR sensors |
| Medical and biotechnological | • Recognition of tenderness<br>• Early diagnosis of tumour<br>• Detection of fever without touching |
| Consumer optics | • Night vision systems<br>• Thermal cameras<br>• Motion control systems |

## 4.2 Fabrication and molding of ChGs in the form of different devices for sensing applications

The importance of ChGs based optical devices is engendered because of their excellent characteristics like small optical losses, extensive transparency in the mid-infrared region, satisfactory mechanical behavior and excellent physicochemical properties. In this section, we will cover the details concerning the fabrication and molding of ChGs in the form of three main optical elements for their potential sensing applications.

### 4.2.1 Infrared optical fibers

The thermal drawing tower method is used to pull the single-index fibers of ChGs from a core-only perform. The double-crucible [19], rod-in-tube [25] or extrusion [26] methods are generally employed for the fabrication of the double-index fibers or preforms of ChGs. The description of some widely used methods is discussed in the subsections.

*4.2.1.1 Thermal drawing tower method*

The facility of the suitable temperature dependence of viscosity in ChGs makes it possible to obtain them in the form of fibers by using the thermal drawing tower method. The general description of the apparatus used in the thermal drawing tower process is shown in figure 4.1. The main component of this apparatus is a drawing tower which consists of a drawing furnace, a diameter controller, and a rotary drum. In this method, the single-index fibers of ChGs are pulled from a core-only preform having desirable geometry in terms of length and diameter. Firstly, the casting of the preform is done inside an ampoule [quartz tube] in the presence of a vacuum. In the next step, it is heated in a well-organized atmosphere of a drawing tower. In the drawing furnace, the softening of the subjacent component of the preform is governed by the thermal gradient of a furnace and the confined heating of glass above its glass-transition temperature. Thus, the drops from its lower part fall in the downward direction and are collected on a rotating drum for pulling into a fiber.

*4.2.1.2 Rod-in-tube method*

The rod-in-tube technique has two major steps in the production of ChG preforms [27]. Generally, melt casting is employed for the rod (core glass) fabrication while rotational casting is used for the tube (cladding glass) preparation as shown in figure 4.2. In the melt casting approach employed for rod fabrication, the basic materials are batched into a sealed vessel under an inert atmosphere by creating a vacuum. This vessel can bear well the vapor pressure of the raw materials. In the rotational casting approach employed for the manufacturing of the tube, the raw material (i.e., glass) is heated inside a cylindrical chamber of a quartz tube above its

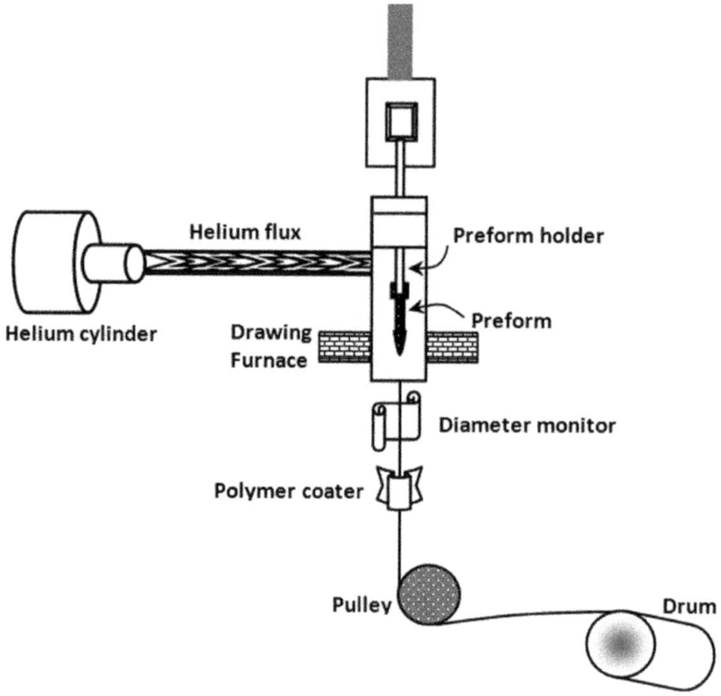


**Figure 4.1.** Synthesis of single index ChG fiber using the thermal drawing tower method.

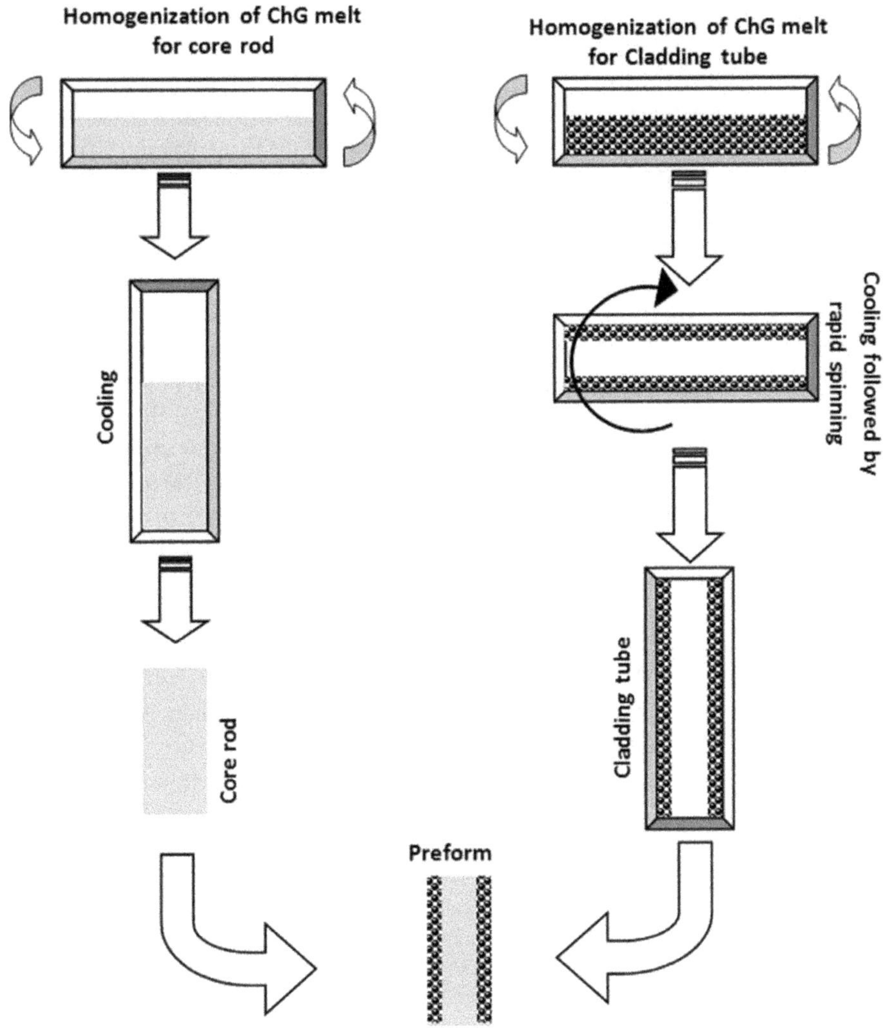


**Figure 4.2.** Synthesis of double index ChG fiber using the rod-in-tube method.

melting point. The glass melt is mounted horizontally in the chamber. The mounted chamber has rotational motion having an elevated rotary speed so that the glass melt is cooled below its glass transition temperature. The shape of the tube is achieved under the influence of the centrifugal force appearing due to the rotational (or spinning) motion. In the final step, the rod (core glass) is placed inside the tube (cladding glass) for the formation of the preform.

*4.2.1.3* In situ *core/cladding casting method*

Le Coq and his research group built up an innovative procedure for the synthesis of double-index ChGs fibers [28]. After the preparation of each glass individually, the core and clad materials are inserted in a silica setup. This arrangement is shown in figure 4.3. The fabrication of the preform is accomplished by a sequence of two independent vacuum distillations, followed by the *in situ* casting of preform through sequential rotation of the setup. The steps involved in the method demonstrated by Le Coq *et al* for the fabrication of preform of TAS (TeAsSe) ChG are shown in figure 4.3. The samples of TAS ChGS are widely used for the manufacturing of the double-index fibers or preforms because of their optimized conditions for fiber drawing. In the initial step, two samples of TAS glasses having suitable but different compositions are synthesized. These samples play the roles of the core and cladding materials. The preform manufacturing of both samples is achieved by employing a unique setup made of silica, as shown in figure 4.3(a). In this setup, the core/clad materials are introduced to acquire the preforms by a progression of two disengaged vacuum distillations pursued by the *in situ* casting of the preform through the sequential circulatory motion of the assembly (see figure 4.3(b) and (c)). The dotted lines indicate the sealing for closing the system (see figure 4.3(a)). The melting of core/clad materials is completed by inserting the sealed ampoules in a furnace with the proper precautions to avoid the amalgamation of both melts (see figure 4.3(d)). The next two steps involve the casting of core/clad materials (see figure 4.3(e) and 4.3(f)).

### 4.2.2 Infrared optical lenses

ChGs based lenses are an outstanding substitute in the specific appropriations for the more costly and product price-driven materials such as Ge, ZnSe and ZnS. Further, they can be utilized in permutation with multiple-element lens assemblies

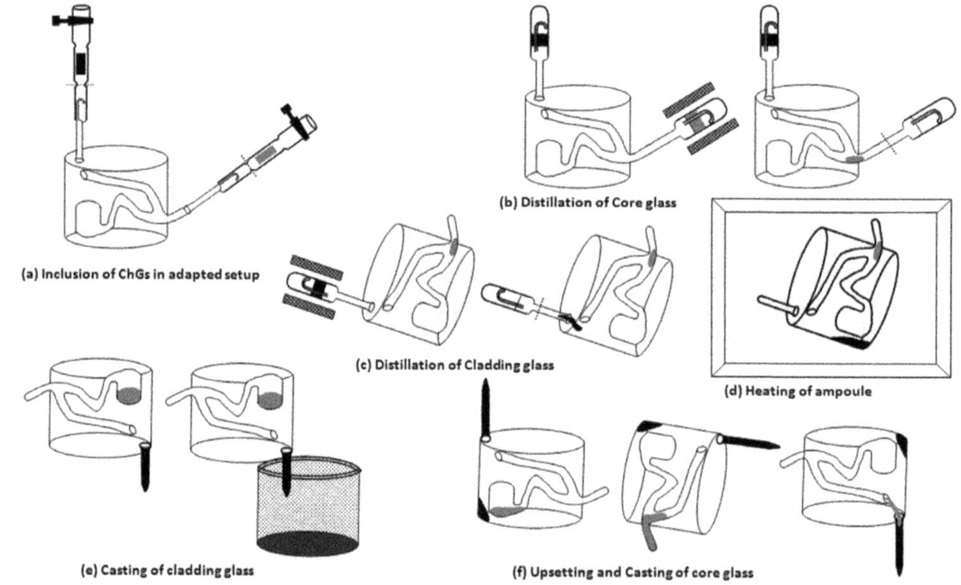


**Figure 4.3.** Synthesis of double index ChG fiber using *in situ* core-cladding casting method (recreated from [28], copyright (2003), with permission from Elsevier).

with customary IR materials. Their optical applications include IR optical lenses and IR optical fibers. ChGs demonstrate good propagation constant, low thermal variation in the index of refraction $\mu$ (i.e., $d\mu/dT$), good dispersion and wide bandwidth. ChGs are the perfect optical material for accomplishing chromatic aberration correction and reducing thermal defocusing of IR optical lenses having the wavelength domain of 2–12 μm.

Hilton *et al* [29] invented a scheme for the fabrication of the preformed lens of ChG by enclosing the sample within the mold. The lens has a top surface and a bottom surface. An appropriate amount of a chosen ChG is set down within the mold and on the apex surface of the preformed lens. In the subsequent step, the softening and melting of ChG sample on the peak location of the preformed lens is achieved by heating the mold in such a manner so that it is tied to the apex face of the preformed lens. A lens surface is formed in the melted ChG sample glass to achieve the shape of a molded lens which is tied to the top surface of the preformed lens. The removal of the molded lens and preformed lens assembly from the mold is the final step.

Figure 4.4 shows a comprehensive representation of the press molding process. The pressing machinery has two molds that have been manufactured with the preferred surfaces (aspheric, spherical, diffractive, plano, etc). The molds are positioned into a sleeve with a ChG glass preform. Molding of the sample preform into the chosen profile is completed by heating and applying pressure to the system in a controlled way so that one can achieve the preferred shaping. The mechanical blueprint and thermal investigation of the molds, sleeves, and preforms play a significant role in achieving the perfect molding of a sample in the form of a lens of high quality.

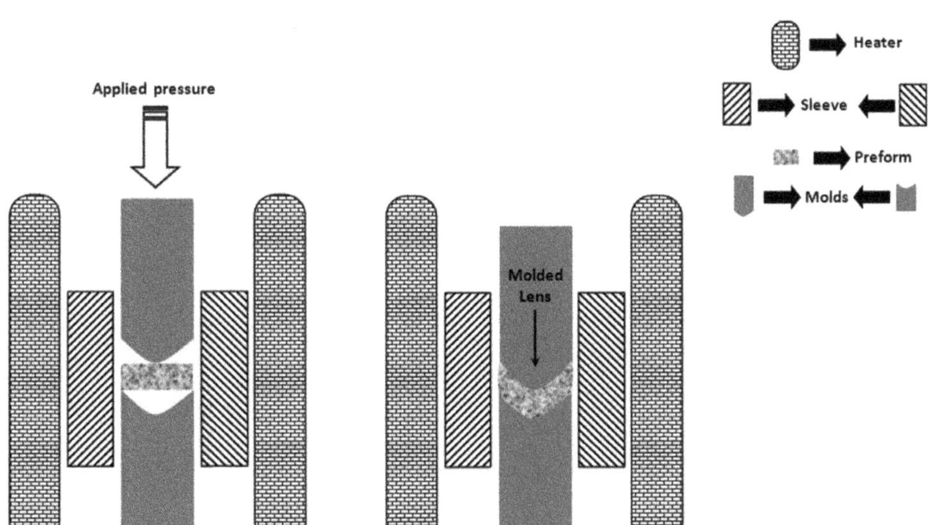


**Figure 4.4.** Description of the press molding process for fabrication of the preformed lens of ChG.

### 4.2.3 Thin film membranes

In the beginning, PLD technique was invented for the deposition of thin films of magnetic and superconducting materials. This technique consists of a highly energetic non-equilibrium process. In this process, a focused laser beam and the sample material play the role of an exterior energy source and the target, respectively. The thin films of the sample are, therefore, deposited due to the interaction between the source and target [30]. The self-governing is another interesting feature of this technique that allows the absorption of the laser irradiation by the plasma to regulate further interaction with the target [31]. The standard procedure of the PLD technique is shown in figure 4.5.

## 4.3 Description of some principals behind the sensing applications

### 4.3.1 Attenuated total internal reflection

The principle of attenuated total internal reflection can be understood by considering the propagation of a light beam of wavelength λ inside of an optical fiber having an index of refraction $\mu_1$ and its reflection from an interface of the fiber and an adjacent sample of refractive index $\mu_2$ [32, 33]. This causes the appearance of the completely reflected beam inside the optical fiber and the evanescent wave in the sample, as shown in figure 4.6. The exponential decay of the strength '$E$' of the electric field having dependence on the distance '$x$' from the fiber exterior is also shown in figure 4.6.

Considering the concept of a skin-depth $S_d$, we can express the exponential decay of the strength $E$ as follows:

$$E(x) = E_0 \exp\left(-\frac{x}{S_d}\right) \tag{4.1}$$

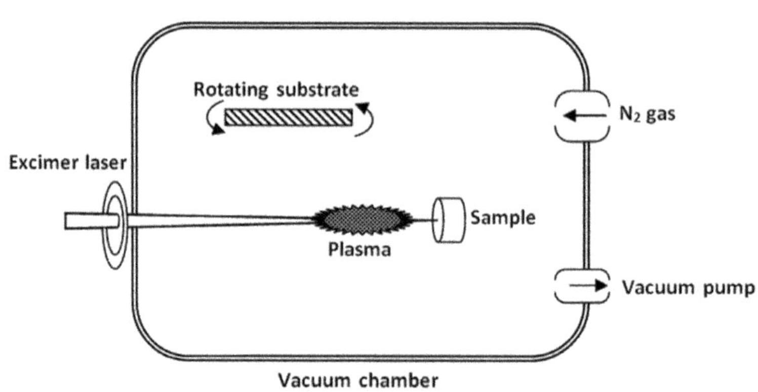


**Figure 4.5.** Synthesis of thin-film membranes of ChGs using PLD technique.

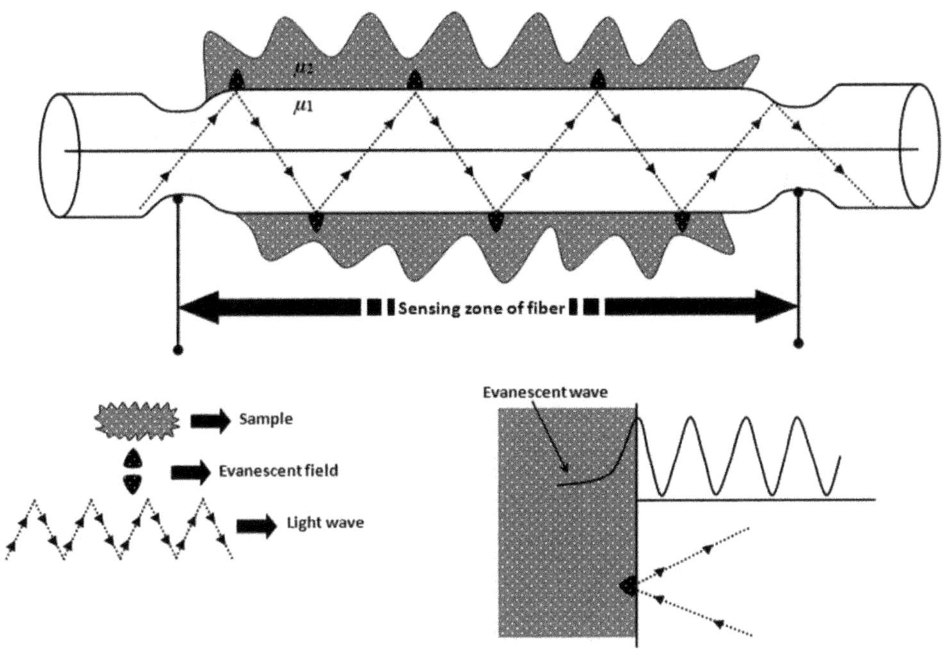


**Figure 4.6.** Illustration of the principle of ATIR used in optical fibers.

The (4.1) following relation can be used to determine the parameter:

$$S_d = \frac{(\lambda/2\pi)}{\sqrt{\mu_1^2 \sin^2 i - \mu_2^2}} \tag{4.2}$$

The (4.2) skin-depth is shown in figure 4.6. When we consider the absorption of the evanescent wave by the specimen at a specific wavelength (say $\lambda_p$), the coefficient of reflectivity ($R$) is less than unity [i.e., $R(\lambda_p) < 1$]. This event is known as attenuated total internal reflection.

### 4.3.2 Fiber evanescent wave spectroscopy

The general evanescent wave is a concept of electro-magnetics and it represents an oscillating electric and/or magnetic field that does not transmit as electromagnetic radiation but whose energy is geographically focused in the locality of the supply having charges and currents of oscillating tendency. The combination of an evanescent wave and an optical fiber is found useful for different kinds of chemical sensing applications [34, 35] in the IR spectroscopic technique based on the principle of attenuated total internal reflection [32, 33]. This technique is known as fiber-optic, evanescent-wave spectroscopy (FEWS).

Keeping in mind the aforesaid general definition of the evanescent wave, we can now define the fiber evanescent wave. In an optical fiber, a light wave transmits in the core as well as in the cladding. The transmission of the light rays in cladding is known as the evanescent wave, which is decayed quickly with an increase in distance from the fiber-axis. In the absence of cladding, the penetration of the evanescent wave in the outside region of the fiber is as shown in figure 4.7. If there is a specimen outside the fiber, which absorbs the evanescent wave, the power of the transmitting light is decreased. In this way, the sensing of the sample can be possible by tracing the corresponding absorption peak in FTIR spectrometer.

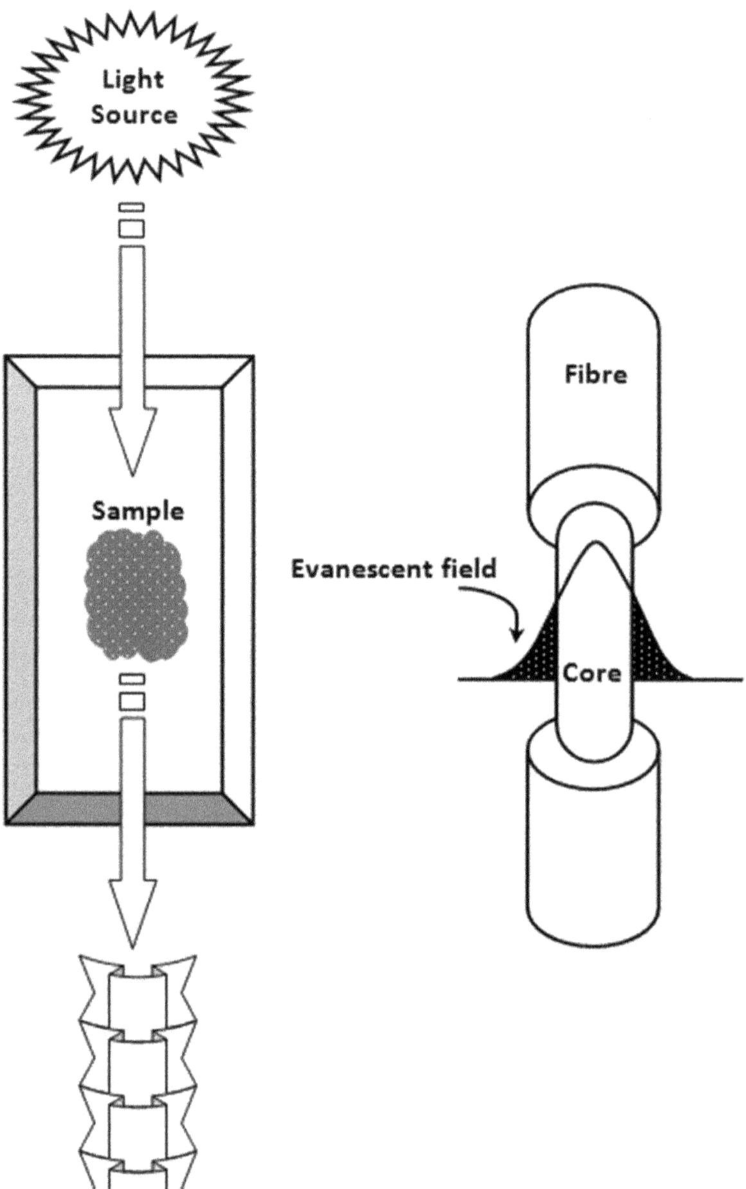


**Figure 4.7.** Diagram showing the working principle of FEWS technique.

For sensing, firstly a cladding from a segment of a suitable IR optical fiber is converted in the form of a small strip for the leaking of the evanescent-wave. This segment of the IR transmitting fiber plays the role of a sensing element and it is placed in a small vessel, inside an IR spectrometer (see figure 4.7). The sample that is to be identified is also placed in the same vessel and it causes the ATIR of the light wave. Consequently, the chemical contents of the sample can be detected through the corresponding absorption peaks [36].

### 4.3.3 Thermal imaging

Thermal imaging is a technique of improving the visibility of things in an atmosphere or local region which we are incapable of seeing or visualizing because of darkness. This method utilizes the IR radiation released from the objects for their detection by generating a thermal image based on that information. Thermal IR imaging is also known as non-visible imaging since the IR spectrum is invisible to the human eye.

The thermal imaging is realized by an apparatus that is commonly known as a thermo-graphic instrument. It usually consists of a thermal pointer that examines the temperature of a definite dot. In some cases, a thermal scanner is employed for the scanning of the huge exterior region of an object. The output is the thermal profile having an optical and thermal appearance in the form of an electrical signal of small time-interval [37]. The block diagram of a general thermo-graphic instrument is shown in figure 4.8. The signals coming from the thermal scanner are further fed into a screen for the processing and to access it in the form of predictable digital display or a video display [37].

## 4.4 Some exclusive examples of sensing applications of ChGs based sensors

### 4.4.1 Application in bio-sensing and food security

One of the applications of Te and/or Se rich ChGs based sensors is related to the IR fingerprinting of biological species (e.g., bio-molecules, bio-tissues, and

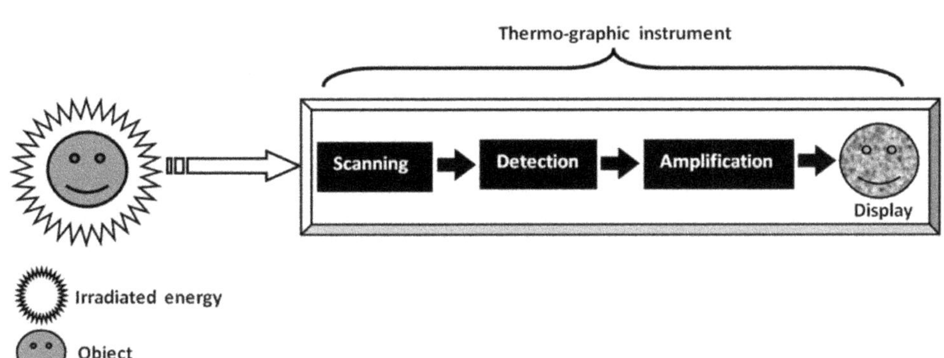


**Figure 4.8.** Block diagram of a thermo-graphic instrument based on the principle of thermal imaging.

micro-organisms) [38]. Such IR fingerprinting is very useful for collecting information about the structural modifications of biological species and the differentiation between specific bio-samples. These fibers allow the transmission of the optical signals expertly because their transparency range is high and it covers the sensing of bio-molecules having fingerprint region between 6 and 15 μm. Further, IR fibers based on ChGs have the potential capability to deal with the biochemical sensing of adjustments that are connected to the decisive stages in progressive biological science.

Keirsse *et al* [38] successfully demonstrate an application of ChGs fibers for bio-sensing by using mouse liver tissues as test materials. For the experiment, the test samples were the livers of famished and nourished rats. They examined the whole liver tissues with the ChG fiber under the almost authentic clinical circumstances for understanding the metabolic adjustments connected to starvation [38].

This group performed similar experiments to show that the metabolic amendment in test animals can be distinguished by analyzing their serum or some other biological fluids [39]. For this, they used the serum samples by extracting the blood from a normal mouse and an overweight mouse. By using principle component analysis and recording the FEWS spectra, this group was able to show the two separate and distinct zones for differentiating between both mice (i.e., the normal one and the overweight one).

This group conducted another experiment to monitor the bio-sensing of a bacterial bio-film consisting of protein molecules by employing an optical fiber of a TAS system (see figure 4.9). The aim of this experiment was to observe the possible interaction of bio-film with the ChGs [40]. They used a sensor keeping in mind two main tasks: this first was real-time sensing and the second was space-time *in situ* scrutinizing of the dispersal of bacterial biofilm. To accomplish this objective, they used a remote analysis approach by executing the ChG optical fiber directly in the cell-culture dish, as illustrated in figure 4.9. The gelose on this dish plays the role of the solid substrate. When the tapered sensing zone of the TAS fiber touches the gelose outside the edge of the bacterial biofilm, the bacteria come into contact with its sensing zone and the two major bands (i.e., amide I and amide II bands) of the protein IR spectrum are increased as a result of the increase in the bio-mass due to mass migration.

### 4.4.2 Early cancer diagnostics

A research group of the University of Nottingham investigated the biocompatibility of optical fibers made by using Sb–Se–Ge containing ChG for the early diagnosis of cancer [41–44]. The main focus of their research was on evolving a skin cancer recognition arrangement by employing mid-infrared (MIR) spectroscopy with a clear intention of carrying the machinery to the stratum of exact *in vivo* diagnostics. The functioning of the MIR probe can be made more responsive and cautious, and highly precise results can be achieved if the utmost part of the light passes through the cancer-affected tissues of the patient. To achieve this condition, the super-continuum generation (SCG) plays a vital and significant role. Selenium and sulfur

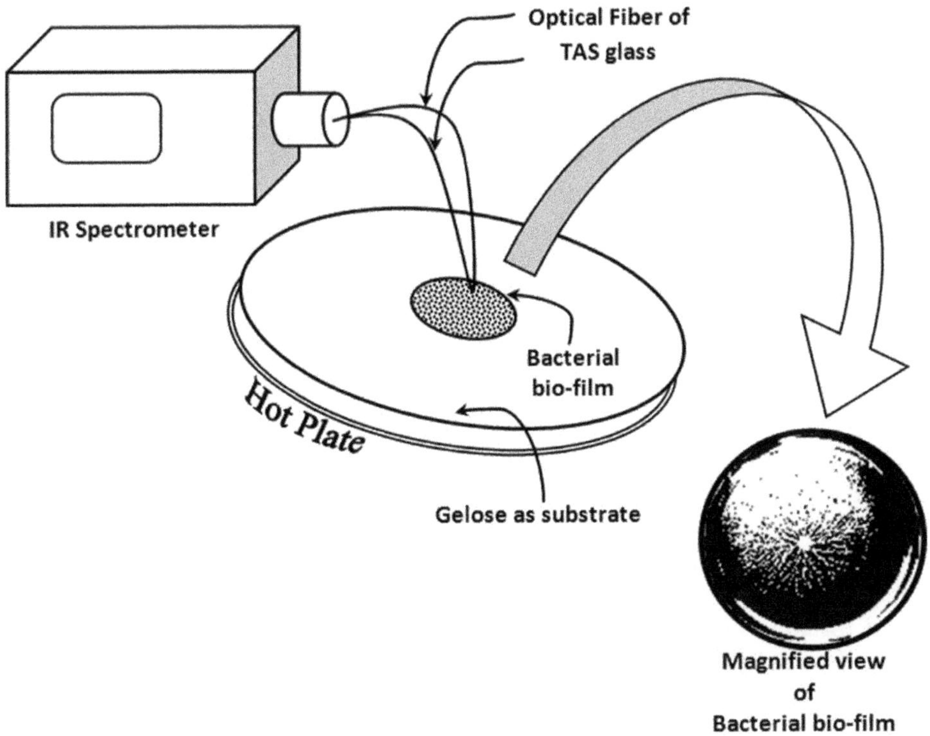


**Figure 4.9.** Bo-sensing of a bacterial biofilm by with the help of the optical fiber of a TAS system.

rich ChGs demonstrate excellent SCG in their corresponding optical waveguides and fibers [45] as well as being suitable materials for utilization in fiber optics for various biomedical applications [46].

Seddon and her group developed highly pure MIR optical fibers of Se–Ge–Sb system having low losses as an effective solution [43, 44]. They observed that these new optical fibers have an extremely large value of Kerr coefficients, low optical losses and excellent optical non-linearity (> $10^3$ times as compared to silica glasses). The setup for what was developed by this group for analysis of skin cancer is shown in figure 4.10. The main components of this setup were ChG optical fiber and Re-doped ChG fiber laser during the detection of cancer. These components make possible the broadband MIR molecular sensing by pumping SCG. These researchers are still working on improving the technology for offering superior and more prospective ways of fast diagnosis of skin cancer at the local general practice clinic rather than in a far situated specialized hospital, so that the patient feels comfortable.

### 4.4.3 Monitoring of pollutants in groundwater

Groundwater is one of our key sources of water. Regrettably, groundwater is hazardous and unfit for human health if it is susceptible to pollutants. The main

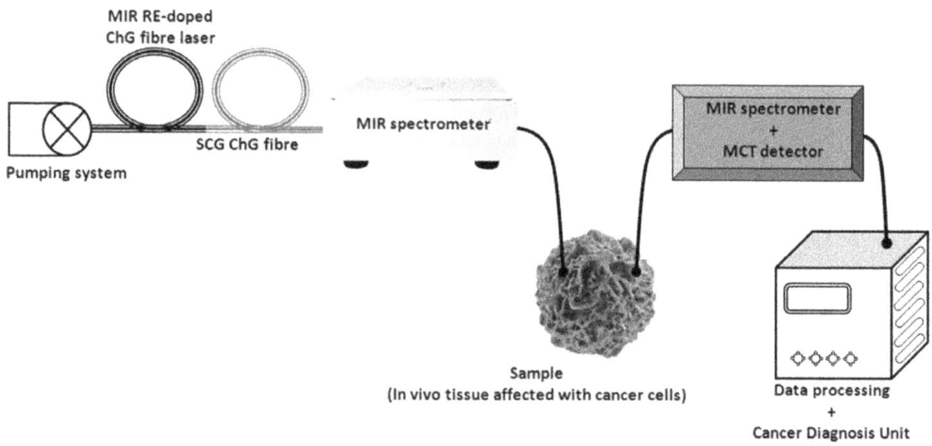


**Figure 4.10.** System consisting of MID RE-doped ChG fiber laser and SCG ChG fiber for early cancer diagnostics.

source of groundwater contamination is the mixing of man-made waste of different products like diesel fuel, oil, road salts, and other chemicals. ChG optical fiber can be used for the estimation of the organic pollutants in the groundwater available under different locations (e.g., soils, artesian basin, aquifers, ocean surfaces, and marine sands). An active research group [21] of Naval research Laboratory, USA used a site characterization and analysis penetrometer system (SCAPS) having a *truck*-mounted cone penetrometer system with the combination of a very long ChG optical fiber and a handy FTIR unit to experiment with the detection of the gasoline contamination in groundwater (see figure 4.11). This group was able to demonstrate the detection limit of ~130 ppm of marine gasoline. Some researchers of three different countries (France, Austria, and Germany) jointly demonstrated preliminary but very important pilot-scale tests in the groundwater of aquifers situated at three different locations by using an optical fiber of TAS ChG [47]. They were able to detect different organic pollutants with noticeable sensitivity of detection at these three sites. The results of their studies are summarized in table 4.2.

### 4.4.4 Night vision systems for surveillance assignments

Increasing worry over accidents, particularly during the night, forces the necessity of auto night vision systems all over the world. Furthermore, there has been tremendous demand in the market for using this kind of arrangement in military vehicles to improve functionality of driving. In the past, optical lenses made of polished germanium were used commonly for many years in thermal imaging. These lenses were found useful for an extensive range of security and business applications. The major drawbacks of these lenses are the high cost and limitations of developing new commercial systems. For the fabrication of the optical elements in the IR optics, the reasonably straightforward synthesis of ChGs permits a reduction in the manufacturing cost as compared to that of Ge-based devices. The development of a novel ChG is reported by a research group for near-IR as well as mid-IR and long-

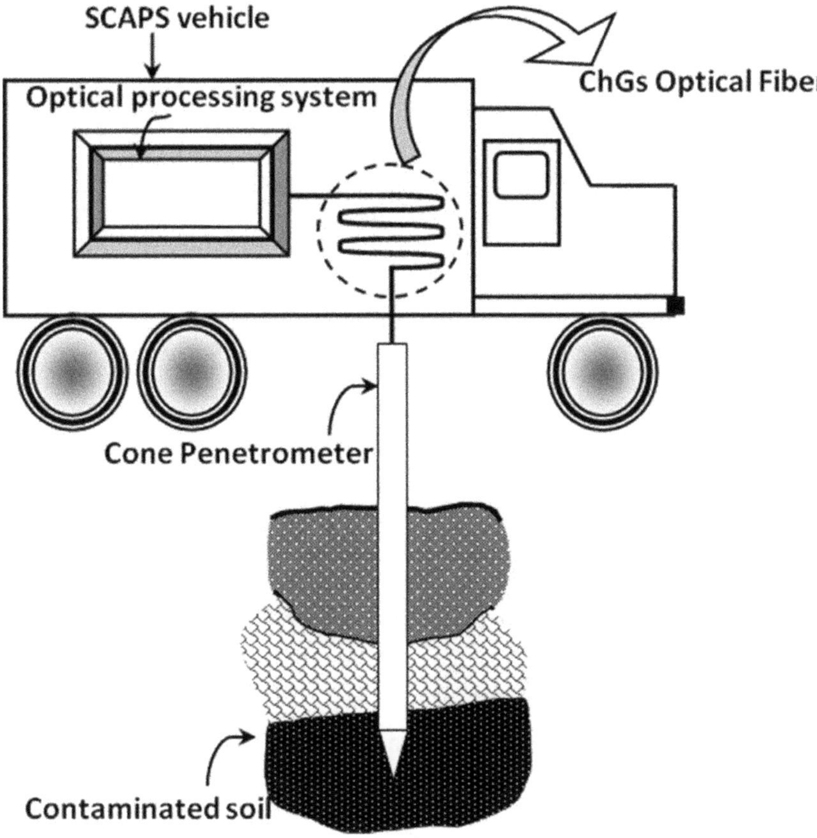


**Figure 4.11.** SCAPS system consisting of the optical fiber of TAS ChG for monitoring of pollutants in groundwater.

**Table 4.2.** Details of the pilot-scale tests in the ground water by Michel *et al* [47].

| Location | Position of absorption band (in $cm^{-1}$) | Corresponding detected pollutants | Limit of detection | Type and identification of vibrational modes |
|---|---|---|---|---|
| Artificial aquifer in Vienna | 910 | $C_2Cl_4$ | 1 mg $dm^{-3}$ | Stretching mode of C-Cl |
| Industrial uncultivated land in Paris | 1255 | $C_2HCl_3$ | 1 μg $dm^{-3}$ | |
| Artificial aquifer in Munich | 910<br>1036 | $C_2Cl_4$<br>$C_6H_4Cl_2$ | 1 mg $L^{-1}$ | |

IR applications. This new material is known as GASIR-1 (Umicore IR Glass) and the credit for its fabrication goes to engineers of Umicore Electro-Optic Materials [48]. Table 4.3 tabulates the comparison of different physical properties of GASIR ChG and germanium. One can realize easily the benefits of the former material as compared to Ge. For instance, there is a minute alteration in the refractive index with changes in temperature in GASIR which further causes a little variation of focal length. This allows trouble-free mechanical designing of the optical lenses of GASIR due to excellent opto-thermal stability. Such a facility of passive athermalization can reduce cost effectively.

The growing number of road accidents and rising safety concerns have drawn attention towards the global auto night vision system market. According to the company, one recent example of the application of this ChG is in the night-vision camera integrated into BMW's 7 series automobiles (see figure 4.12). The manufacturer companies of four-wheelers are using such a night-vision system having a 'far infrared' camera consisting of a GASIR lens that can identify people, animals or strange things during a night journey (see figure 4.12). ChGs based lenses are useful for military applications in weapon-mounted night vision devices [48]. Army soldiers can use such advanced night-vision systems to keep them safer and improve their capability during their missions in the night (see figure 4.13).

### 4.4.5 Monitoring of global warming

We all are aware of the universal problem of global warming and the fact that carbon dioxide has a major contribution to the greenhouse effect. Since $CO_2$ has two characteristic absorption bands at 4.3 and 15 μm, highly developed IR optical devices are essential for its remote detection [49]. ChGs based optical fibers, especially Te-based glass fibers are the finest solution for this purpose because of their wide transparency in the IR region.

Charpentier *et al* [50, 51] developed an optical fiber-based system of $GeSe_4$ ChG for monitoring the presence of carbon dioxide as shown in figure 4.14. They found that this fiber is appropriate for transmitting IR signals and it can be employed as a gas sensor in real-time to check leakage of $CO_2$ gas in a remote $CO_2$ geological storage zone. The figure 4.14 depicts the experimental arrangement used by this

**Table 4.3.** Comparison of various physical properties of GASIR ChG and germanium [48].

| Physical property | GASIR ChG | Germanium |
|---|---|---|
| Transmission domain (μm) | 0.8–15 | 2–15 |
| Density (gm $cm^{-3}$) | 4.40 | 5.32 |
| Young's modulus | 18 | 103 |
| Coefficient of thermal expansion ($\mu K^{-1}$) | 17 | 5.9 |
| Temperature coefficient of refractive index ($\mu K^{-1}$)[a] | 55 | 400 |
| Refractive index at 10 μm | 2.5 | 4.0 |
| Maximum working temperature (°C) | 240 | 80 |

[a] At 20 °C and 10.6 μm.

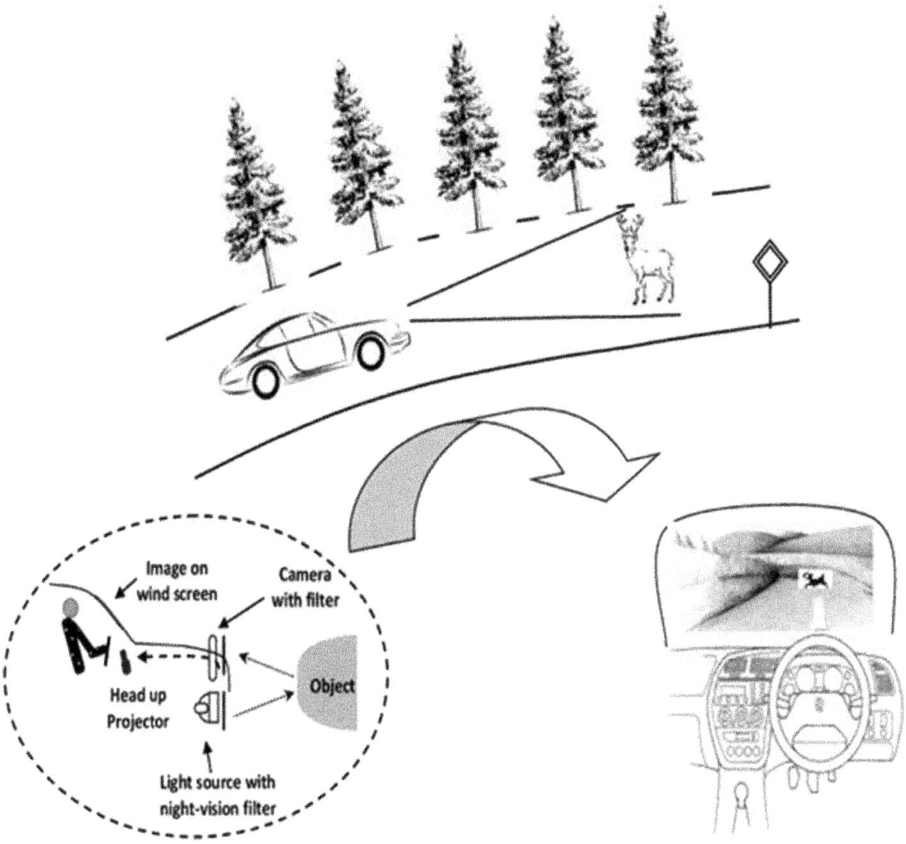


**Figure 4.12.** Four-wheeler with night-vision accessory consisting of ChG lens for safety purposes during night journeys.

group. The two collinear optical fibers of $GeSe_4$ ChG are enclosed in a chamber of cylindrical geometry. The sliced iron tubes are used for the alignment. Their other ends are linked with an FTIR spectrometer. The IR light beam starts from the FTIR spectrometer and enters the chamber through the input fiber. The output fiber controls this beam which returns it to the detector. To conduct the concentration-dependent experiment, argon gas is mixed with $CO_2$ gas. It plays the role of a purge gas too. A $CO_2$ gas analyzer is used to determine its precise concentration in the gas flux. This group performed another experiment of fluorescence absorbance for demonstrating the mid-IR optical sensing of $CO_2$ gas by employing a $Dy^{3+}$ doped ChG fiber of the quaternary GaGeSbS system [51].

### 4.4.6 Other significant applications

Infrared optical fibers, electrode arrays, thin films of ChGs play an efficient role as chemical sensors for the detection of inorganic anions, heavy metal cations as well as gases [52–61]. Natale *et al* [52] found that $Cl^-$ and $F^-$ inorganic ions can be detected

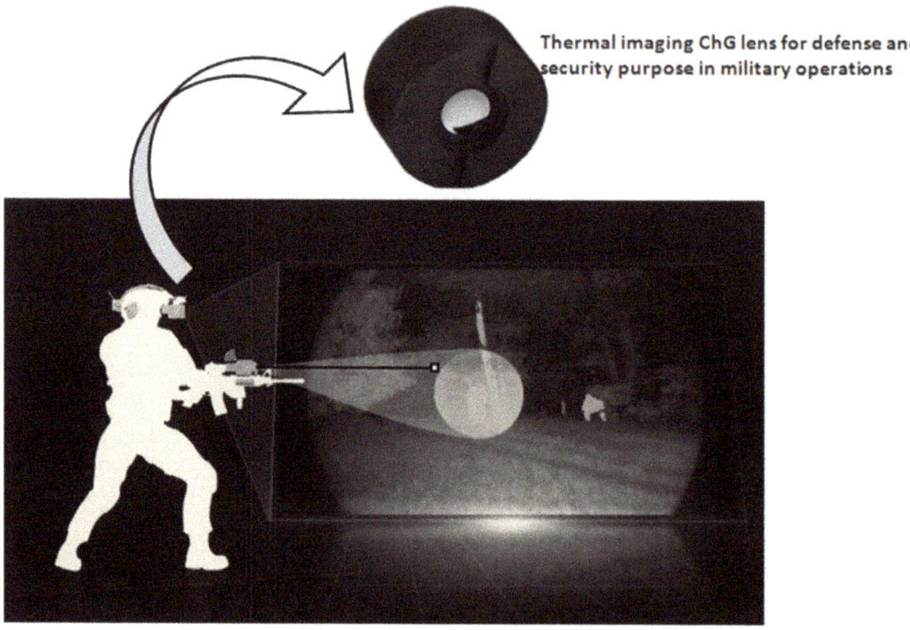


**Figure 4.13.** Advanced night-vision systems consisting of ChG lens for military operations.

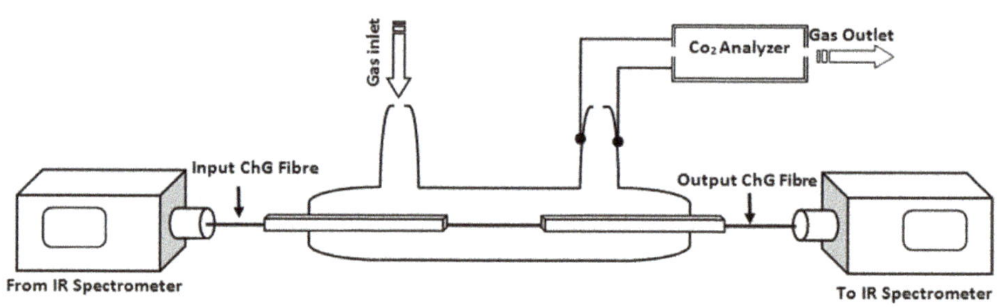


**Figure 4.14.** Schematic experimental set-up of ChG fiber for sensing of $CO_2$ gas.

satisfactorily by using electrode arrays of ChGs. Legin and his research group performed a series of experiments [53, 54] for the determination of the presence of a small amount (~μg) of heavy elements (Cu, Fe, Cr, Cd, and Hg) in natural and wastewaters by using ChGs chemical sensors. They verified the accuracy of their experimental results by demonstrating a good agreement with the corresponding results of atomic absorption spectroscopy. Mourzina *et al* [55] developed a novel thin-film Pb micro-sensor with the support of ChGs. Marian *et al* [56] used a ternary Ge–As–Te system for the fabrication of gas sensors having good sensitivity to detect the small concentrations of $NO_2$ gas. Cali *et al* [57] employed a chalcogenide membrane for the sensing of the sulfide species in the specimen. Spelthahn *et al* [58] developed a sensor array with heavy metal-sensitive ChG membranes for the simultaneous finding of the four ions $Ag^+$, $Cu^{++}$, $Cd^{++}$, and $Pb^{++}$ in solution. Dai and his group [59] designed a unique image-guided FTIR sensor with the help of ChG optical fibers for the sensing of the gases released into the atmosphere by the fuel combustion process. Mariac *et al* [60] employed ion-selective electrodes of Fe(III):ChG by adding Fe in a particular ChG ternary glass of a parent Se–Ge–Sb system and developed ChG membrane for the sensing of iron(III) in seawater. Tsiulyanu and Ciobanu [61] chose a ChGs of quaternary As–Ge–S–Te system for developing AC operating gas sensors and demonstrated their sensing ability at ambient temperature.

## 4.5 Conclusions

Chalcogenide glasses based optical fibers are very stable and so the sensors or sensing devices made from them demonstrate excellent durability. It is generally found that using ChGs based fibers for sensing applications below 100 °C is completely safe for management, storing, and freight purposes. At a temperature roughly equal to 100 °C, it is suggested to operate under fume hood surroundings. ChGs fibers have high resistance against atmospheric moisture and devitrification. The products designed by using optical devices of ChGs have excellent refractive index uniformity and stability. They are primarily employed for scrutinizing equipment and systems, marine/ship, fire protection/police, sensors and IR products. ChGs based optical elements can be planned to substitute many germanium applications when the price is a major concern, as well as benefits of relatively high index, lower density, good thermal stability properties and moldable glass properties. The excellent chromatic rectification makes it feasible to utilize their lenses instead of ZnSe for specific applications. The application of ChGs of a Te–As–Se (TAS) system is extensively acknowledged by diverse research teams and their optical fibers are found to be extremely suitable for MIR sensing in different branches of sciences (e.g., food, environmental, and biological sciences).

The processing of ChGs based devices can be utilized in distinctive visual machining practices including milling, sparkling and diamond flexing, and can be coated with IR coating samples having customary anti-reflective character. Well-liked assortments of ChGs including Se–Ge–As, Se–Ge–Sb, and $As_2M_3$ ($M = S, Se$) can be proposed.

In conclusion, we see that the development of various ChGs based optical elements (e.g., optical fibers, lenses, thin films, etc) with the aim of innovation and an extensive range of applications is an on-going research activity. This field is still developing and new, potential as well as interesting sensing applications of ChGs are coming each year.

## References

[1] Mehta N 2006 *J. Sci. Indust. Res.* **65** 777–86

[2] Kolobov A V, Fons P, Frenkel A I, Ankudinov A L, Tominaga J and Uruga T 2004 *Nat. Mater.* **3** 703–8

[3] Wuttig M and Yamada N 2007 *Nat. Mater.* **6** 824–32

[4] Matsunaga T *et al* 2011 *Adv. Funct. Mater.* **21** 2232–9

[5] Wuttig M and Salinga M 2012 *Nat. Mater.* **11** 270–1

[6] Luckas J, Krebs D, Grothe S, Klomfa J, Carius R, Longeaud C and Wuttig M 2013 *J. Mater. Res.* **28** 1139–47
[7] Kumar A and Mehta N 2018 *J. Mater. Chem.* C **6** 2747–59
[8] Mehta N and Kumar A 2013 *Recent Patents Mater. Sci.* **6** 59–67
[9] Mehta N 2015 *Rev. Adv. Sci. Eng.* **4** 173–82
[10] Sharma A and Mehta N 2016 *Mater. Lett.* **178** 178–80
[11] Srivastava A, Pal S K and Mehta N 2020 *Mater. Chem. Phys.* **241** 122326
[12] Pal S K, Srivastava A and Mehta N 2019 *J. Alloys Compd.* **806** 660–7
[13] Srivastava A, Chandel N and Neeraj Mehta N 2019 *J. Environ. Chem. Eng.* **7** 103032
[14] Churbanov M F 1995 *J. Non-Cryst. Solids* **184** 25–9
[15] Slusher R E, Lenz G, Hodelin J, Sanghera J S, Shaw L B and Aggarwal I D 2004 *J. Opt. Soc. Am.* B **21** 1146–55
[16] Sanghera J S, Shaw L B and Aggarwal I D 2002 *C.R. Chim.* **5** 873–83
[17] Bureau B *et al* 2004 *J. Non-Cryst. Solids* **345/346** 276–83
[18] Zhang B *et al* 2015 *J. Am. Chem. Soc.* **98** 1389–92
[19] Sanghera J and Aggarwal I D 1999 *J. Non-Cryst. Solids.* **256&257** 6–16
[20] Seddon A B 1995 *J. Non-Cryst. Solids* **184** 44–50
[21] Sanghera J, Shaw L B, Busse L, Nguyen V, Pureza P, Cole B, Harbison B B and Aggarwal I D 2000 *Fiber Integrated Opt.* **19** 251–74
[22] Le Coq D, Boussard-Pledel C, Fonteneau G, Pain T, Bureau B and Adam J L 2003 *Mater. Res. Bull.* **38** 1745–54
[23] Kim W H *et al* 2016 *J. Non-Cryst. Solids* **431** 8–15
[24] Shiryaev V S and Churbanov M F 2017 *J. Non-Cryst. Solids* **475** 1–9
[25] Sanghera J S and Aggarwal I D 1997 *J. Non-Cryst. Solids* **213&214** 63–7
[26] Furniss D and Seddon A B 1999 *J. Non-Cryst. Solids* **256&257** 232–6
[27] Nishii J, Yamashita T and Yamagishi T 1989 *Appl. Opt.* **28** 5122–7
[28] Le Coq D, Boussard-Pledel C, Fonteneau G, Pain T, Bureau B and Adam J L 2003 *J. Non-Cryst. Solids* **326** 451–4
[29] Hilton A 2010 *Chalcogenide Glasses for Infrared Optics* (New York: McGraw-Hill)
[30] Saenger K L 1993 *Processing of Advanced Materials* ed J A Mc-Geough and M Datta (London: Chapman & Hall) p 1
[31] Chrisey D B and Hubler G H 1994 *Pulsed Laser Deposition of Thin Films* (New York: Wiley)
[32] Harrick N J 1987 *Internal Reflection Spectroscopy* (New York: Wiley)
[33] Mirabella F M 1993 *Internal Reflection Spectroscopy* (New York: Marcel Dekker)
[34] Lew A, Depeursinge C, Cochet F, Berthou H and Parriaux O 1984 Single-mode fiber evanescent wave spectroscopy *Proc. SPIE 0514, 2nd Int. Conf. on Optical Fiber Sensors: OFS'84 (12 November 1984)* pp 71–4
[35] Ruddy V, MacCraith B D and Murphy J A 1990 *J. Appl. Phys.* **67** 6070–4
[36] Connelly M C 2005 *Fiber Sensors* (Limerick: Elsevier)
[37] Kruegle H and Abram F 2006 Low-light-level cameras, thermal infrared imagers *CCTV Surveillance* 2nd edn ed H Kruegle (Amsterdam: Elsevier) ch 19
[38] Keirsse J, Boussard-Pledel C, Loreal O, Sire O, Bureau B, Turlin B, Leroyer P and Lucas J 2003 *J. Non-Cryst. Solids* **326&327** 430–3
[39] Hocde S, Loreal O, Sire O, Boussard-Pledel C, Bureau B, Turlin B, Keirsse J, Leroyer P and Lucas J 2004 *J. Biomed. Opt.* **9** 404–7

[40] Keirsse J, Bousard-Pledel C, Loreal O, Sire O, Bureau B, Leroyer P, Turlin B and Lucas J 2003 *Vib. Spectrosc.* **32** 23–32
[41] Seddon A B 2011 *Int. J. Appl. Glass Sci.* **2** 177–91
[42] Seddon A B 2013 *Phys. Status Solidi* B **250** 1020–7
[43] Seddon A B, Napier B, Lindsay I, Lamrini S, Moselund P M, Bang O and Farris M 2018 *Analyst* **143** 5874–87
[44] Seddon A B 2017 Biomedical applications in probing deep tissue using mid-infrared (MIR) supercontinuum optical biopsy *Deep Imaging in Tissue and Tissue-Like Media with Linear and Nonlinear Optics* ed R R Alfano and L Shi (Singapore: Pan Stanford Publishing printed in GB) ch. 8, pp 231–97
[45] Sanghera J S, Aggarwal I D, Shaw L B, Florea C M, Pureza P, Nguyen V Q, Kung F and Aggarwal I D 2006 *J. Optoelectron. Adv. Mater.* **8** 2148–55
[46] Sanghera J, Shaw L B, Busse L E, Talley D and Aggarwal I D 1999 *Proc. SPIE* 3596 178–87
[47] Michel K, Bureau B, Boussard-Pledel C, Jouan T, Adam J L, Staubmann K and Baumann T 2004 *Sens. Actuators* B **101** 252–9
[48] Calvez L 2017 *C.R. Phys.* **18** 314–22
[49] Wang G, Nie Q, Wang X, Shen X, Chen F and Zhang X 2011 *J. Appl. Phys.* **110** 043536
[50] Charpentier F, Bureau B, Troles J, Boussard-Pledel C, Pierres K M L, Smektala F and Adam J L 2009 *Opt. Mater.* **31** 496–500
[51] Starecki F *et al* 2015 *Sens. Actuators* B **207** 518–25
[52] Natale C D, Davide F, Brunink J A J, D'Amico A, Vlasov Y G, Legin A V and Rudnitskaya A M 1996 *Sens. Actuators* B **34** 539–42
[53] Legin A V, Bychkov E A and Vlasov Y G 1995 *Sens. Actuators* B **24&25** 309–11
[54] Legin A V, Vlasov Y G, Rudnitskaya A M and Bychkov E A 1996 *Sens. Actuators* B **34** 456–61
[55] Mourzina Y, Schoning M J, Schubert J, Zander W, Legin A V, Vlasov Y G, Kardos P and Luth H 2000 *Sens. Actuators* B **71** 13–8
[56] Marian S, Tsiulyanu D and Liess H D 2001 *Sens. Actuators* B **78** 191–4
[57] Cali C, Taillades G, Pradel A and Ribes M 2001 *Sens. Actuators* B **76** 560–4
[58] Spelthahn H, Kirsanov D, Legin A, Osterrath T, Schubert J, Zander W and Schoning M J 2012 *Phys. Status Solidi* A **209** 885–91
[59] Maric M, Sohail M, Veder J P and Marco R D 2015 *Sens. Actuators* B **207** 907–17
[60] Dai X, Liu X, Liu L, Zhu B and Fang Z 2015 *Sens. Actuators* B **220** 414–9
[61] Tsiulyanu D and Ciobanu M 2016 *Sens. Actuators* B **223** 95–100

# Chapter 5

# Advanced dynamic and static calibration methods for optical imaging sensors

**Jin Li and Zilong Liu**

The internal-parameter calibration of optical imaging sensors (i.e., optical cameras) is a prerequisite for the three-dimensional position reconstruction (i.e., 3DPR positioning) of an object based on optical images. The accurate calibration of optical imaging sensors with easy operation is very valuable for 3DPR application. Here, we introduce an efficient dynamic and static calibration method (SCM), enabling accurate measurements for the internal-parameter of an optical imaging sensor. The SCM adopts a structured aperture focal plane with a known focal length of a normal optical system and provides two-dimensional (2D) equally spaced sub-beams. The 2D grids with the structured prior position in the SCM are used as calibration targets. The SCM-based method has potential use as a standardization measurement method for optical imaging sensor assembly.

## 5.1 Introduction

High-resolution optical cameras have become essential tools in non-contact observation applications, such as mapping and environmental monitoring [1–7]. The photogrammetry constitutes one of the most significant functions of an optical camera. In a photogrammetry mission, an optical camera can utilize its images to obtain a position of the observed object (called 'positioning'). For example, in the remote sensing field, the GeoEye-1 satellite can precisely position an object within 5 m of its true location on the surface of the Earth with CE90 (Circular Error of 90%) [8, 9]. The Pleiades-1B provides a high geo-location accuracy (3 m @ CE90) [10, 11]. The successful implementation of the above positioning function directly depends on the measurements of the external and internal geometric parameters of the camera. Here, the camera geometric calibration is the measurement of its internal parameters. Internal geometric calibration is a prerequisite for realizing the image position reconstruction function of cameras.

doi:10.1088/978-0-7503-2707-7ch5  

Currently, many optical camera calibration methods are being developed in different application cases. In the computer vision field, many approaches to calibrate the internal parameters of optical cameras exist for different computer vision applications [12–22]. A famous calibration method in the computer vision field is self-calibration methods [23–27]. For example, a self-calibrating bundle adjustment is developed for camera calibrations in [28, 29]. In the optics field, optical camera calibration approaches have been developed to solve the calibration problem from the optical aspects [30–36]. In the optical field, a famous calibration method is based on diffractive optical elements (DOEs) for wide-angle geometric camera calibration [37, 38]. The DOEs can generate multiple 2D diffraction grids which can be used to calibrate cameras for photogrammetry.

In the remote sensing field, camera calibration is performed in on-ground and on-orbit conditions [39–48]. Therefore, the calibration methods for remote sensing cameras accordingly include on-ground calibration methods and on-orbit calibration methods. On-ground calibration approaches widely use laboratory measuring angles (LMAs) [49–52]. The LMA method utilizes a precision turntable to the different field of views (FOVs) measurements to complete the calibration of the camera. Recently, several advanced on-ground calibration approaches [53, 54] have displayed multiple advantages, including easy operation and high accuracy. These new methods provided new investigation direction for on-ground calibrations of remote sensing cameras. For the on-orbit calibration of remote sensing cameras, the famous calibration methods are ground control point (GCP) methods [55–59]. Another on-orbit calibration method is based on 180-degree satellite maneuvers (180-SM) [60–62]. Recently, several new on-orbit self-calibration methods have been developed to push the current on-orbit calibration technology for remote sensing cameras [63–65].

The above-mentioned optical camera calibration methods can be summarized into two categories: (1) dynamic calibration; and (2) static calibration. Dynamic calibration approaches (e.g. LMA) mainly adopt a movement mechanism to control an uncalibrated camera to capture predefined targets. In dynamic calibrations, a precision turntable is an ideal movement mechanism, where an uncalibrated optical camera is placed on a precision turntable to capture the light of single-pixel illumination, produced by a collimator. At different rotation angles of the turntable, multiple positions from different FOVs can be obtained to realize the calibrating of the camera. Another dynamic calibration example is the on-orbit calibration of remote sensing cameras using 180-degree satellite maneuvers (180-SM). Static calibration approaches are based on the static working of optical cameras. For example, crossed-phase DOEs are used as the calibration target while the uncalibrated camera works at the static case. Another static calibration method based on a micro-electromechanical system (MEMS) integration method, which is based on a point-source focal plane, integrated on the focal-plane of an un-calibrated camera, is used for calibrating a smart camera [66]. This method can implement the self-calibration and has high calibration accuracy because this method does use the extra predefined targets (e.g. DOEs).

Alternatively, we introduce another efficient static solution using a micro-scaleplate of light (mSOL) [67] in this chapter. The mSOL is a normal optical system with a known focal length and prior self-luminescence targets. The mSOL can be used as an infinite target, simulating ground objective, to calibrate a camera. The mSOL is scalable, which can be configured with different scales. In particular, the mSOL can be designed into a microstructure. In this case, it can easily integrate into an optical camera system for the camera self-calibration, without the increase of development difficulty. This method has the merits of high accuracy, easy operation, and less required auxiliary equipment. This method provides a novel calibration concept for optical camera calibrations.

The rest of this paper is organized as follows. Section 5.2 introduces the principle of image positioning and camera calibration. In section 5.3, the dynamic calibration principle is introduced. Section 5.4 shows the static calibration method. Section 5.5 analyzes the future development direction. Section 5.6 concludes the dynamic and static calibration method.

## 5.2 Principle of camera calibrations

In this section, section 5.2.1 introduces the position determination principle using images obtained by the calibrated camera. In section 5.2.2, the camera calibration principle is outlined. Section 5.2.3 illustrates the camera model for the calibration application. Finally, section 5.2.4 explains the camera distortion model in the calibration process.

### 5.2.1 Position determination principle using optical cameras

The image position determination (i.e. positioning) process is that images obtained by an optical camera are used to calculate the position of an object in the world coordinate system (i.e. object space coordinate system). Figure 5.1 shows the positioning principle and the coordinate transforms flow from the image to the object position. There are three important coordinate systems as follows:

1. Camera coordinate system ($O_c - X_cY_cZ_c$): the origin is the center of the optical system of a camera. In the pinhole camera model, the origin is the aperture center. The $Z_c$ is the optical axis of the camera. $X_c$ and $Y_c$ are the other two axes of the camera system.
2. Image plane coordinate system ($O_i - X_iY_iZ_i$): the origin $O_i$ is the intersection of the optical axis and the image plane. $X_i$ and $Y_i$ are parallel to $X_c$ and $Y_c$, respectively.
3. World coordinate system ($O_w - X_wY_wZ_w$): This coordinate system defines the position of observation objects.

In the image positioning processing, the images coordinate system is converted to the world coordinate system. The transform between the image coordinate system and the camera coordinate system is completed by the internal parameters. Accordingly, the transform between the camera coordinate system and the world coordinate system is completed by the external parameters. Let $M_{i-c}$ express the

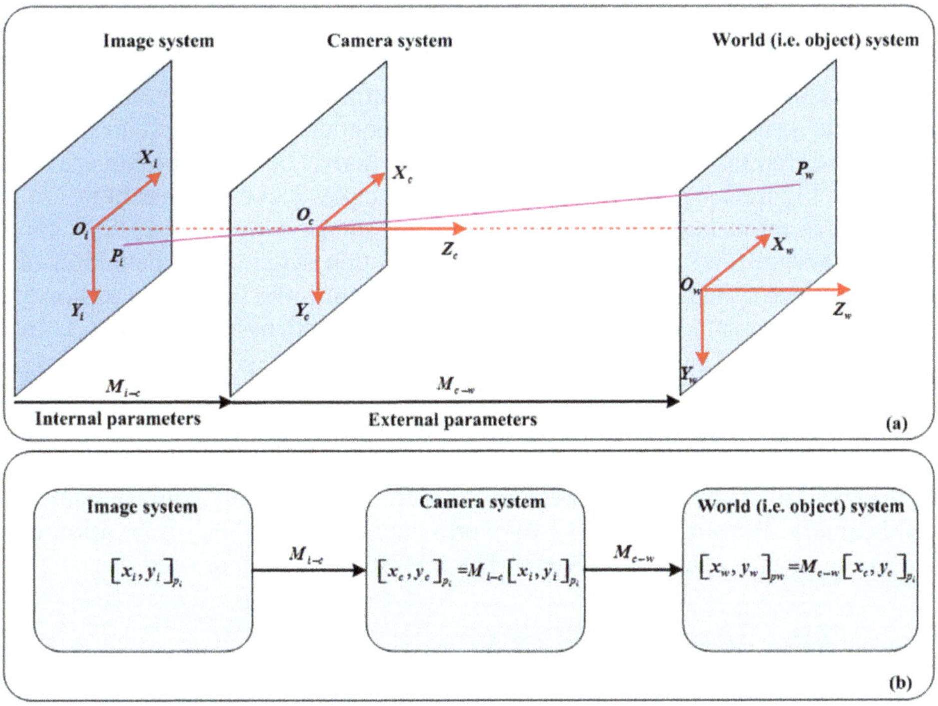


**Figure 5.1.** Position determination principle of an optical imaging sensor (a) and working flow (b).

internal parameters matrix. Let $M_{c-w}$ express the external parameters matrix. Let the position of an observation object point be $(x_i, y_i)$ in the image coordinate system. In the camera coordinate system, the position of the object point can be expressed as:

$$\begin{bmatrix} X_C \\ Y_C \\ Z_C \end{bmatrix} = M_{i-c} \begin{bmatrix} X_i \\ Y_i \\ 1 \end{bmatrix} \tag{5.1}$$

In the world coordinate system, the position of the object point can be expressed as:

$$\begin{bmatrix} X_w \\ Y_w \\ Z_w \end{bmatrix} = M_{c-w} \begin{bmatrix} X_C \\ Y_C \\ Z_C \end{bmatrix} = M_{c-w}M_{i-c} \begin{bmatrix} X_i \\ Y_i \\ 1 \end{bmatrix} \tag{5.2}$$

In equation (5.2), the $M_{i-c}$ and $M_{c-w}$ are known, $(X_i, Y_i)$ is measured. The $(X_w, Y_w, Z_w)$ is obtained by the $M_{i-c}$, $M_{c-w}$, and $(X_i, Y_i)$. Thus, the $M_{i-c}$ and $M_{c-w}$ are prerequisites of the image positioning. In an actual position mission, the coordinate systems may be complicated and more coordinate systems may be included. The beginning of the coordinate system transformation is the image coordinate system and the final coordinate system is the world coordinate system. The intermediate coordinate systems are determined based on the actual positioning mission.

### 5.2.2 Camera calibration principle

The camera calibration process is the determination of the coordination transform matrix $M_{i-c}$ using a predefined object coordination system. The predefined object coordination system is defined by the user. This coordinate system has a known position observed object point and the transform matrix between the camera system coordinate. Figure 5.2 shows the calibration principle. The camera calibration also involves three important coordinate systems, i.e. image coordinate system, camera coordinate system, and predefined object coordination system. The definition of the image coordinate system, the camera coordinate system is the same as in section 5.2.1. The predefined object coordination system is denoted by $(O_p - X_pY_pZ_p)$. In the $O_p - X_pY_pZ_p$ system, the origin $O_p$ locates on the optical axis; the $X_p$ and $Y_p$ are parallel to $X_i$ and $Y_i$, respectively.

The predefined coordination system is designed by the user. The observation object point is known. The transform matrix (denoted by $M_{p-w}$) is also designed by the user and is also known. The predefined object is imaged on the image plane of an optical camera. We still use $(x_i, y_i)$ to express the position of an observation object point in the image coordinate system. The imaging can be expressed as:

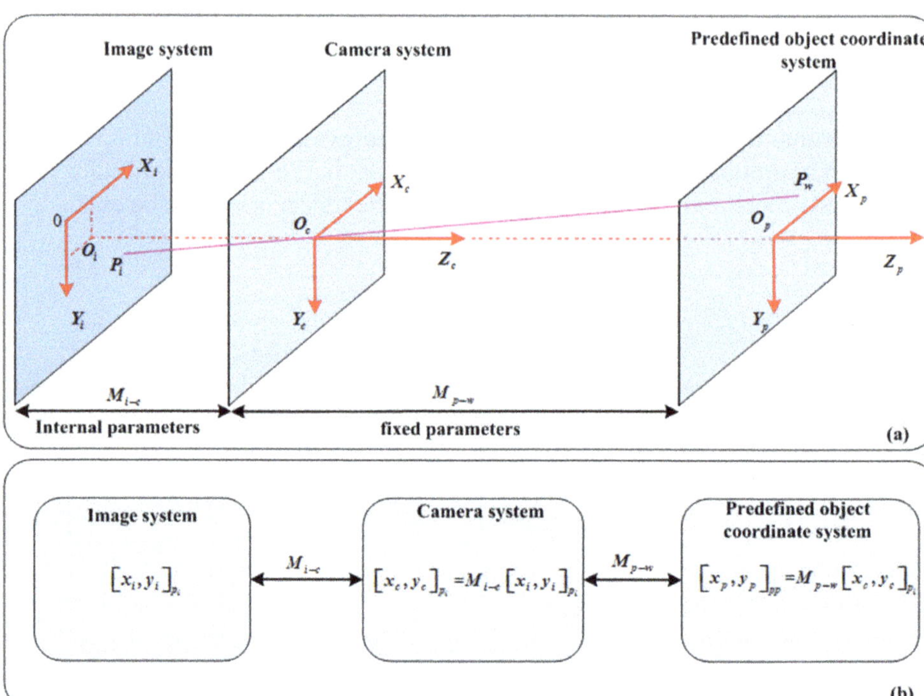


**Figure 5.2.** Calibration principle of an optical imaging sensor (a) and working flow (b).

$$\begin{bmatrix} X_p \\ Y_p \\ Z_p \end{bmatrix} = M_{c-p} \begin{bmatrix} X_C \\ Y_C \\ Z_C \end{bmatrix} = M_{c-p} M_{i-c} \begin{bmatrix} X_i \\ Y_i \\ 1 \end{bmatrix} \tag{5.3}$$

In equation (5.3), the ($X_p$, $Y_p$) and $M_{c-p}$ are known, ($X_i$, $Y_i$) is measured by the camera. The internal parameter $M_{i-c}$ is known. Thus, the predefined parameters, ($X_p$, $Y_p$) and $M_{c-p}$, are prerequisites of the image camera calibration. In other words, the calibration methods are how to design a proper predefined object coordinate system.

### 5.2.3 Camera calibration model

From section 5.2.2, the camera calibration is implemented using the geometric relationship between a 2D image and a 3D object. In the geometric relationship model, the camera is a pinhole model. Figure 5.3 shows the pinhole model of a camera and the coordinate system for the camera calibration. The pinhole model regards the optical system of a camera as a single small aperture.

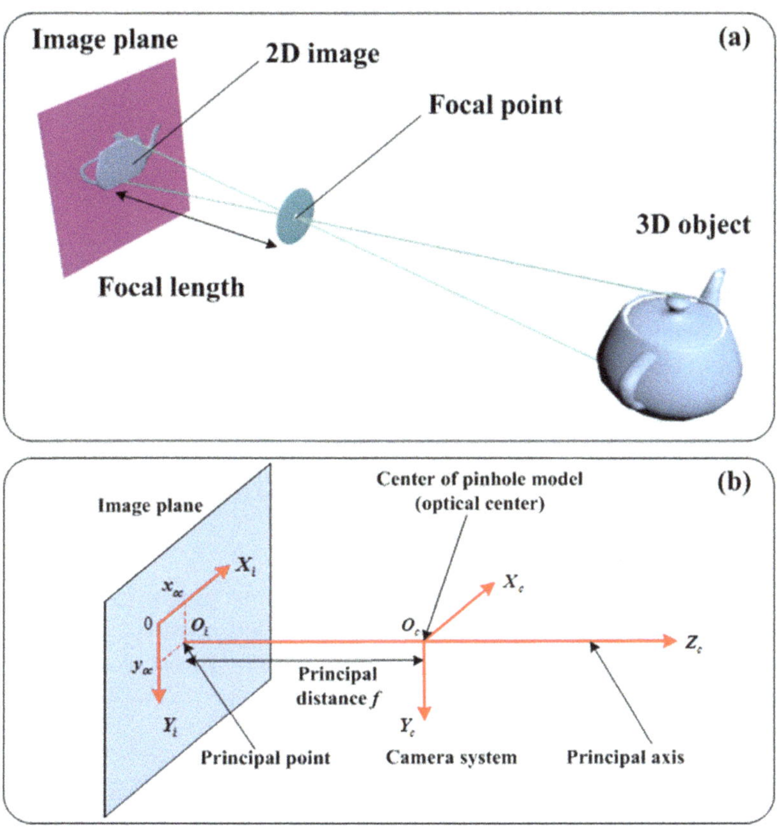


**Figure 5.3.** Pinhole imaging model used for the camera calibration (a) and the coordination system (b).

Camera calibration operation determines the geometric parameters including the principal distance and principal point. In an ideal system, the origin of the image coordinate system is the principal point (see figure 5.1). The internal parameter matrix $M_{c-p}$ can be expressed as:

$$M_{c-p} = \begin{bmatrix} f & 0 & 0 \\ 0 & f & 0 \\ 0 & 0 & 1 \end{bmatrix}, \tag{5.4}$$

where $f$ is the principal distance. In the actual system assembly, the origin of the image coordinate system is different from the principal point (see figure 5.3). Equation (5.4) can be rewritten as:

$$M_{c-p} = \begin{bmatrix} f & 0 & x_{oc} \\ 0 & f & y_{oc} \\ 0 & 0 & 1 \end{bmatrix} \tag{5.5}$$

Substituting equation (5.4) into equation (5.3), the calibration equation is expressed as:

$$\begin{bmatrix} X_p \\ Y_p \\ Z_p \end{bmatrix} = \begin{bmatrix} f & 0 & x_{oc} \\ 0 & f & y_{oc} \\ 0 & 0 & 1 \end{bmatrix} M_{i-c} \begin{bmatrix} X_i \\ Y_i \\ 1 \end{bmatrix} \tag{5.6}$$

The above geometric imaging relationship for equations (5.4–5.6) is constructed under the pinhole model.

### 5.2.4 Distortion model in camera calibration

In equation (5.6), the distortion of an optical system is not considered. In the actual system, optical distortion still exists. When considering distortion, equation (5.6) can be rewritten as:

$$\begin{bmatrix} X_p \\ Y_p \\ Z_p \end{bmatrix} = \begin{bmatrix} f & 0 & x_{oc} \\ 0 & f & y_{oc} \\ 0 & 0 & 1 \end{bmatrix} M_{i-c} \left( \begin{bmatrix} X_i \\ Y_i \\ 1 \end{bmatrix} + \begin{bmatrix} X_i^{\text{distortion}} \\ Y_i^{\text{distortion}} \\ 1 \end{bmatrix} \right), \tag{5.7}$$

where $(X_{\text{i}}^{\text{distortion}}, Y_{\text{i}}^{\text{distortion}})$ is the position offset due to the optical distortion. In the camera calibration, the radial distortion model is widely used [14, 68, 69]. Figure 5.4 shows the two classical radial distortion examples. The position offset is calculated by the following [70]:

$$\begin{cases} X_i^{\text{distortion}} = X_i(k_1 r^2 + k_2 r^4 + k_3 r^6) \\ Y_i^{\text{distortion}} = Y_i(k_1 r^2 + k_2 r^4 + k_3 r^6) \end{cases} \tag{5.8}$$

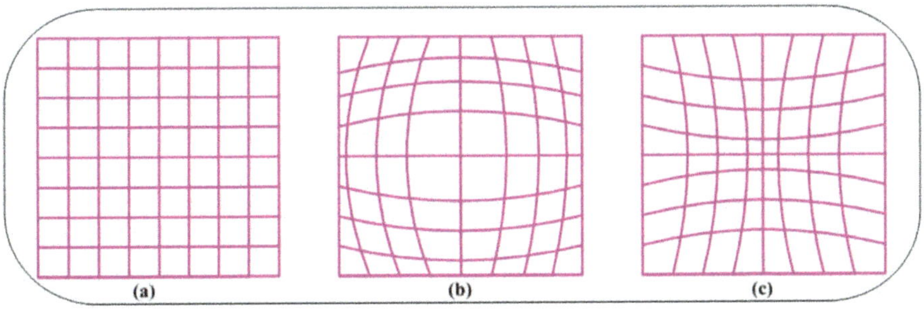


**Figure 5.4.** Two classical radial distortion cases of optical lens, no distortion (a), positive radial distortion (b), and negative radial distortion (c).

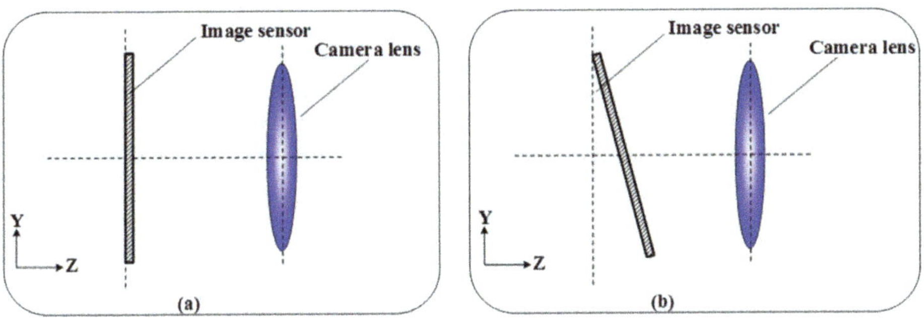


**Figure 5.5.** A classical tangential distortion case of optical lens, no distortion (a) and tangential distortion (b).

with

$$r^2 = X_i^2 + Y_i^2, \tag{5.9}$$

where $k_1$, $k_2$, and $k_3$ are the radial distortion coefficients of a lens.

In an actual system, other distortion maybe exists. For example, figure 5.5 also shows a classical tangential distortion example. The tangential distortion is that the image plane has a slant angle. Accordingly, the position offset is calculated by the following [71]:

$$\begin{cases} X_i^{\text{distortion}} = 2 \times p_1 \times X_i \times Y_i + p_2 \times (r^2 + 2 \times X_i^2) \\ Y_i^{\text{distortion}} = p_1 \times (r^2 + 2 \times Y_i^2) + 2 \times p_2 \times X_i \times Y_i \end{cases} \tag{5.10}$$

with

$$r^2 = X_i^2 + Y_i^2, \tag{5.11}$$

where $p_1$ and $p_2$ are the tangential distortion coefficients of a lens.

## 5.3 Dynamic calibration approaches

In this section, section 5.3.1 introduces the general principle of dynamic calibration. Section 5.3.2 shows the calibration model used for the dynamic process.

### 5.3.1 The principle of the dynamic camera calibration

As with the aforementioned dynamic calibrations, dynamic camera calibration approaches mainly adopt a movement mechanism to control an uncalibrated camera to capture predefined targets. The motion way includes rotation or shift to capture the targets of the predefined coordinate system at different positions. Figure 5.6 shows the two motion ways for the dynamic calibration of an optical camera. A camera may be rotated and the calibration is achieved based on the angle relationship. A camera may be shifted and the calibration is achieved based on the shift position relationship. The 180-SM calibration method is that the camera is the shift case.

In dynamic calibrations, a precision turntable is widely used as an ideal movement mechanism. Figure 5.7 shows the dynamic calibration configuration constituting a turntable, a collimator, and an uncalibrated optical camera. The uncalibrated optical camera is fixed to the turntable and a collimator is placed in the front of the camera. The light of single-pixel illumination produced by a collimator is acquired by the camera to obtain an observation position image. At different rotation angles of the turntable, multiple observation positions from different FOVs can be obtained to realize the calibrating of the camera.

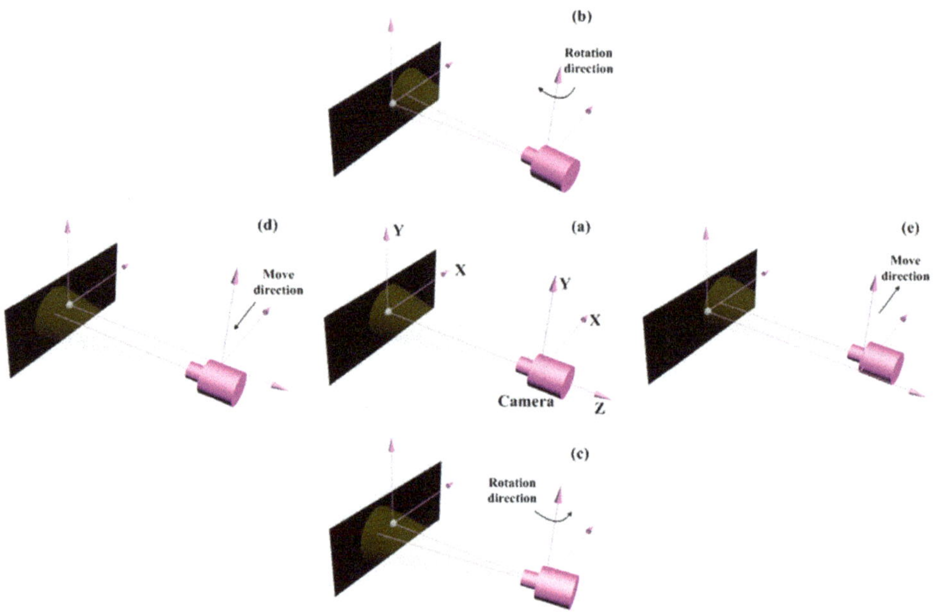


**Figure 5.6.** Two dynamic calibration approaches, the camera is located on the original position (a); the camera is rotated around $y$-axis with 10 degrees (b); the camera is rotated around $y$-axis with −10 degrees (c); the camera moves along the negative direction of $x$-axis (d); the camera moves along the positive direction of the $x$-axis (d).

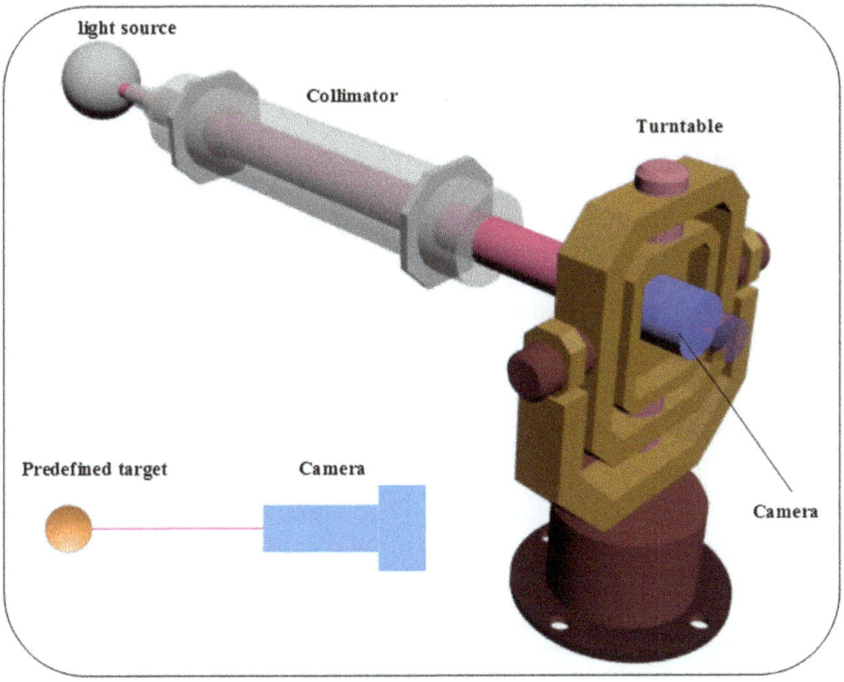


**Figure 5.7.** Calibration configurations of the dynamic calibration method.

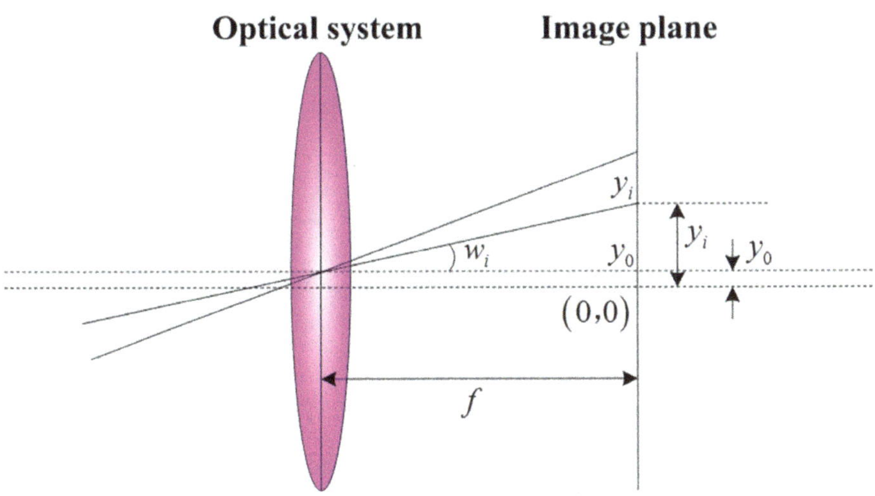


**Figure 5.8.** Imaging model of the dynamic calibration method.

### 5.3.2 Calibration model used for the dynamic calibration

The imaging model in the horizontal direction is shown in figure 5.8. Here, we use $w_i$ to express a rotation angle at the rotation time $i$ ($i = 1,2,3,\ldots,N$); $f$ is the principal length of the uncalibrated camera. Let $y_0$ be the horizontal position of the principal point. In the horizontal direction, the ideal image position is expressed as

$$y_i = y_0 + f \times \tan(w_i) \tag{5.12}$$

In the actual system, the distortion exists. Thus, the image position is

$$y_i = y_0 + f \times \tan(w_i) + \Delta y \tag{5.13}$$

Based on the optical distortion theory, an optical system has minimal distortion at the position of the principal point. Thus, the principal distance and the horizontal position of the principal point can be obtained by the optimizing problem as:

$$\min_{y_0, f} \sum_{i}^{I} (y_i - y_0 - f \tan w_i), \tag{5.14}$$

A least square method is widely used to solve equation (5.14). The principal distance and the horizontal position of the principal point can be expressed as follows:

$$f = \frac{\left(\sum_{i=1}^{N} y_i \tan^2 w_i \cdot \sum_{i=1}^{N} \tan^3 w_i\right) - \left(\sum_{i=1}^{N} y_i \tan w_i \cdot \sum_{i=1}^{N} \tan^4 w_i\right)}{\left(\sum_{i=1}^{N} \tan^3 w_i\right)^2 - \left(\sum_{i=1}^{N} \tan^2 w_i \cdot \sum_{i=1}^{N} \tan^4 w_i\right)}, \tag{5.15}$$

$$y_0 = \frac{\left(\sum_{i=1}^{N} y_i \tan^2 w_i \cdot \sum_{i=1}^{N} \tan^2 w_i\right) - \left(\sum_{i=1}^{N} y_i \tan w_i \cdot \sum_{i=1}^{N} \tan^3 w_i\right)}{\left(\sum_{i=1}^{N} \tan^3 w_i\right)^2 - \left(\sum_{i=1}^{N} \tan^2 w_i \cdot \sum_{i=1}^{N} \tan^4 w_i\right)} \tag{5.16}$$

Accordingly, the vertical position of the principal point is calculated in the same way.

### 5.3.3 Dynamic calibration with multi-aperture MEMS light lead-in devices

To further improve the dynamic calibration accuracy, a multi-aperture MEMS light lead-in device (MEMS-LLAD) is used as the optical focal plane to geometrically calibrate cameras. Figure 5.9 shows the proof-of-concept calibration setup. The MEMS-LLAD is composed of multiple apertures, where each aperture can pass through lights while other areas cannot. Figure 5.9(a) shows a multiple aperture example. The aperture pattern can be designed by the user. As shown in figure 5.9(b), the calibration system included an optical camera, a MEMS-LLAD, a collimator, and a high-accuracy turntable. The MEMS-LLAD is installed on the focal plane of the collimator. The collimator can generate the collimated beam.

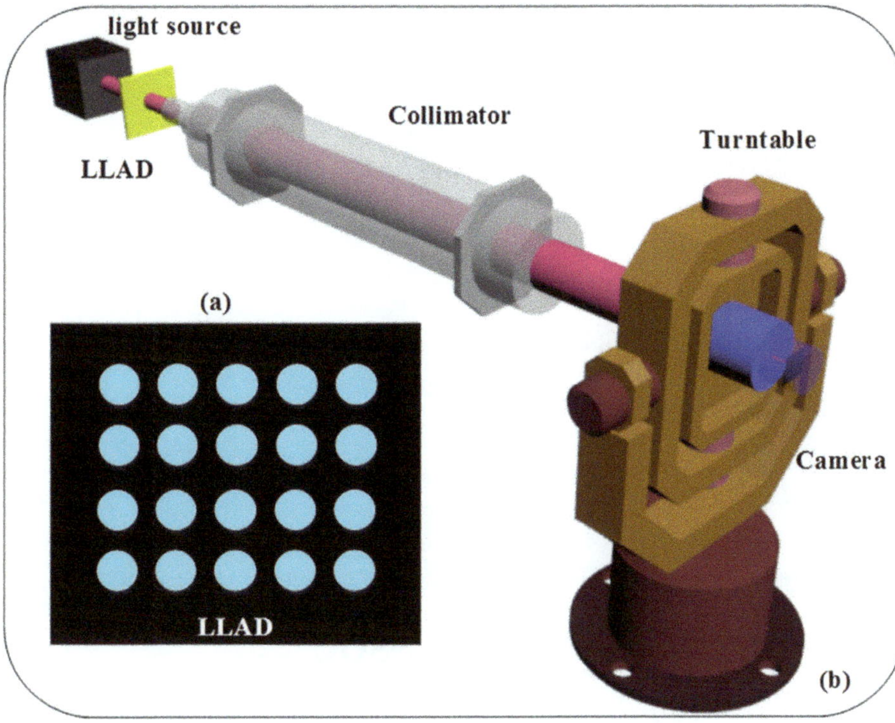


**Figure 5.9.** Dynamic calibration using multi-aperture MEMS, a multiple aperture example (a); the calibration configuration (b).

When the MEMS-LLAD is illuminated by lights, the multiple collimated beams are produced. Since the collimated beam is parallel, a collimator can provide an infinite target for the uncalibrated camera. Different from the dynamic single aperture method, the multiple-aperture method can obtain multiple points at each rotation angle of the high-accuracy turntable. A multi-image matrix formed by the multiple points can decrease the random error of the dynamic calibration system. Moreover, the proposed method does not increase the additional volume and power consumption of the calibration system. The proposed method adopts this new construction and implementation method based on the MEMS-LLAD to guarantee the reliability for high-accuracy calibration of optical cameras.

Currently, the MEMS-LLAD is fabricated by MEMS machining technologies. A MEMS digital micromirror device (DMD) has a potential of the dynamic camera calibration. We have adopted a DMD to implement digital grayscale generation equipment (DGGE) [72]. Figure 5.10 shows the DGGE principle. The DGGE is composed of a light source, a detector, a DMD controller, and optics components. We have found that the DMD is a very powerful device to dynamically produce single or multiple apertures. We can revise the DGGE to achieve a new dynamic camera calibration setup for future works.

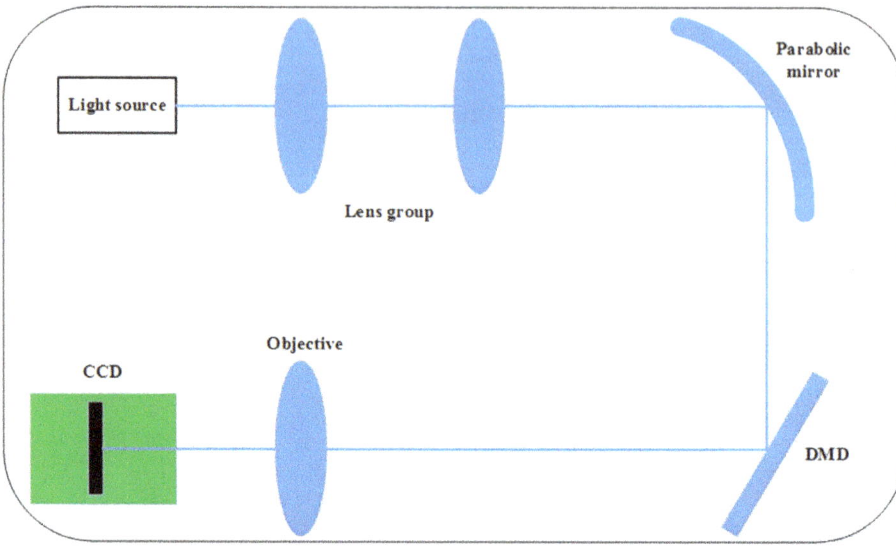


**Figure 5.10.** Schematic diagram of digital grayscale generation equipment (DGGE).

## 5.4 Static calibration principle with mSOL

In this section, the structure is as follows. In section 5.4.1, the general principle of static calibration methods is described. Section 5.4.2 introduces the static calibration principle with DOEs. Section 5.4.3 demonstrates the calibration configurations of the static calibration with mSOL. Section 5.4.4 explains the calibration theory for the static calibration with mSOL. In section 5.4.5, the position extraction approach used for static calibration with mSOL is introduced. Section 5.4.6 shows the application example using the static calibration method.

### 5.4.1 Static calibration general principle

Different from dynamic calibration methods, static calibration approaches are performed under the static working conditions of uncalibrated optical cameras. The dynamic calibration methods usually adopt a simple predefined coordinate system in conjunction with a dynamic camera coordinate system. The static calibration method only needs a specific predefined coordinate system in connection with a fixed camera coordinate system. To successfully complement the calibration task, the static calibration methods need the known information of a predefined target coordinate system. Figure 5.11 shows the static calibration principle. An uncalibrated camera is placed directly in the front of a predefined coordinate system with a specific target. The known information on predefined targets is used to calculate the internal parameters of the uncalibrated camera. Let $[X_p, Y_p, Z_p]$ be the position of a predefined target. $[X_I, Y_I, Z_I]$ expresses the position of the predefined target image. The transform matrix between the predefined target position and its image position is denoted by $D_T$. The relationship between the predefined target information and the image positions can be expressed as:

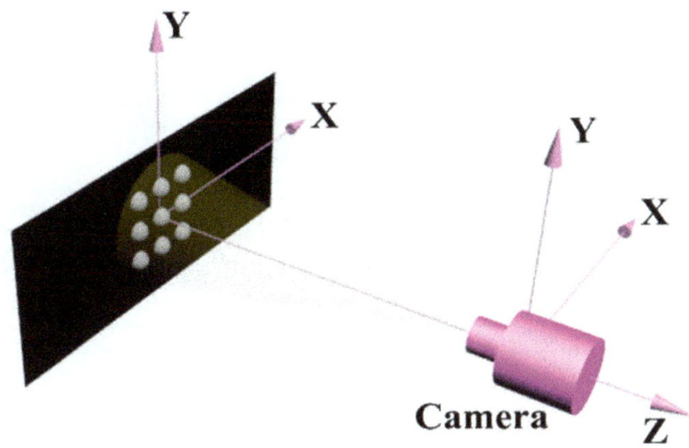


**Figure 5.11.** General principle of static calibration methods.

$$\begin{bmatrix} X_p \\ Y_p \\ Z_p \end{bmatrix} = D_T \begin{bmatrix} X_I \\ Y_I \\ Z_I \end{bmatrix} \tag{5.17}$$

The internal parameters are included in $D_T$. By solving equation (5.17), the internal parameters are calculated. Thus, a static calibration method is how to design the predefined targets. In section 5.4.2, the specific predefined coordinate system is implemented by the DOEs with the known propagation directions. In section 5.4.3, the specific predefined coordinate system is implemented by the mSOLs with the known distance and position information.

### 5.4.2 Static calibration principle with DOEs

A typical static calibration method for geometric camera calibration is based on diffractive optical elements (DOEs). The DOEs in conjunction with a laser beam illumination can generate multiple 2D diffraction grids to calibrate cameras for photogrammetry [37, 38]. Figure 5.12 shows the calibration principle using a DOE. The calibration configuration is composed of a laser, two lenses (L1 and L2), a DOE, and an uncalibrated camera. The laser beam is expanded and collimated by the beam expander. The beam expander is composed of two lenses. The collimated beam illuminates the DOE. The diffractive beam is captured by the uncalibrated camera. The DOE can produce multiple sub-beams with the known propagation directions. The direction of the diffracted sub-beams can be expressed as:

$$\begin{pmatrix} X \\ Y \\ Z \end{pmatrix} = \begin{pmatrix} \lambda f_x \\ \lambda f_y \\ \sqrt{1 - \lambda^2 \left(f_x^2 + f_y^2\right)} \end{pmatrix}, \tag{5.18}$$

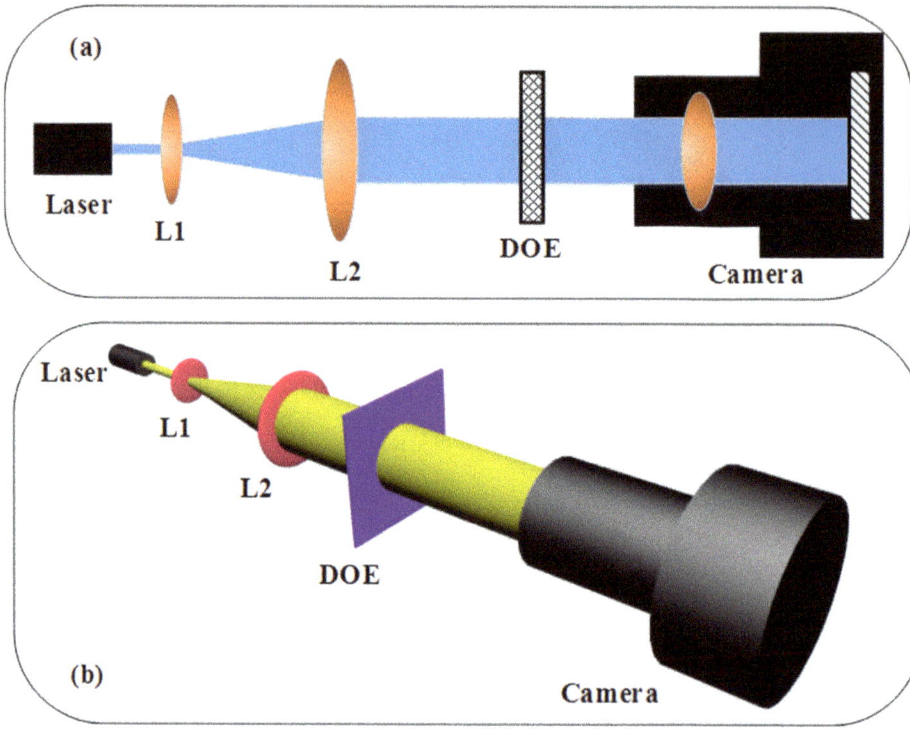


**Figure 5.12.** Schematic diagram of the DOE calibration for cameras, 2D optical configurations (a) and 3D optical configurations (b).

where $\lambda$ is the beam wavelength, $f_x$ and $f_y$ are the spatial frequency of the DOE. Equation (5.18) can provide a known knowledge of a calibration target. Thus, the DOE can be used to complete the camera calibration. The DOEs can be fabricated by E-beam lithography technology [36]. The DOEs can be implemented using twin-crossed 1D phase grating [37].

### 5.4.3 Calibration configurations with mSOL

Alternatively, an mSOL-based calibration method is developed [67] to achieve a predefined target. Figure 5.13 shows the calibration principle using the mSOL. The mSOL is defined as a normal optical system with a known focal length. The mSOL is composed of predefined self-luminescence 2D grids. From the mSOL, the outgoing lights are parallel. The mSOL can be utilized as a 2D infinite grid target to calibrate a camera. In the remote sensing imaging field, the mSOLcan simulate ground objective points. So, the mSOL is very suitable for the remote sensing camera calibration.

The mSOL is designed into an independent model. Figure 5.13 shows the calibration configurations of an mSOL. The mSOL is composed of an area-array light source (LS), a structured aperture focal plane (SAFP), and a normal optical system (NOS). The NOS is a known focal length. The SAFP adopts a multi-aperture structure with the predefined position information. Let the total number of micro-apertures on the SAFP

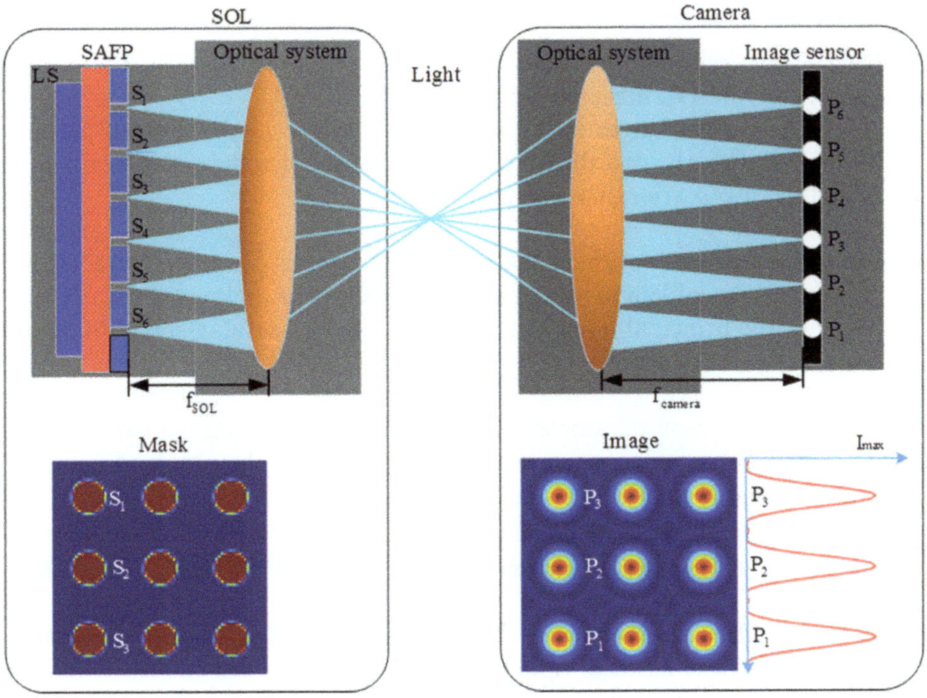


**Figure 5.13.** Calibration principle using the mSOL.

be $N$. We use $S_i$ ($i = 1, 2, \ldots, N$) to denote a micro-aperture in the SAFP. The distance between $S_{i-1}$ and $S_i$ is known, which can be designed by the user. To generate the parallel lights, the SAFP is located on the focal plane of the normal optical system. An area-array light source is used to illuminate the SAFP to produce multiple visual point light sources. These visual point light sources are located on the focal plane of the NOS. The outgoing lights from the NOS are parallel. The mSOL can be used as a predefined target because the position of multiple visual point-light sources and the focal length of the NOS are known.

In the calibration processing, an uncalibrated camera is located directly in front of the mSOL. The outgoing lights from the mSOL enter into the uncalibrated camera and optical point images are formed by the camera optical system. The image sensor of the uncalibrated camera can collect the optical energy of image points and convert the optical point images into digital point images. The mSOL has the visual point sources of $N$. The $N$ optical grids are produced on the image plane of the camera. Accordingly, a digital image with $N$ grids is obtained by the uncalibrated camera. A point extraction method can be used to obtain the position information in the image coordinate system. Let $P_i$ express an image grid in the image coordinate system. The $P_i$ of the observed grids in conjunction with predefined apertures (i.e., positions and focal length) of SAFP can be used to extract the internal parameters of the uncalibrated camera. The calculation equations of internal parameters will be shown in section 5.4.4. The point extraction algorithm will be explained in section 5.4.5.

### 5.4.4 Calibration theory

To calculate the internal parameters of an uncalibrated camera, the following parameters of the mSOL are used as:

1. $f_{\text{SOL}}$: the optical length of the mSOL, which is *a priori* a known parameter.
2. $(L_{sx}(i), L_{sy}(i))$: the coordinate of $i$th structured micro-aperture of the mSOL, which is also *a priori* a known parameter.

To calibrate the internal parameter of a camera, camera parameters are used as:

1. $f_{\text{camera}}$: the focal length of the mSOL, which is an unknown parameter.
2. $(u_0, v_0)$: the principal point of the uncalibrated camera, which is an unknown parameter.

Besides, the grid position parameter extracted from the images is used as:

1. $(L_{px}(i), L_{py}(i))$: the coordinate of $i$th image grid of the observed image in an ideal optical system of a camera.
2. $(\hat{L}_{px}(i), \hat{L}_{py}(i))$: the coordinate of $i$th image grid of the observed image in an actual optical system of a camera, which is also a measured parameter.

To construct the calibration equation using the mSOL, we first consider an uncalibrated camera is an ideal optical system. As the aforementioned calibration theory, the ideal uncalibrated camera can be considered as a pin-hole model. In the pin-hole model, the origin of the image coordinate system is located on the optical axis of the optical system of the ideal uncalibrated camera. Thus, the coordinate position of the principal point is (0, 0) in the camera coordinate system. Under the ideal optical system condition, the position of grids observed by the uncalibrated camera can be expressed as:

$$\begin{cases} \dfrac{L_{px}}{f_{\text{camera}}} = \dfrac{L_{sx}}{f_{\text{SOL}}} \\ \dfrac{L_{py}}{f_{\text{camera}}} = \dfrac{L_{sy}}{f_{\text{SOL}}} \end{cases} \tag{5.19}$$

The matrix form of equation (5.19) can be expressed as:

$$\begin{bmatrix} \hat{L}_{px} \\ \hat{L}_{py} \\ 1 \end{bmatrix} = \begin{bmatrix} L_{px} \\ L_{py} \\ 1 \end{bmatrix} = \begin{bmatrix} f_{\text{camera}} & 0 & 0 \\ 0 & f_{\text{camera}} & 0 \\ 0 & 0 & 1 \end{bmatrix} \begin{bmatrix} \dfrac{L_{sx}}{f_{\text{SOL}}} & \dfrac{L_{sy}}{f_{\text{SOL}}} & 1 \end{bmatrix}^T \tag{5.20}$$

Second, the actual uncalibrated camera is considered to construct the calibration equation. An actual uncalibrated camera has multiple errors existing in the design, fabrication, and assembly process. In the actual assembly of an optical camera, the origin of the image coordinate system deviates from the optical axis of the uncalibrated camera. Thus, the coordinate position of the principal point is not equal to (0, 0) in the camera coordinate system. As the aforementioned definition in section 5.4.4, $(u_0, v_0)$

is the principal point of the uncalibrated camera in the actual optical system. In the actual optical system condition, equation (5.20) can be revised into:

$$
\begin{aligned}
[\hat{L}_{px}\ \hat{L}_{py}\ 1]^T &= \begin{bmatrix} L_{px} \\ L_{py} \\ 1 \end{bmatrix} + \begin{bmatrix} u_0 \\ v_0 \\ 1 \end{bmatrix} \\
&= \begin{bmatrix} f_{\text{camera}} & 0 & u_0 \\ 0 & f_{\text{camera}} & v_0 \\ 0 & 0 & 1 \end{bmatrix} \begin{bmatrix} \dfrac{L_{sx}}{f_{\text{SOL}}} & \dfrac{L_{sy}}{f_{\text{SOL}}} & 1 \end{bmatrix}^T
\end{aligned} \tag{5.21}
$$

Moreover, the optical system of an actual uncalibrated camera has optical distortion. The optical distortion is mainly produced by the optical design of the lenses used for the camera optical system. We use $\Delta(x, y)$ to express the position deviation between the actual and ideal position of a grid image because of the camera distortion. When the camera distortion is considered, the calibration equation of equation (5.19) can be revised as:

$$
\begin{aligned}
[\hat{L}_{px}\ \hat{L}_{px}\ 1]^T &= \begin{bmatrix} L_{px} \\ L_{py} \\ 1 \end{bmatrix} + \begin{bmatrix} u_0 \\ v_0 \\ 1 \end{bmatrix} + \Delta(x, y) \\
&= \begin{bmatrix} f_{\text{camera}} & 0 & u_0 \\ 0 & f_{\text{camera}} & v_0 \\ 0 & 0 & 1 \end{bmatrix} \begin{bmatrix} \dfrac{L_{sx}}{f_{\text{SOL}}} \\ \dfrac{L_{sy}}{f_{\text{SOL}}} \\ 1 \end{bmatrix} + \Delta(x, y)
\end{aligned} \tag{5.22}
$$

To solve equation (5.22), the position deviation, $\Delta(x,y)$, should be expressed based on a camera distortion model. To construct the expression of $\Delta(x, y)$, we use $(d_x, d_y)$ to denote an undistorted position in normalized image coordinates. The normalized image coordinates are calculated based on the grind image coordinates and the focal length of the optical system of the camera and the mSOL. When a radial distortion is considered, the deviation position, $\Delta(x,y)$, can be expressed as

$$
\begin{aligned}
\Delta(x, y) &= \begin{bmatrix} \Delta L_{px} \\ \Delta L_{py} \\ 0 \end{bmatrix} = \begin{bmatrix} dx \\ dy \\ 0 \end{bmatrix} K \\
&= \begin{bmatrix} \dfrac{L_{px}}{f_{\text{camera}}} \\ \dfrac{L_{px}}{f_{\text{camera}}} \\ 0 \end{bmatrix} K = \begin{bmatrix} \dfrac{L_{sx}}{f_{\text{SOL}}} \\ \dfrac{L_{sy}}{f_{\text{SOL}}} \\ 0 \end{bmatrix} K
\end{aligned} \tag{5.23}
$$

where $K$ is the distortion parameters. In the camera calibration, the radial distortion model is widely used. The $K$ of the radial distortion model is expressed as

$$K = k_1 r^2 + k_2 r^4 + k_3 r^6, \tag{5.24}$$

where $r$, $k_1$, $k_2$, and $k_3$ are the coefficients to represent the distortion of the optical system of the uncalibrated camera. The distortion coefficients can be calculated as

$$\begin{aligned} r^2 = x^2 + y^2 &= \left(L_{px}/f_{\text{camera}}\right)^2 + \left(L_{py}/f_{\text{camera}}\right)^2 \\ &= \left(L_{sx}/f_{\text{SOL}}\right)^2 + \left(L_{sy}/f_{\text{SOL}}\right)^2 \end{aligned} \tag{5.25}$$

In the combination of equations (5.22–5.25), the camera calibration equation can be expressed as:

$$\begin{bmatrix} \hat{L}_{px} \\ \hat{L}_{px} \\ 1 \end{bmatrix} = \begin{bmatrix} f_{\text{camera}} & 0 & u_0 \\ 0 & f_{\text{camera}} & v_0 \\ 0 & 0 & 1 \end{bmatrix} \begin{bmatrix} \dfrac{L_{sx}}{f_{\text{SOL}}} \\ \dfrac{L_{sy}}{f_{\text{SOL}}} \\ 1 \end{bmatrix} + \begin{bmatrix} \dfrac{L_{sx}}{f_{\text{SOL}}} \\ \dfrac{L_{sy}}{f_{\text{SOL}}} \\ 0 \end{bmatrix} (k_1 r^2 + k_2 r^4 + k_3 r^6) \tag{5.26}$$

Based on equation (5.26), the internal parameters of an uncalibrated camera can be calculated by solving the following optimization problem:

$$(\hat{u}_0, \hat{v}_0, \hat{f}_{\text{camera}}) = \min_{m} \left\| \begin{bmatrix} \hat{L}_{px} - u_0 \\ \hat{L}_{py} - v_0 \end{bmatrix} - f_{\text{camera}} \begin{bmatrix} \dfrac{L_{sx}}{f_{\text{SOL}}} \\ \dfrac{L_{sy}}{f_{\text{SOL}}} \end{bmatrix} (1 + k_1 r^2 + k_2 r^4 + k_3 r^6) \right\|^2 \tag{5.27}$$

### 5.4.5 The position extraction approach of the predefined target images

In order to accurately determine the position ($\hat{L}_{px}$, $\hat{L}_{py}$) in equation (5.27), an efficient position extraction algorithm of the predefined target images is necessary. The camera calibration widely uses two point-extraction algorithms. The first method is based on a centroid extraction method. The centroid extraction method firstly labels connected components in a binary image. Then, the set of properties of the connected components is measured. Finally, the centroid of a point image is extracted. The second method is based on a correlation operation technology. The principle of the extraction method based on the correlation operation technology is shown in figure 5.14. The correlation operation between a standard point image and an observed point image is performed to obtain the position information of each point. The correlation operation can be expressed as:

$$C(m, n) = \sum_{i=0}^{L} \sum_{j=0}^{L} P_1(m + i, n + j) P_0(i, j), \tag{5.28}$$

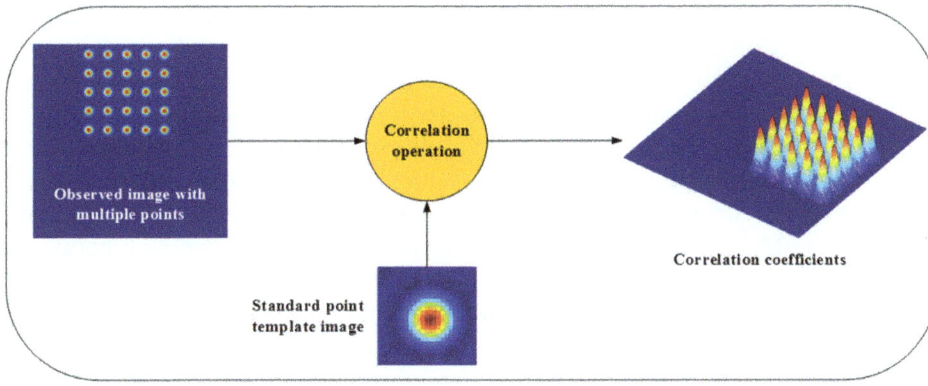


**Figure 5.14.** The position extraction algorithm based on the correlation operation technology.

where $P_0(i, j)$ is extracted template image. We use a centroid method to extract the position of each image point from the correlation value. The centroid value is calculated by the following equation:

$$x = \frac{\sum_{i=1}^{10}\sum_{j=1}^{10} m \times C(m, n)}{\sum_{i=1}^{10}\sum_{j=1}^{10} C(m, n)}, \; y = \frac{\sum_{i=1}^{10}\sum_{j=1}^{10} n \times C(m, n)}{\sum_{i=1}^{10}\sum_{j=1}^{10} C(m, n)} \quad (5.29)$$

The accurate position of an image point of mSOL can be calculated using equation (5.29). Here $(x, y)$ is equal to $[\hat{L}_{px} \;\; \hat{L}_{py}]^T$ in equation (5.27).

To further improve the extraction accuracy of point images, the optical aberrations of the imaging process can be corrected. In the calibration process, vibration, system errors, and extraction errors can produce optical aberrations existing in the point image captured by the uncalibrated camera. First, we extract an array template from the measured image. Second, multiple image points are used as the observed points to invert the point spread function (PSF) using sparse total variation technologies [73–75]. The reconstructed PSF is used to reconstruct the high-resolution image to correct the optical aberrations.

### 5.4.6 Applied examples

The mSOL can be configured with different scales for different application cases. The mSOL can be designed into a microstructure for an integration system. In this case, the micro-mSOL can be easily integrated into an optical camera system for the camera self-calibration, without the increase of the development complexity. Generally, the mSOL can be directly placed in the front of an uncalibrated camera to complete the calibration processing. Here, we take an optical camera with a three-mirror-anastigmatic (TMA) [76–80] optical structure as an example. A TMA optical structure usually two types: view-off-axis-type TMA and view-coaxis-type TMA.

To implement the integration between an mSOL and an uncalibrated camera, a feasible integration method is that the mSOL is installed directly on the truss of the uncalibrated camera. Under the general application case, the mSOL can be directly placed in the front of an uncalibrated TMA-based camera.

Figure 5.15 shows the application example of mSOL for calibrating a camera using a TMA-based optical system. The optical system of the camera is a view-coaxis-type TMA structure. The calibration installation configuration between the mSOL and the TMA camera should follow a rule. That is the alignment between the TMA camera and the mSOL is ensured. The calibration procedure is performed as the following steps. In the first step, the TMA camera works for obtaining an image with multiple grids. In the second step, the grid position in the obtained image is extracted. Third, equation (5.27) can be applied to measure the internal parameters of the TMA camera. In the fourth step, multiple measurements are performed using the three steps. Finally, the average measure can be used as the final calibration result of the TMA camera.

## 5.5 Discussion and future development directions

Fundamentally, the dynamic or static calibration method applies a predefined coordinate system to reconstruct the internal parameters of an uncalibrated camera. The dynamic calibration approaches can fully utilize the motion information in conjunction with the images to measure the internal parameters. The static calibration approaches use the predefined coordinate system with specific

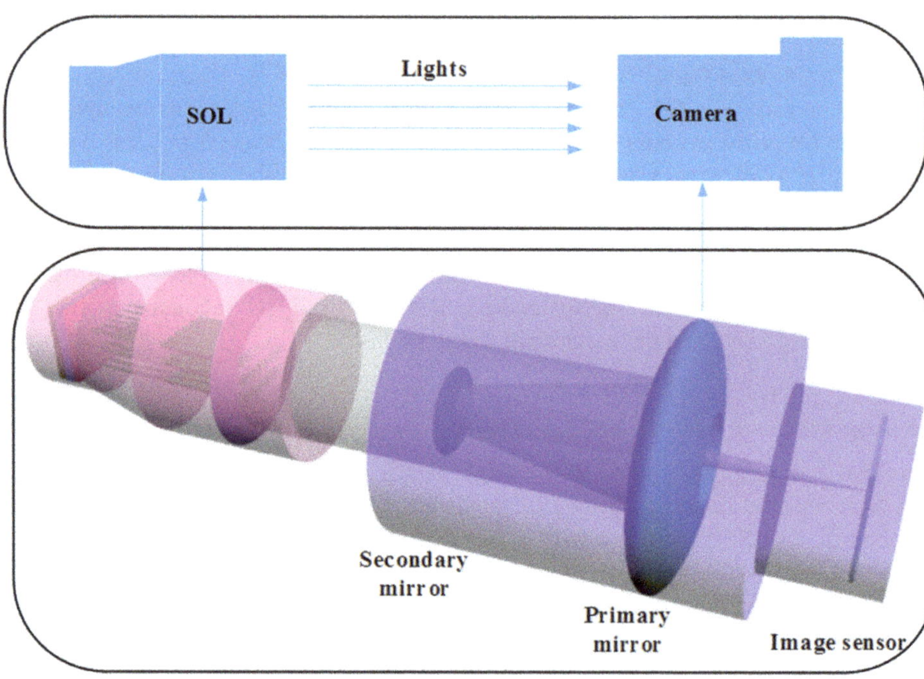


**Figure 5.15.** Calibration example with mSOL.

functionality to complete the calibration. Two approaches have different advantages and disadvantages. The use of both methods should be determined based on the actual calibration mission. In the future, there are several possible development directions. First, the combination of the dynamic and static methods may improve the calibration accuracy. Second, the construction of a perfect predefined coordination system is a promising direction. Third, the use of the calibrated system to calibrate the uncalibrated cameras is the new development direction. Fourth, the introduction of new components and new principles is a very interesting research direction.

## 5.6 Conclusion

The geometric calibration is a necessary step in a photogrammetry mission of optical imaging sensors. Two important geometrical calibration methods (i.e. static and dynamic calibrations) for optical image sensors have been introduced. Dynamic calibration approaches widely adopt a movement mechanism (e.g. turntable) to rotate an uncalibrated camera to obtain multiple positions from different FOVs. Static calibration approaches use the advance calibration components/systems (e.g. DOE) to complete the calibration under the static working of optical cameras. In this chapter, we demonstrate a valuable static calibration method using an mSOL, enabling direct generation of 2D predefined grids to complete the camera calibration. The mSOL-based method has several merits, such as high accuracy, easy operation, and less required auxiliary equipment. Two calibration methods have different application cases based on the user.

Acknowledgements: we would like to acknowledge the encouragement of Professor Yanyan Liu and Professor Guonig Li. We would like to acknowledge the assistance of our research team.

## References

[1] Laliberte A S, Goforth M A, Steele C M and Rango A 2011 *Remote Sens.* **3** 2529–51

[2] Govil K, Welch M L, Ball J T and Pennypacker C R 2020 *Remote Sens.* **12** 166

[3] Li J, Xing F, Sun T and You Z 2015 *Opt. Express* **23** 6187–208

[4] Kong L, Zhao Y, Dong L, Jian Y, Jin X, Li B, Feng Y, Liu M, Liu X and Wu H 2013 *Opt. Express* **21** 17464–71

[5] Ehlers M, Gähler M and Janowsky R 2003 *ISPRS J. Photogramm. Remote Sens.* **57** 315–26

[6] Endres F, Hess J, Sturm J, Cremers D and Burgard W 2013 *IEEE Trans. Rob.* **30** 177–87

[7] Tauro F, Olivieri G, Petroselli A, Porfiri M and Grimaldi S 2016 *Environ. Monit. Assess.* **188** 1–11

[8] Fraser C S and Ravanbakhsh M 2009 Georeferencing from Geoeye-1 imagery: early indications of metric performance *ISPRS Hannover Workshop (Hannover, June 2009)* ed C Heipke, K Jacobsen, S Müller and U Sörgel

[9] Aguilar M A, Aguilar F J, Mar Saldaña M D and Fernández I 2012 *Photogrammetric Eng. Remote Sens.* **78** 247–57

[10] Brian C 2014 Pleiades 1B and SPOT 6 image quality status after commissioning and 1st year in orbit *Proc. of the Joint Agency Commercial Imagery Evaluation (JACIE) Workshop (Louisville, March 2014)*

[11] Gaudin-Delrieu C, Lamard J L, Cheroutre P, Bailly B, Dhuicq P and Puig O 2017 The high resolution optical instruments for the Pleiades HR Earth observation satellites *Int. Conf. on Space Optics—ICSO (Toulouse, October 2008)* ed J Costeraste, E Armandillo and N Karafolas 105660C
[12] Zhang Z 2000 *IEEE Trans. Pattern Anal. Mach. Intell.* **22** 1330–4
[13] Zhang Z 2004 *IEEE Trans. Pattern Anal. Mach. Intell.* **26** 892–9
[14] Weng J, Cohen P and Herniou M 1992 *IEEE Trans. Pattern Anal. Mach. Intell.* **10** 965–80
[15] Heikkila J 2000 *IEEE Trans. Pattern Anal. Mach. Intell.* **22** 1066–77
[16] Herrera D, Kannala J and Heikkilä J 2012 *IEEE Trans. Pattern Anal. Mach. Intell.* **34** 2058–64
[17] Wei G Q and De M S 1994 *IEEE Trans. Pattern Anal. Mach. Intell.* **16** 469–80
[18] Maoling Q, Songde M and Yi L 2000 *Acta Autom. Sin.* **26** 43–55
[19] Tsai R 1987 *IEEE J. Robot. Autom.* **3** 323–44
[20] Ying X and Hu Z 2004 *IEEE Trans. Pattern Anal. Mach. Intell.* **26** 1260–71
[21] Wang L L and Tsai W H 1991 *IEEE Trans. Pattern Anal. Mach. Intell.* **13** 370–6
[22] Heikkila J and Silven O 1997 A four-step camera calibration procedure with implicit image correction *Proc. of IEEE Computer Society Conf. on Computer Vision and Pattern Recognition (Puerto Rico, June 1997)* ed D Plummer and I Torwick pp 1106–12
[23] Maybank S J and Faugeras O D 1992 *Int. J. Comput. Vision* **8** 123–51
[24] Faugeras O D, Luong Q T and Maybank S J 1992 Camera self-calibration: theory and experiments *European Conf. on Computer Vision (Santa Margherita Ligure, May 1992)* ed G Sandini pp 321–34
[25] Hartley R I 1993 Euclidean reconstruction from uncalibrated views *Joint European-US Workshop on Applications of Invariance in Computer Vision (Azores, October 1993)* ed J L Mundy, A Zisserman and D Forsyth pp 235–56
[26] De M S 1996 *IEEE Trans. Robot. Autom.* **12** 114–20
[27] Caprile B and Torre V 1990 *Int. J. Comput. Vision* **4** 127–39
[28] Fraser C S 2001 Photogrammetric camera component calibration: a review of analytical techniques *Calibration and Orientation of Cameras in Computer Vision (Washington, June 2001)* ed A Gruen and T S Huang pp 95–121
[29] Lipski C, Bose D, Eisemann M, Berger K and Magnor M 2010 Sparse bundle adjustment speedup strategies *WSCG Short Papers Post-conference Proc. (Plzen, February 2010)* ed V Skala pp 85–8
[30] Hong Y, Ren G and Liu E 2015 *Opt. Express* **23** 23992–4003
[31] Ricolfe-Viala C, Sanchez-Salmeron A J and Valera A 2012 *Opt. Express* **20** 27691–6
[32] Lin P D and Sung C K 2007 *Opt. Express* **15** 3012–22
[33] Wei Z and Liu X 2015 *Opt. Express* **23** 18897–914
[34] Yılmaztürk F 2011 *Opt. Express* **19** 18164–74
[35] Ricolfe-Viala C and Sanchez-Salmeron A J 2011 *Opt. Express* **19** 10769–75
[36] González-Aguilera D, Rodríguez-Gonzálvez P, Armesto J and Arias P 2011 *Opt. Express* **19** 2676–93
[37] Bauer M, Grießbach D, Hermerschmidt A, Krüger S, Scheele M and Schischmanow A 2008 *Opt. Express* **16** 20241–8
[38] Thibault S, Arfaoui A and Desaulniers P 2011 *Opt. Lett.* **36** 4770–2
[39] Fraser C S 1997 *ISPRS J. Photogramm. Remote Sens.* **52** 149–59
[40] Lichti D D and Kim C 2011 *Remote Sens.* **3** 1014–28

[41] Hieronymus J 2012 Comparison of methods for geometric camera calibration *Int. Archives of the Photogrammetry, Remote Sensing and Spatial Information Sciences (Melbourne, August 2012)* ed M Shortis and J Mills pp 595–9
[42] Li J, Liu F, Liu S and Wang Z 2017 *IEEE Sens. J.* **17** 1663–72
[43] Luhmann T, Piechel J and Roelfs T 2013 *Thermal Infrared Remote Sensing: Sensors, Methods, Applications* ed C Kuenzer and S Dech (Dordrecht: Springer) pp 27–42
[44] Remondino F and Fraser C 2006 *Int. Arch. Photogramm. Remote Sens.* **36** 266–72
[45] Honkavaara E, Ahokas E, Hyyppä J, Jaakkola J, Kaartinen H, Kuittinen R, Markelin L and Nurminen K 2006 *ISPRS J. Photogramm. Remote Sens.* **60** 387–99
[46] Lichti D D, Kim C and Jamtsho S 2010 *ISPRS J. Photogramm. Remote Sens.* **65** 360–8
[47] Beyer H A 1990 *ISPRS J. Photogramm. Remote Sens.* **45** 17–32
[48] Barazzetti L, Mussio L, Remondino F and Scaioni M 2011 Targetless camera calibration *Int. Archives of the Photogrammetry, Remote Sensing and Spatial Information Sciences (Trento, March 2011)* ed F Remondono and S EI-Hakim p 8
[49] Kornus W, Lehner M, Blechinger F and Putz E 1996 *Int. Arch. Photogramm. Remote Sens.* **31** 90–8
[50] Yuan F, Qi W J and Fang A P 2014 Laboratory geometric calibration of areal digital aerial camera *IOP Conf. Series: Earth and Environmental Science (Beijing, April 2013)* ed G Huadong 012196
[51] Yuan F, Qi W, Fang A, Ding P and Xiujuan Y U 2013 Laboratory geometric calibration of non-metric digital camera *MIPPR 2013: Remote Sensing Image Proc., Geographic Information Systems, and Other Applications (Wuhan, October 2013)* ed J Tian and J Ma 89210A
[52] Clarke T A and Fryer J G 1998 *Photogramm. Rec.* **16** 51–66
[53] Li J and Liu Z 2019 *IEEE Trans. Circuits Syst. Video Technol.* **30** 2550–8
[54] Li J and Liu Z 2017 *Microsyst. Nanoeng.* **3** 1–7
[55] De Lussy F, Greslou D, Dechoz C, Amberg V, Delvit J M, Lebegue L, Blanchet G and Fourest S 2012 Pleiades HR in flight geometrical calibration: location and mapping of the focal plane *Int. Archives of the Photogrammetry, Remote Sensing and Spatial Information Sciences (Melbourne, August 2012)* ed M Shortis and J Mills pp 519–23
[56] Fourest S, Kubik P, Lebègue L, Déchoz C, Lacherade S and Blanchet G 2012 Star-based methods for Pleiades HR commissioning *Int. Archives of the Photogrammetry, Remote Sensing and Spatial Information Sciences (Melbourne, August 2012)* ed M Shortis and J Mills pp 513–8
[57] Mian O, Lutes J, Lipa G, Hutton J J, Gavelle E and Borghini S 2015 *Int. Arch. Photogramm. Remote Sens. Spat. Inf. Sci.* **40** 397
[58] Mulawa D 2004 *Int. Arch. Photogramm. Remote Sens. Spat. Inf. Sci.* **35** 1–6
[59] Chiang K W, Tsai M L and Chu C H 2012 *Sensors* **12** 9161–80
[60] Greslou D, de Lussy F, Amberg V, Dechoz C, Lenoir F, Delvit J M and Lebègue L 2013 Pleiades-HR 1A&1B image quality commissioning: innovative geometric calibration methods and results *Earth Observing Systems XVIII (California, August 2013)* ed J J Butler, X Xiong and X Gu 886611
[61] Delvit J M, Greslou D, Amberg V, Dechoz C, Delussy F, Lebegue L, Latry C, Artigues S and Bernard L 2012 Attitude assessment using Pleiades-HR capabilities *Int. Archives of the Photogrammetry, Remote Sensing and Spatial Information Sciences (Melbourne, August 2012)* ed M Shortis and J Mills pp 525–30

[62] Jovanovic V M, Bull M A, Smyth M M and Zong J 2002 *IEEE Trans. Geosci. Remote Sens.* **40** 1512–9
[63] Li J and Tian S F 2017 *Sci. Rep.* **7** 1–9
[64] Li J and Zhang Y 2017 *IEEE Photonics Technol. Lett.* **29** 1939–42
[65] Li J, Xing F, Chu D and Liu Z 2016 *Sensors* **16** 1176
[66] Li J, Zhang Y, Liu S and Wang Z 2016 *Remote Sens.* **8** 893
[67] Li J and Liu Z 2018 *Sens. Actuators* A **273** 98–106
[68] Zhang G, He J and Yang X 2003 *Opt. Laser Technol.* **35** 457–61
[69] Hartley R and Kang S B 2007 *IEEE Trans. Pattern Anal. Mach. Intell.* **29** 1309–21
[70] Prescott B and McLean G F 1997 *Graph. Model Image Process.* **59** 39–47
[71] Wang J, Shi F, Zhang J and Liu Y 2008 *Pattern Recognit.* **41** 607–15
[72] Liu Z, Jiang Y, Li Y, Li J, Li Z, Zhang S, Lian Y and Liu R 2019 *IEEE Access* **7** 137552–9
[73] Li J, Liu Z and Liu F 2017 *Opt. Express* **25** 4018–37
[74] Chan T F and Wong C K 1998 *IEEE Trans. Image Process.* **7** 370–5
[75] Marquina A 2009 *SIAM J. Imag. Sci.* **2** 64–83
[76] Gu Z, Yan C and Wang Y 2015 *Opt. Express* **23** 1398–405
[78] Korsch D 1977 *Appl. Opt.* **16** 2074–7
[79] Meng Q, Wang H, Wang K, Wang Y, Ji Z and Wang D 2016 *Appl. Opt.* **55** 8962–70
[80] Zhang X, Xu S, Ma H and Liu N 2017 *Opt. Express* **25** 12867–83

**IOP** Publishing

Advances in Modern Sensors
Physics, design, simulation and applications
**G R Sinha**

# Chapter 6

# Smart and wearable sensors used in numerous modern applications and their significance

**N Hema and Suma Dawn**

Smart systems play an important role in today's world. Due to technological evolution and introduction of Industry 4.0, technological invention also needs upgradation. This revolution has triggered development of smart and connected devices. This chapter introduces what smart sensors are, and what properties they should possess to be called smart devices. Smart systems are built from smart sensors. This chapter details how sensors are designed to be smart sensors for smart application. Many popular examples of smart applications currently in use are discussed, such as: smart city, smart home, smart agriculture, smart water meter and many other popular applications. This chapter also introduces the smart factory, which is again a developing technology due to Industry 4.0. This chapter offers the knowledge to develop your own smart system by giving exposure of the required sensor and its properties. These properties need to be programmed in a processor that is interfaced with a sensor, which converts a simple sensor into a smart sensor.

Due to evolution in smart sensors technology, most of the health monitoring devices have advanced, especially wearable sensors. These wearable sensors are integrated into wearable objects and they should be portable, consume less power, be user friendly and so on. These wearable smart sensors have a gyroscope and accelerometer, wearable electrodes to measure the electric pulses, temperature sensor, altimeter, proximity sensor and biochemical sensors as common sensors in design of smart wearable devices. Various vital health measuring smart wearable devices are discussed with their architecture and the typical smart sensor used in them.

## 6.1 Introduction

Sensors are basically used by various automated and embedded systems developers where their job is to collect the physical environmental input continuously and send

doi:10.1088/978-0-7503-2707-7ch6  

it to the controller unit to process the sensor data further. As technology has evolved in recent years, IoT-based applications have become part of our daily lives. In such IoT applications, sensors are not only required to capture the data, but also to do more functionality with the goal of reducing data communication and power consumption. Solution(s) to these problems is based on the development of smart sensors.

Smart sensors are an integration of sensors and actuators with a processor and a communication module for specific tasks. These smart sensors are the device that not only takes input from the physical environment but at the sensor level itself, some processing of these data is done using built-in computational resources such as microprocessors and controllers to perform certain predefined tasks. These may be dependent on detection of certain specific inputs. The data and operations are processed before transmitting the intermediary output on to the next level processing through communication. As defined by the IEEE 1451 standard, smart sensors are sensors with small memory and standardized physical connection to enable communication with the processor and data network. Smart sensors have properties such as self-calibration, communication, computation, multi-sensing, and cost improvement.

Smart sensors are mainly used in smart devices and in smart factories. Smart devices that we use in our day-to-day life such as Amazon Echo, Smart Watch, Smart TV, Philips Hue bulbs, enable you to control both intensity of light and the color of your lights; Ecobee4 allows you to control air temperature with voice commands, NetGear Arlo Q is a truly useful tool for home security, EcovacsDeebot N79S is used for the automated vacuum cleaner, smart parking, smart traffic control and so on. This smart autonomous device saves a significant amount of time in our busy routine lives. Details of such smart devices will be discussed in this chapter.

Smart factories have various industrial smart sensors that are connected for optimization, transparency, productivity and agility. The need for this smart factory is to connect the physical objects, sharing the data between physical objects, increasing resource efficiency and to increase productivity. The benefits of such smart factories have production which is faster, cheaper, better usage of product, improved recall process of product, decreased environmental impact and better smart supply chain. Smart sensors not only warrant additional accuracy and automation in the process of data collection of the existing physical environment but also enable transmission of more accurate data that are not constrained with many noisy data amongst the less faulty recorded and documented information. These devices enable accurate and robust monitoring and controlling procedures that can be made use of in numerous environments, including smart grid systems, reconnaissance in the battlefield zone, and exploration of science applications, amongst others. Some of the popularly known smart factories are Whirlpool, Siemens, Hirotec, Hewlett-Packard, and Ocado. Even food industries are also gearing up for smart automation. Details of such smart sensors and smart factories are discussed in this chapter.

When it comes to vital sign monitoring of health parameters, then wearable sensors play an important role. Wearable sensors, especially devices that monitor body temperature, blood pressure, respiration, glucose, heart rate, oxygen saturation, electrocardiogram, and others are progressively in-line with health monitoring requirements. These also provide an understanding of the status of activity/fitness level for tracking self-health. At the clinical level, these devices can be used for providing more data to technicians, doctors, patients and other stakeholders in the health industry with the potential for early diagnostic, assessment and treatment guidance.

## 6.2 Smart sensors properties

Typically, sensors perform a specific function of input by sensing the physical variation in the characteristics of a system, in response to some physical stimulus. This stimulus could be variation in the temperature, variation in light condition, variation in resistance or changes in the chemical composition of gas. Gas sensors are available for checking different chemicals composition like a methane gas sensor, a carbon dioxide sensor, carbon monoxide sensor, and oxygen sensor. Therefore, gas sensors are able to measures any changes in these chemical compositions in the environment and send this signal to the device connected.

In addition to sensing features, smart sensors should have embedded intelligence to make the decision to filter the undesired data for efficient communication. Therefore, a smart sensor (or intelligent sensor) requires system-on-chip-like microcontrollers with the sensing module, interfacing, signal processing and intelligence through decision making coding. The design goal of smart sensor systems is to provide data to the user with increased reliability and integrity.

The smart sensor intelligence mainly includes testing, identification, validation or adaptation functions, all on self, for sensor's data. Figure 6.1 gives typical smart sensor structure. Sensors used in the smart sensor system can be passive or active sensor, analog or digital sensor or scalar or vector sensor.

The modern definition defined by Yurish [1] of smart or intelligent sensors can be formulated now in the following way: '*Smart sensor is an electronic device, including sensing element, interfacing, signal processing and one- or several intelligence*

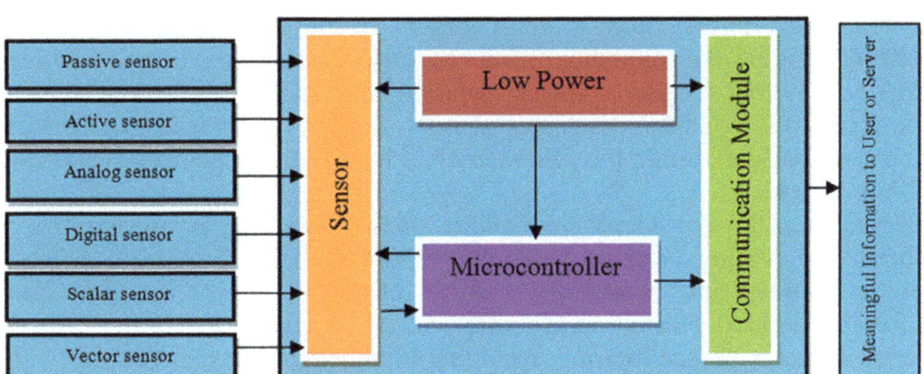


**Figure 6.1.** Smart sensor systems architecture.

*functions as self-testing, self-identification, self-validation or self-adaptation*'. The '*intelligence*' is the keyword in this definition. Among all the properties, self-adaptation is an important and new function of smart sensors and sensor systems in the current scenario. Adaptive algorithms are used for self-adaptation smart sensors systems. These adaptive algorithms are based on the precision measurements of parameters of frequency-time.

While designing smart sensors, they should have certain capabilities before sending meaningful information to the user or server. The following are the main property on any typical smart or intelligent sensor [2]:

i. self-calibration;
ii. reliability or self-health assessment;
iii. self-healing;
iv. compensated measurements;
v. self-adaptability for exchange accuracy for speed and conversely;
vi. Self-adaptability for low power consumption.

### 6.2.1 Self-calibration

Sensors data plays an important role in defining the accuracy of the system developed. Calibration of such sensors is important, it is the process of correcting systematic errors while reading data. The process involves the comparison of two sensors' data: one from a known sensor measurement (standardized units) and another from the unknown sensor to adjust the correction parameters in order to provide an accurate measurement [3].

Smart sensors based devices once installed in-field are exposed to environmental noise and hardware failures. This exposure leads to the mismatch between initial or factory calibration conditions with in-field conditions making the calibration process challenging. Due to this reason, in-field sensor calibration schemes become impractical, time consuming, and difficult to achieve [3].

Design of reconfigurable systems needs collaboration of smart sensors which are capable of handling various sensor inputs. The reconfiguration system cannot afford spending more time in their calibration and hence requires self-calibration or autocalibration sensors. Since a smart sensors system has a microcontroller unit (MCU) in it, it has the capability of running various algorithms for self-calibration. Rivera *et al* [4] has proposed artificial neural networks (ANN) for autocalibration methodology for nonlinear intelligent sensors. Možek *et al* [5] has designed self-learning and self-calibration for smart sensors using a digitally controlled closed-loop calibration system.

Pereira *et al* [6] considered probability density function as an adaptive self-calibration algorithm for smart sensors to minimize the calibration time, increase level of accuracy and reduce the number of calibration points. The chapter 'Calibration and self-calibration of smart sensors' by Michiel Pertijs in [7] discusses different forms of self-calibration, including interfacing of a smart sensor with an actuator to generate local calibration signal, and interfacing of additional sensors to reduce the cross-sensitivity.

### 6.2.2 Reliability or self-health assessment

Smart systems are equipped with various sensors to sense the surrounding environment to make decisions based on these sensors' input and system behavior. Unfortunately, sensors are susceptible to faults due to physical wear and tear or aging. These faults lead to failure of the smart system. Reliability of the sensor plays an important role in a smart sensor system. Detecting of these faults and diagnosing a fault at sensor-level itself, is very important [8]. Due to hardware fault there can be various sensor errors. Elhoushi *et al* [9] classify these errors into deterministic or stochastic.

### 6.2.3 Self-healing

Wearable sensors have to be flexible and this flexibility has problems that lead to scratching and other damage, which potentially destroys their electronic functionality. Such wearable sensors need to be self-healing. Self-healing materials are an emerging class of smart materials, and hence smart sensor. These sensors are capable of automatic or stimulated based repair of their damage under external stimuli, such as heat, light, solvents, etc [12].

Huynh and Haick define the term 'self-healing sensor' [13]. The self-healing sensor idea was generated to mimic human skin wound healing. However, the healing mechanism of human skin is more complicated and the effort has been made to create simple self-healing sensors. Usually these sensors are made of chemiresistors. A chemiresistor based self-healing sensor is a resistor that responds to changes in its resistance. The self-healing skin sensor material has synthetic polyurethane [13].

Recent nanotechnology has improved application of metal oxides. In biosensor application, metal oxide based sensors are popular. These sensors effectively interact with target molecules depending on the surface of the metal oxide. Also, to improve the sensitivity of the sensor, tailoring of the surface properties of metal oxide is important [14].

Organic semi-conducting molecules or polymers are used in the organic electronic sensors that are used in wearable devices. These organic sensors have self-healing polymers due to their nature of insulation [15]. Most of the self-healing sensors are based on supramolecular chemistry where molecular interactions constitute hydrogen bonding, metal-ligand coordination, pi-pi stacking, and so on. Many searches include different types of substance used for the self-healing. For example, anticorrosion coatings of self-healing use rare earth compounds [16]. Microvascular-based self-healing materials are used for large-scale fracture damage [17].

Sensors are designed to identify mechanical deformation, which is required in the future of artificial intelligence products. The design of these sensors requires them to be sensitive, versatile with regard to strain, and self-healing. Hence, they are developed using the hierarchical structure design of combining metal–ligand chemistry [18]. Wearable sensors like fit bands that track heart rate or steps by the user are popular. To monitor sweat while exercising, a headband has been designed by South Korean scientists. Sweat can be measured to obtain potassium and sodium ions to monitor the health of the person. These headbands have been designed with flexible threads that can heal on their own within 20 s. These threads are made of carbon fiber electrodes coated with a citric acid-based polymer, and can measure electrolyte concentrations within sweat. Self-healing is possible due to hydrogen bonding that is incorporated in the polymer [19].

The most two important errors that commonly occur in sensors are stuck error (bias) and drift error. Bias error always returns a constant value, whereas drift error returns values which continually drift positive (or negative), and these two faults are hard to detect [10]. Different fault detection methods are proposed depending upon the type of the sensor used in the system to detect the failure of sensors and hence failure of the system. Various diagnosis methods including sensor/faulty component isolate are adapted in the smart sensor itself to make the sensor data more reliable. It is always efficient to have passive sensors as a backup to replace faulty sensors and improve self-health assessment. Smart wearable body sensors used in real-time applications should have fault detection and diagnosis properties in them.

Most of the common approaches [8] to identify faults are: (a) correlation tests are applied on the sensor readings. Here, more than one sensor reading is required from the same type. (b) Check for predefined suspicious patterns; if identified then mark them as uncertain input. (c) Apply a structural model to verify that the uncertain sensors reacted to a fault and to diagnose the root cause of the fault. Bayesian belief network (BBN)-based sensor fault detection and identification is also proposed by Mehranbod *et al* [11]. BBN is proposed for detection and identification of different sensor fault types like bias, drift and noise.

### 6.2.4 Compensated measurements

To increase the reliability of smart sensors the silicon manufacturer also has to design multi-layered sensors and sensor arrays that are able to provide internal compensation. Depending on the availability of different sensing material we may have different types of smart sensors with reduced error and cost. Sensors are designed to measure the physical changes in the environment [20].

A MEMS accelerometer can be operated in either a closed-loop or an open-loop mode. In general, maximum linearity and bandwidth can be obtained in closed-loop mode and maximum dynamic range is obtained in open-loop mode [21]. Some of the good quality, expensive and factory calibrated sensors may have some offset error. But offsets do not change the sensitivity of the output; due to this characteristic, most of the errors appear as a simple shift of the offset point and an error-free sensor output is still obtained. The process of obtaining error-free sensors output is the auto-zero calibration technique, and this can be powerful to implement in applications [22, 23].

Multi-sensor systems may require frequent recalibration, which is generally costly and time consuming. This cost is due to the large number of reference sampling and also due to limited availability. Instead of using auto-calibration, recalibration of the sensor system using a full set of calibration samples, drift correction and calibration update can be used [24].

For dynamic assessment of a real-time smart system, compensation algorithms can be directly programmed in hardware, embedding them in a data acquisition system (DAQ). Different compensation algorithms can be used based on the capability to facilitate the readjustment of sensor reading systems for easy measure of any variable and eliminate the systematic errors. Some of the compensation algorithms that are used in DAQ are: (i) Piecewise Linear Interpolation Compensation Algorithm; (ii) Polynomial Progressive Compensation Algorithm; (iii) Improved Polynomial Progressive Compensation Algorithm; and (iv) Artificial Neural Network Compensation Algorithm [25].

The current design of smart sensors has both sensor and dedicated interface electronics as a single system-in-package [20]. To obtain different performance characteristics, these sensors need to be configured in different modes. One such example is design of a smart temperature sensor, in which a normal temperature sensor IC built using CMOS technology has inaccuracies of less than ±2 °C. This accuracy is the best accuracy obtained so far in a CMOS temperature sensor. With a similar concept, smart wind sensors and smart Hall sensors can be designed.

### 6.2.5 Self-adaptability: exchange accuracy for speed and vice versa

To refer to a sensor as smart it needs to have self-adaptability properties within it. Self-adaptability is based on adaptive algorithms used in the measuring systems. Due to changing properties of a signal or measuring condition, measuring procedure is changed automatically and this is called adaptive measurements [26]. This adaption of changing measurement is due to increasing the accuracy, decreasing measuring time and reduction of power consumption. Different adaptation algorithms are designed based on the need of the smart sensor system. For example, a pressure sensor system for gas pipelines and antilock braking system (ABS) for automobiles has adaptive measurements based on the different sensing elements. A functional block diagram was designed by Kirianaki *et al* [27].

The adaptive algorithm resides in the sensor system which requires computing power (microcontroller or embedded computer). Such self-adaptive sensors systems are developed by integrating the sensor with UFDC and universal sensors or transducers interfaces (USTI) which is the minimal possible hardware and lowest cost. In the gas pipe, quick measurement of pressure of the gas is very important and this is an example of a self-adaptive smart pressure sensor system [26].

#### *6.2.5.1 Self-adaptability: low power*

The success of a smart sensor system based application depends on the how long the batteries are going to last. The demand for ultra-low power sensors is designed to reduce power consumption. New sensors to reduce power consumption by 50%–60% over older solutions have been designed [28]. Sensors are designed to do some computation and data analysis internally. This is called smart hardware ‘digital block’. An advanced embedded pedometer is an example of a hardwired step counter that works independently inside the sensor without CPU intervention. A good real-world example is industrial predictive maintenance, where the sensors can categorize and identify different malfunctioning states in the equipment before waking the microprocessor to react.

Designing smart sensor systems for harsh commercial environments places unique demands on developers to reduce power consumption. While smart sensor systems for consumer and industrial applications may seem similar on the surface, developers must deal with significant differences in power, connectivity and robustness requirements. One way to achieve this is to keep the processor in sleep mode and for wakeup only when it is necessary for the device to transmit and receive information with restricted connection time. For communication, a low power wireless system is used like Bluetooth Low Energy [29].

In addition to the above mentioned smart sensor properties, we also need self-adapting software. Several IoT frameworks are also designed to include various requirements to achieve different objectives in a changing environment, and the requirements must be dynamically satisfied at runtime [30].

## 6.3 Smart sensors types

Sensors simply convert physical variables into electrical signals. They have needed to evolve into something more sophisticated to act as smart sensors. The following functionalities are required in smart sensors [31]:

(i) low cost;
(ii) physically small;
(iii) wireless or wired connection;
(iv) self-identification and self-validation;
(v) very low power;
(vi) robustness;
(vii) self-diagnostic and self-healing;
(viii) self-calibrating;
(ix) data pre-processing.

The following are some of the popular sensors used by many smart sensors' applications [32]:

A. *Proximity sensors*: measure the location of the neighbouring objects without any physical contact.
B. *Position sensors:* detect the existence of objects in a restricted area by detecting their motion.
C. *Motion detection sensors:* used to detect all the kinetic and physical movements present in the location.
D. *Velocity sensors*: compute the rate of modification in constant location capacity and location values at known intervals.
E. *Temperature sensors:* helpful in measuring heat energy.
F. *Pressure sensors:* sense the expanse of force and convert it into appropriate signals.
G. *Chemical sensors:* used to quantify the chemical configuration of the setting.

H. *Humidity sensors*: quantify air temperature as well as moisture and signal the humidity in the setting.
I. *Water quality sensors:* used for ion observations in addition to temperature, PH, turbidity, conductivity, and measurement of dissolved oxygen.
J. *Gyroscope sensors*: sense any slope or angular movements in the object by quantifying angular velocity.
K. *Optical sensors*: are beneficial in sensing electromagnetic vitalities like light.
L. *Gas sensors:* used to monitor changes of the air quality and detect the presence of various gases.
M. *Smoke sensors:* sense smoke (airborne particulates and gases), and its level.
N. *Level sensors:* determine the level or amount of fluids, liquids or other substances that flow in an open or closed system.
O. *Image sensors:* used to convert optical images into electronic signals for displaying or storing files electronically.

Depending on the application requirement we can have different smart sensors systems. For example, the **smart temperature sensor**, as shown in figure 6.2, is the one which has a temperature sensor LM 35 (analog sensor) interfacing circuit with a processor that has some communication capability. These sensors not only read data from the sensor but also do pre-processing before sending them further.

Now instead of analog, we have a digital temperature sensor like the DS1621, which generates 9 bits temperature data. The operating voltage of this sensor is from 2.7 to 5.5 V, then the user can define thermostatic settings by using the value of resistors R1 and R2 from 4.7 to 10 KΩ to design a **digital smart temperature sensor.**

An accelerometer sensor generates the magnitude and direction of the acceleration. One of the popular ADXL335 accelerometer sensors provides three axes (*X*, *Y*, and *Z*) values in analog voltage. To develop a **smart accelerometer sensor** an interfacing circuit is designed.

The use of these sensors includes a wide variety of applications such as: navigation systems in aircraft and missiles; tablets and digital cameras to keep images on screen upright; in drones for fight stabilization; to detect vibration in cars, machines, buildings; process control systems and safety installation. Also, they are used to study the behavioral patterns of animals.

Another very popular sensor is the gas sensor MQ-2. It measures and detects concentration of different gases. Gas sensor MQ-2 provides the concentration of

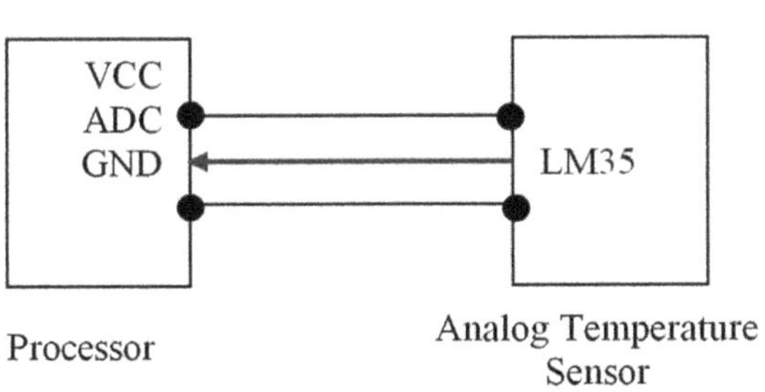


**Figure 6.2.** Analog smart temperature sensor.

LPG, propane, and hydrogen in analog voltage. To design a **smart gas sensor** we need an interfacing circuit.

Applications of gas sensor include: air quality monitoring; leakage detection of toxic gases; manhole and sewage treatment; automotive exhaust and alcohol breath test. All smart sensors have to withstand harsh environments and need to be robust while acquiring the data. Different algorithms for *self-scrutiny*, *self-characterization*, and *self-corroboration* are programmed inside these processors of smart sensors to withstand harsh environments and provide more accurate and timely data.

## 6.4 Smart sensor applications

Various sensors are used in several areas of application for depiction of smart as well as varied IoT system requirements. Some of the applications of smart sensors are discussed in the next session.

### 6.4.1 Smart cities

This includes smart parking, noise handling, traffic management, waste management, smart lighting and building structural health monitoring. A variety of sensors are used for measuring various metrics that include:

(1) **Smart parking system:** this requires monitoring of parked vehicles and smartly parking the vehicles to reduce traffic congestion and also from an economic interest. Due to recent IoT and smart system evolution, drivers can efficiently identify the suitable parking spaces with smart parking services. Infrared sensors, ultrasonic, optical sensors, cameras, acoustic, piezoelectric, smart energy-efficient sensors, mobile sensors, accelerometers, gyroscope sensors, light sensors and RFID are some of the common sensors used by various research works on smart parking between 2001 and 2016 [33]. These sensors with lower power device communication devices like Bluetooth form the smart parking system. Here sensors capture the environment facts and microcontrollers make the decision about the free space and inform the users via communication.

(2) **Building structural health:** this system monitors the physical conditions of the ingredients and volume of vibrations of nearby structures such as buildings, monuments, roads, bridges, etc, to predict the health and strength in structures. Fiber optic, strain gauge, wind velocity and direction, temperature and humidity, vibration, tension force, acceleration, speed and axle meters, digital camera, temperature sensors, pressure, displacement sensors, and vibration sensors [34] are some commonly used sensors to measure structural health. Any damage due to physical damage or overloading load of the structure are measured in real time and any serious measurements are to be notified via wireless communication.

(3) **Smart noise handling:** sound monitoring aims to avert any form of mishappening in crowded zones, gunshots, screams, horns, and road accidents in real time. Different kinds of acoustic, meteorological, $CO_2$, and

temperature sensors are used, along with low power wireless communication devices to build a smart noise handling system [35].

(4) **Smart traffic management:** road congestion monitoring can be implement using sensors like infrared, RFID, motion, camera, smart phone, and radars [36]. Currently Google Maps uses the smart phone and direction–distance sensors data which estimate speed of the vehicle to predict the traffic congestion.

(5) **Smart waste management:** to detect full trash containers, trash segregation and to pick up trash from different trash containers, optimized routes can be proposed. Again, depending on the complexity of waste management one can have various sensors in the system with low power communication [37].

(6) **Smart street light system:** depending on the intensity of the daylight, streetlights can be turned on/off, which can reduce human intervention as well as save electricity. Adaptive and intelligent lighting systems can be explored for public places as well [38]. Some of the common sensors used are light sensors, temperature sensors and power meter sensors.

### 6.4.2 Smart environment

Environmental monitoring sensors also play an important role in the economy for wellbeing in any country. Various smart environment monitoring applications are available, like air pollution detection, forest fire detection and catastrophy warning. Many sensors such as light, temperature, humidity, chemical, gyroscope, biosensors, accelerometer, smoke, optical and others can be widely used to create smart environment sensors.

(1) **Forest fire detection:** various sensors like smoke, temperature and humidity sensors capture forest fires, and adaptive neuro-fuzzy inference system (ANFIS) is used to calculate the true presence of fire and generate a fire alert. This adaptive algorithm measures the changing rate of smoke, the changing rate of temperature and humidity to check the presence of fire and its potential spread [39].

(2) **Air pollution detection:** temperature and humidity sensors, AM2301 and PM2.5 are specific sensors used to monitor and control the emissions of harmful and toxic gases of factories and vehicles within a certain vicinity. Increase in pollutant levels can be handled in a timely way so as to protect the environment to a greater extent [40].

(3) **Natural disaster warning system:** various parameters like earth vibrations, moisture in the soil, density etc, are monitored for the possibility to identify danger well before a landslide or earthquake. Various sensors like weather, horizontal and vertical liquid level, vibration, $CO_2$, temperature, dust, and microphone are used by the natural disaster warning system. The horizontal and vertical liquid level sensors are used for flood identification and vibration sensors measure the ground vibration to detect a landslide/earthquake. The $CO_2$, temperature, dust, and microphone sensors are used for noise pollution. Generally, to build smart sensors for modern devices,

various sensors like weather, horizontal and vertical liquid level, vibration, $CO_2$, temperature, dust, and microphone are integrated on single boards with GPRS modules, and rechargeable batteries with integrated wireless communication [41].

#### *6.4.2.1 Smart water management system*

Water monitoring is required to check water quality, its ground level, pollution level and leakages in containers to save water. Temperature, humidity, water meter, pressure, occupancy, and water quality sensors are some common sensors used in a smart water management system. A smart sensors system built with these sensors can be used for applications in the proper management of water. The following are some of the specific water management systems:

(1) **Underground water leakages:** for this, water pressure and acoustic sensors are used to detect tanks and pipes or underground pipe leakages to avoid damage caused by them [42].
(2) **Flood monitoring:** horizontal and vertical liquid level sensors are used to monitor the level of water and its variations from time to time in rivers, dams or reservoirs, which can save lives and crops from damage.
(3) **Smart water meters:** real-time smart water meters monitor water usage of the user, and they are an essential and key component of the water management system. This system controls the valves of a pump to not waste water and will make users mindful of their water consumption and help them to reduce their daily water usage. Also, any abnormal water usage is alerted to reduce water loss [43].

#### *6.4.2.2 Smart security system*

Light, gyroscope, biosensor, chemical, magneto, accelerometer, temperature, proximity, position and infrared sensors are used to build smart security systems in the surroundings. Some of the applications providing security are discussed in the following:

(1) **Smart perimeter control:** whenever there is any entry in an unauthorized area, an alert and action can be implemented in restricted areas. Many variants of biosensors are used to identify whether the person is authentic or not. One such method is the face recognition system [44] and iris detection system and so on.
(2) **Smart explosive detection:** detection of harmful gas levels, radiation levels, and leakages can be done smartly with the use of sensors. Sensors can be attached to the walls or containers at factories, mines and industry, where there is a danger of leakage happening.

#### *6.4.2.3 Smart agriculture*

Temperature, humidity, water quality, chemical, proximity, PH of soil, soil moisture, rainfall, sunshine, acoustic and position sensors are widely used in numerous agriculture applications. The data required for the agriculture application are amount of moisture in the soil, climate conditions and pesticides for the desired crop. In a smart irrigation system, moisture soil sensors are attached to the ground that can detect the moisture in the soil. In addition to soil moisture data, humidity, temperature and crop type decide the appropriate amount of watering to be dispatched. These smart irrigation systems also take real-time precipitation and prediction data into consideration to reschedule the irrigation [45–47].

(1) **Smart crop management:** sensor data of weather conditions in the environment are used to forecast snowfall, ice formation, rain, and drought. Crop management can be done to prevent damage caused by weather, especially for greenhouse vegetation [46]. Smart irrigation uses data from climatic and soil moisture sensors to schedule the irrigation [45, 47]. Currently, drones are also used in smart agriculture for the overall monitoring of the crop.

(2) **Smart water sprinkler system:** spatial temporal sprinkling is used to reduce the required amount of water. This ultimately improves the production of fruit, vegetables and crops, and their quality with limited resources. Indoor farming is in need of such smart water sprinkler systems to automatically schedule crop irrigation.

(3) **Smart compost:** once organic waste segregation is done, the next step is to compost it into manure. The right humidity and temperature are important for the prevention of fungus and other microbial contaminants to reduce into manure. In the current scenario, it is an interesting area of smart agriculture.

#### *6.4.2.4 Smart homes*

Light, gyroscope, biosensor, chemical, magneto, accelerometer, temperature, proximity, position, infrared and camera sensors are integrated to different devices at home to creating a smart environment. Smart homes make decisions on when to switch on/off appliances, and detect intruders by sending alerts which results in a better lifestyle. Some applications in smart homes are as follows:

(1) **Remote control of appliances**: intelligently switching on/off appliances makes life easy and safe by preventing accidents and also saving energy.

(2) **Intrusion detection systems:** due to increase in the nuclear family, houses are left locked for many hours and can attract intruders. The smart home can detect intruders by motion or with biosensors at the entrance and report this to the concerned user. One can remotely watch their home and be assured of the security. Authors in [48] presented an intrusion detection system which senses motion at the door and alerts the owner by telegram message.

(3) **Energy and water saving:** smart water meters are used to control the consumption of water and in turn save electricity.

### 6.4.3 Smart factories

Factories have moved from automation to smart factories due to evolution in Industry 4.0. Automation, the application of artificial intelligence (AI) and cyber-physical systems has combined physical machines and business processes to constitute smart factories. According to Deloitte University Press [49], the 'smart factory is said to be a flexible system that can self-optimize performance across a broader network, self-adapt to and learn from new conditions in real or near-real time, and run entire production processes.'

Smart factories are needed due to evolution of technologies, high competitive market, and high amount of production within minimum timeline and to reduce risk of failure. The advantage of using smart factories is to reducing cost, increasing efficiency, improving quality, improving predictability and improving safety.

Major components of the smart factories include:

- smart machine;
- smart engineering;
- innovation;
- smart devices;
- smart manufacturing process;
- information technology;
- data analysis.

These components should have communication between them such that a smart machine should communicate with smart devices. The smart manufacturing process should be dynamic, automated, real time and efficient. Smart engineering should follow smart design of the product, smart development of the product and smart planning. Information technology should provide smart software application, smart monitoring and control and a smart management process. The main five key characteristics of a smart factory are:

(1) connection;
(2) optimization;
(3) transparency;
(4) proactivity; and
(5) agility.

Some major features of a smart factory include: connectivity, optimization, transparency, proactivity, and agility. **Connection** is the important feature of the smart factory, requiring processes and materials that can be connected to smart devices, smart machines with data and processes. An **optimized** smart factory allows tasks to be executed with minimal human intervention with high reliability. The automated workflows, improved tracking and scheduling, synchronization of assets and minimal energy consumption inherent in the smart factory can increase productivity, uptime, and quality, as well as reducing costs and overall waste. In the smart factory, the data captured are **transparent**; in addition to this, real-time monitoring, handling of required actions on time, alert message generation and real-time tracking also need to be transparent. In a **proactive** system, employees and systems can anticipate and act before issues or challenges arise, rather than reacting afterwards. This feature can include identifying anomalies, identifying and predictively addressing quality issues, restocking and replenishing inventory, and monitoring safety and maintenance concerns. **Agile** flexibility allows the smart factory to adapt to schedule and product changes with minimal intervention in addition to the self-configuring of the equipment and material flows. Smart factories will leverage multiple digital and physical technologies, including a digital twin, digital design, additive manufacturing machines, and autonomous robots.

## 6.5 Case study: smart home surveillance system using a smart camera

Today's surveillance cameras keep recording 24/7, and even provide remote live feed on mobile devices. But these devices do not provide feedback of unusual events happening in the house. If a crime/theft takes place in the owner's absence, rewinding footage will retrieve what happened in the past hours; but no action can be taken when the crime is being committed unless remote live feed is obtained, which is not possible all the time in this busy world.

By ignore this problem, people may catch the criminal with the recorded footage but this process of catching the criminal has a lot of hassle and is a very time and resource consuming task with increasing criminal activities.

In today's India, most millennials live alone in cities and people who stay in families are mostly from nuclear families, with most people working in the family leaving the house empty in the day. Additionally, with the new trend of e-commerce, a lot of things are nowadays delivered to your doorstep. But this delivery is supposed to be collected by someone in working hours which is not always possible. If this problem is solved, it provides a hassle-free delivery of goods to your door step. The proposed system uses machine learning algorithms on the camera feed to detect any unrecognized activity in real time and alert the home owner. The implementation can be done on a microcontroller like Raspberry Pi with various other sensors [48]. The hardware required is Raspberry Pi 3 and PiCAM, and the software required is Raspbian OS, Face Recognition Library, Dlib and Imutis. For this hardware, a proper face recognition algorithm is what makes the camera a smart camera with the help of the processor.

Haar Cascade is a machine learning-based approach where a cascade function is trained from a lot of positive and negative images. It is then used to detect objects in other images. The algorithm needs a lot of positive images (images of faces) and negative images (images without faces) to train the classifier. Then we need to extract features from it. For this, Haar features on the image are shown as in figure 6.3. They are just like our convolutional kernel. Each feature is a single value obtained by subtracting the sum of pixels under the white rectangle from the sum of pixels under the black rectangle.

The PiCAM is used to take snapshots of the environment and surroundings at regular intervals. The shot is made up of multiple frames or images, and each image is processed using machine learning algorithms to detect faces. Haar Cascade, HOG and Linear SVM are the techniques mainly used to detect the face of a person in a frame. Once a face has been detected, 128-d face embeddings are computed through a deep metric neural network and identification and verification is run on this data obtained through extensive computation. Once the face embeddings have been compared to the ones stored in a database and the person is recognized by the

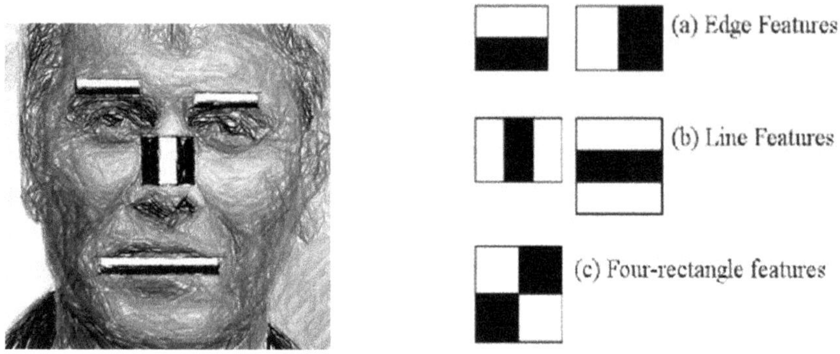


**Figure 6.3.** Sample image for Haar Cascade.

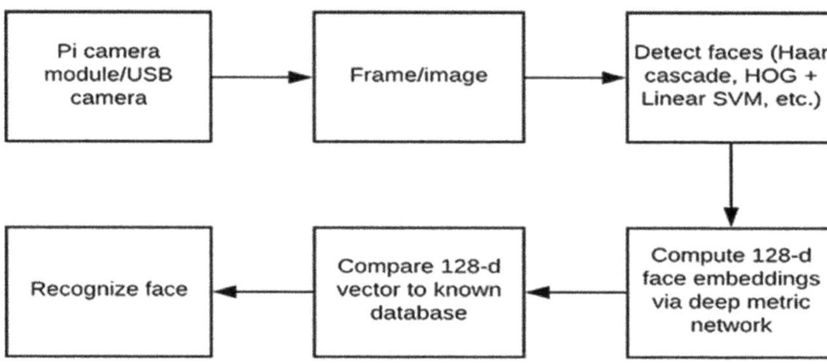


**Figure 6.4.** Implementation of smart surveillance system using a smart camera.

system, the person is let inside. The basic implementation is shown in figure 6.4 of the smart surveillance system simulated in order to ensure better security and safety of homes, offices etc.

## 6.6 Wearable sensors

Wearable sensors are sensor units or small monitoring devices that can be integrated into wearable clothing items or otherwise or desegrated directly with the body so as to monitor it. The monitoring can be done by capturing clinically vital data and processing it to deduce certain conclusions from it. Due to their numerous features, these sensors may be worn on different body parts depending on their requirements and applications, as shown in figure 6.5. The sensors may be used individually or in conjunction with multiple other devices to help in disease or condition prognosis.

Applications of wearable sensors include fitness, disease monitoring, medical prognosis, rehabilitation and physiotherapy monitoring, activity recognition, monitoring during pregnancy, amongst others [65, 72–74, 79, 88, 89, 93, 95]. Some of these aspects can be explored for adults and newborns too [90]. Further, these also have cultural and social aspects. However, their acceptance by individuals and society-at-large is now being seen as these wearable sensors are used in many

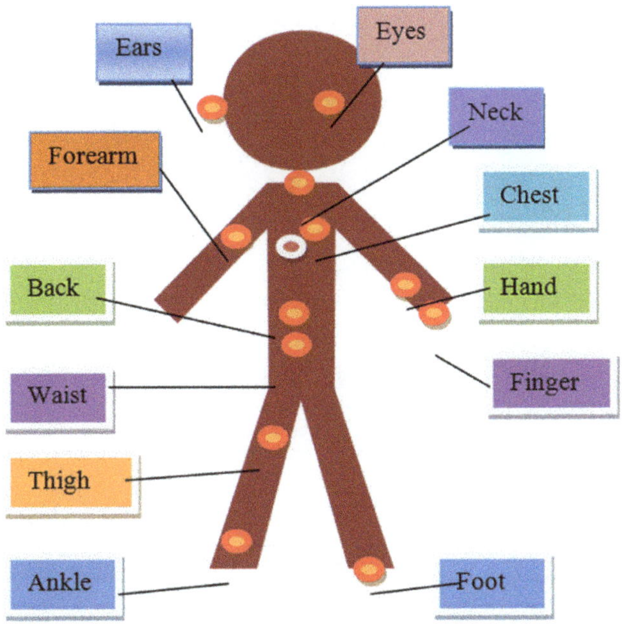


**Figure 6.5.** Bodypart on which the wearable sensors may be attached (courtesy of [56]).

hospitals and are an integral part of disease monitoring in hospitals and caregiving centers [83].

Certain physical aspects of wearable sensors include having a miniature size, not having content that may cause sudden allergic reactions, flexibility and stretchable electronic components. As electronic consumables, there are certain technological perspectives that can make these devices acceptable to various consumers such as individuals, hospitals, healthcare workers, and others. These trends include: (a) perceived benefits in the form of performance expectancy and hedonic benefits; (b) effort expectancy; (c) functional congruence; (d) self-efficacy; (e) social influence; (f) perceived vulnerability such as vulnerability and severity; and (g) perceived privacy risk. These parameters influence the acceptance and adaptation of various healthcare wearable devices [55]. While most of these parameters have a positive effect on individuals intending to adopt the devices, certain factors such as perceived vulnerability and privacy risks have a negative impact.

To able to continuously monitor activity, the sensors should have adequate battery and wireless data transmission capacity as well. These various types of classification for wearable sensors are shown below in figure 6.6.

Wearable sensors may also be categorized as per their certain specific features. This is shown in figure 6.7.

Wearable sensors may be thought of as being designed in three basic stages:

(i) The first types belong to those that were developed as wearable devices decades ago. These devices basically established the wearable sensors. These usually constituted:

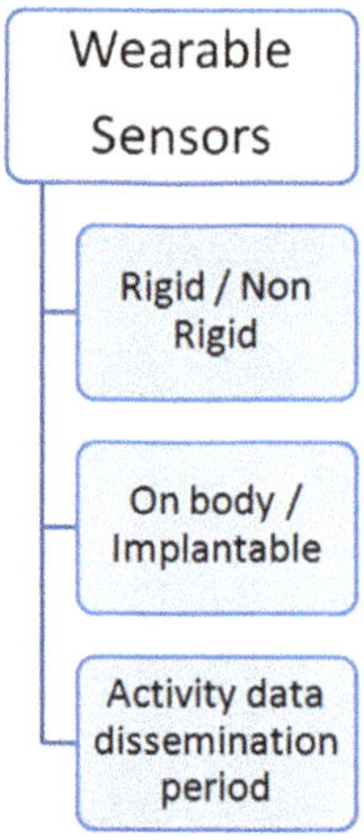


**Figure 6.6.** Types of wearable sensors.

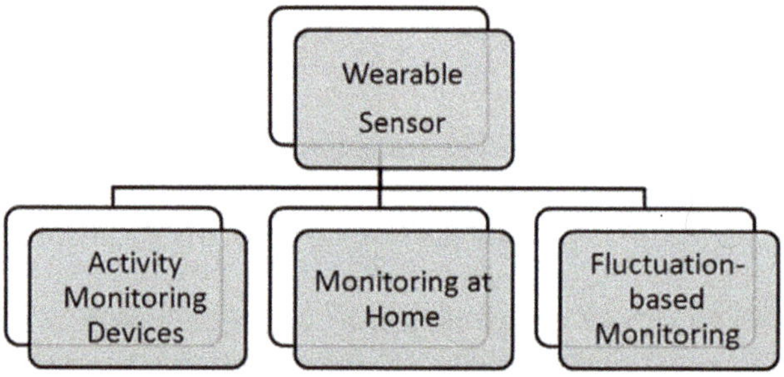


**Figure 6.7.** Classification of wearable sensors as per their specific features.

(a) electrodes for measuring biopotentials, e.g., medical or elite sports;
(b) microphones, e.g., headphones, hearing aids, etc;
(c) early dedicated devices such as GPS, temperature sensors, etc.

(ii) The second type of wearable devices was designed for smartphones and other such devices, i.e., later incorporated as wearable devices. These devices were 'made wearable'. The devices include inertial measuring units such as accelerometers, gyroscope, compass and smartphone interface platforms such as camera, touch inputs, etc.

(iii) These are 'made-for-wearable' sensors as per commercial evaluation. Examples include textiles and footwear for sensors, force/pressure sensors, motion, etc. Properties include low-design consumption, flexible, stretchable, etc.

A study was done by IDTechEx (URL: https://www.idtechex.com/en/research-report/wearable-sensors-2018–2028-technologies-markets-and-players/555), that tracked over 42 different wearable product types for revenue. This distribution is done in figure 6.8.

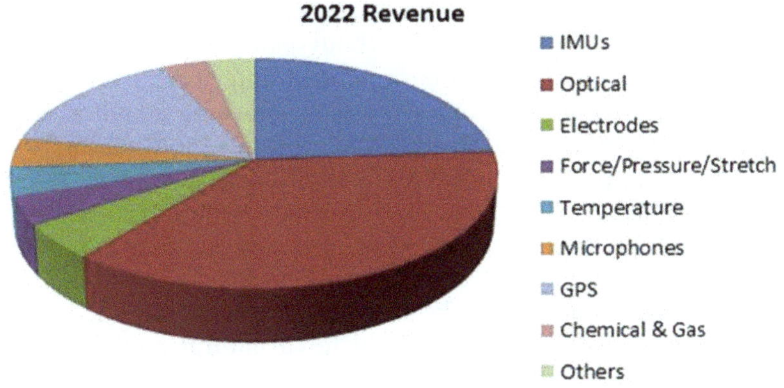


**Figure 6.8.** Revenue distribution nine types of wearable product types available in the market (Courtesy URL: https://www.idtechex.com/en/research-report/wearable-sensors-2018–2028-technologies-markets-and-players/555).

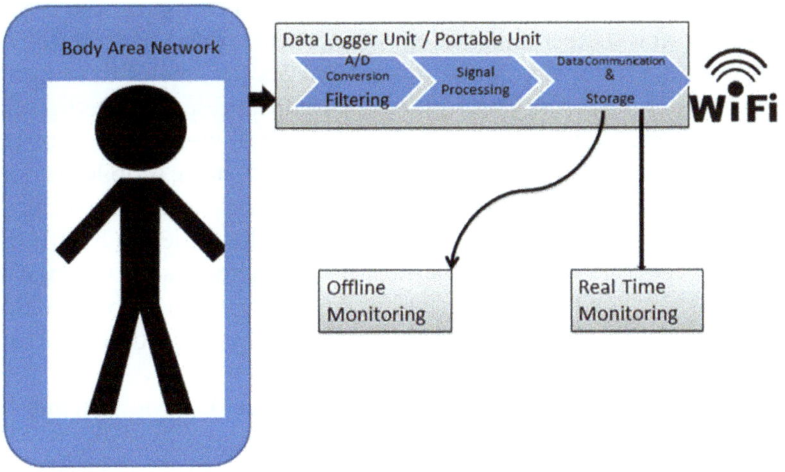


**Figure 6.9.** Generic architecture of wearable health devices systems [50].

The generic architecture of wearable health devices systems is depicted in figure 6.9, as shown by Dias and Cunha [51]. The basic architecture has four parts:

(A) Body-area network; this includes an interconnected network of sensors placed near the body area. These sensors are usually, tiny, low-cost, and low-power in nature. These sensors are interconnected to allow for data aggregation and sending it to external networks for remote processing;

(B) Data logging/portable unit: this is the unit where all the information is gathered, containing the outputs and inputs;

(C) Offline monitoring;

(D) Real-time monitoring.

Some of the popular wearable sensor-based devices include

(i) ADAMM is a piece of wearable (body-patch) technology by Health Care Originals that helps to monitor asthma and present timely intervention more easily. It detects your symptoms—tracking cough rate, respiration patterns, heartbeat, temperature and other symptoms of interest deviation from pre-set pattern renders a vibration or alarm system.

(ii) Valedo back therapy for back-related posture corrections helps in rehabilitation such as recovering arm and leg functionalities, helping with gait and balance and diminishing back pain.

(iii) Quell Relief is another of the new wearable smart healthcare technologies that is drug-free and non-surgical for a natural pain-relief response. It utilizes neurotechnology as a solution to chronic pain crises. It also allows for personalization to body types and for tracking activity, sleep, gait, and pain.

(iv) HealthPatch MD by Vital Connect is a new patch-based sensor set for healthcare professionals. It enables technicians and caregivers to watch over the vital parameters of their patients to recognize the posture of the person and detect falls using ECG, heart rate, heart rate variability, respiratory rate, skin temperature, and accelerometer data inputs.

(v) Proteus Discover by Proteus Digital Health presents a method that makes use of ingestible sensors. Attached is a small wearable sensor patch. The information is connected as an application on a mobile device through which a provider portal can help in measuring medical treatment effectiveness. It also gives an insight into the patient's health patterns.

(vi) Smart Stop by Chrono Therapeutics presents a smart miniature device that is designed for convalescing for the general public in overcoming their chronic addiction. These can be personalized to suit a patient's individual needs

(vii) Google Smart Contact Lenses aim to assist diabetic patients by continuous monitoring of their glucose level in tears.

(viii) Patient movements in bed monitoring susceptibility to pressure ulcers, and many such sensors and wearable devices are also available.

## 6.7 Applications of wearable sensors

Various forms of wearable devices are available for various applications like heart monitoring, sweat monitoring, cancer detection, Alzheimer's and Dementia prediction, wearable artificial kidney, Parkinson's disease and vision biosensors. These applications use various smart sensors with integrated processors for various data analysis and communication.

### 6.7.1 Programmable bio-electric ASIC sensors

Recording of multiple bio-electric signals can be done using application-specific integrated circuit (ASIC) sensors. Electroencephalography (EEG) is used in non-invasive brain–computer interface as it has high temporal resolution and portability.

EEG signals can be measured by attaching 64 electrodes to the subject's scalp using a cap. The placement of the electrodes is as shown in figure 6.10. Spatial and temporal feature selection can be analyzed from the aggregated data.

Classification methods are used to analyze features [52, 53]. EEG-based car control was shown by authors Yu *et al* [54]. These signals are non-stationary in nature and may have motion artifacts that must be reduced before using the data for further processing [78]. Data collection from such devices can be studied by classification of cognitive functions. Short recordings at particular sampling rates using signal processing methods are usually done. Although all 64 channels are usually not used, main electrodes using 12–16 channels are considered.

Personal electrocardiogram (ECG)/Elektro-kardiographie (EKG) devices are also available for determining if the heart-beat rhythm is atrial fibrillation. In an ECG, the heartbeat consists of valves closing between the atria and the ventricles to ensure the unidirectional blood flow. The musculature response of the heart is audible. These devices may not only monitor the cardiac rhythm but also alert patients in case of any abnormal symptoms. Single-lead ECG recording can be taken from Apple Watch, the KardiaBand, and others for reliable recording of the cardiac rhythm. The features include heart rate, heart rate variability, stress, fatigue, heart age, breathing index, mood, etc. Cardiac signals may be monitored by using conductive textile electrodes that have certain additional features such as flexibility, elasticity, and others [66].

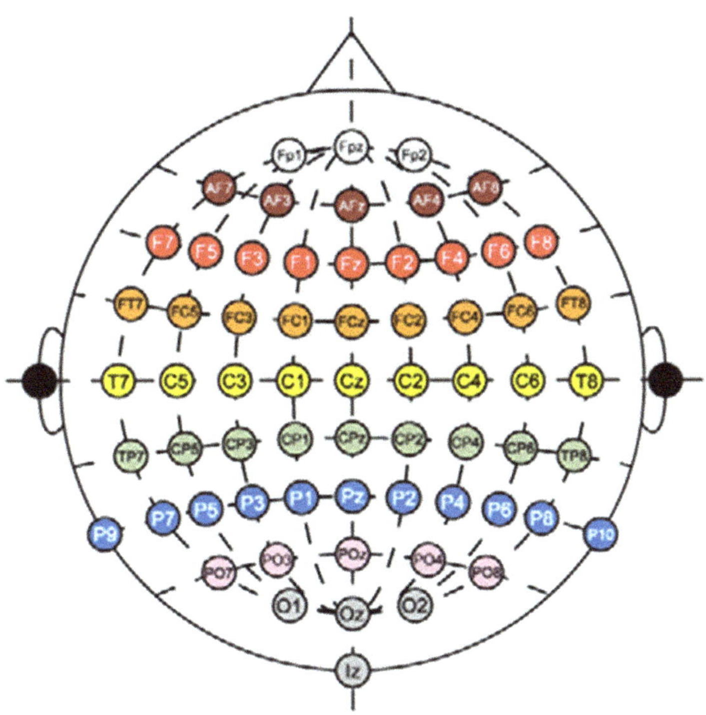


**Figure 6.10.** Spatial location of 64-channels of EEG electrodes.

### 6.7.2 Diabetes wearable medical device

Diabetic patients need to monitor their blood sugar/glucose levels regularly. Such a monitoring system consists of a handheld reader and a sensor worn on the back of the upper arm that is inserted just under the skin. Regulatory systems consist of continuously delivering insulin at set and variable rates throughout the day to mimic the insulin release of a healthy pancreas [55]. Examples include The FreeStyle Libre, Omnipod's DASH System, Eversense wearable, MiniMed 670G system, K'Track Glucose from Pkvitality and others. Their costs vary according to the functionality provided by the systems.

### 6.7.3 Cancer detecting wearable device

Such wearable devices are known to have 'filters' that circulate within the blood for cancer cells. In the normal liquid biopsy procedure, patients are required to test blood and urine samples to look for cancer markers. This is a very complex system. Devices such as '3D-nanopatterned microfluidic chip' are now known to successfully detect cancer markers in the tiniest drop of blood or in a component of the blood called plasma.

Implementation workflow of wearable technology systems was shown in [56]. Data acquisition is through the devices, with storage and transport being performed in big-data repositories. The information is analyzed for feedback and treatment.

### 6.7.4 Wearable sweat-sensor

Every day almost 500–700 ml of hypotonic fluid secreted by the sweat glands is released by the human body for thermoregulation under climate conditions. This secretion by the eccrine gland provides important psychological data such as electrolytes and metabolites. These include sodium and potassium ions and lactate and glucose, respectively. The interaction between blood cells and sweat glands was shown by [59, 60]. Such devices can also help in hydration monitoring, as shown by Butterworth *et al* in [67].

Sweat biomarkers can be used for the diagnosis of genetic conditions, the existence of cystic fibrosis in infants, monitoring of diabetes, detection of ischemia, and forms of skin injuries and diseases [9]. The authors of [9] have also discussed numerous sensors and their metrics used for analyzing sweat data. Enzymatic amperometric sweat based ethanol sensing can also be done for understanding the blood alcohol level. These are usually patch-based sensors [61]. Many more diseases can be analyzed with the proper use of sweat data, as shown by the authors in [58].

### 6.7.5 Wearable peritoneal dialysis device

Renal failure implies that after-metabolism products are not ejected from the body. There exist two forms of renal failure: (a) acute renal failure (ARF) considered to be usually a reversible disease state with adequate treatment; (b) chronic kidney disease (CKD) considered to be irreversible and require kidney transplantation. These treatments require constant medical attention called renal replacement therapy

(RRT) or dialysis, which is an artificial blood purification process. The methods of dialysis can be categorized as: (i) hemodialysis (HD); or (ii) as peritoneal dialysis. Certain details were given by the authors of [62].

The wearable artificial kidney (WAK) has shown the potential to conquer certain demerits of RRT or dialysis technique. Its features include having smaller size and weight for easy mobility, reduced amount of water consumption, reduced cost to facility at-home dialysis service, human physiology based blood purification, and others.

### 6.7.6 Predicting the progress of Alzheimer's and dementia diseases

Alzheimer's disease is a rapidly growing healthcare ecosystem burden [57]. While dementia is a set of symptoms that negatively affect memory [84]. Cognitive, behavioral, sensory, and motor changes should be monitored to understand the prognosis of the disease. Being a neurological or neurodegenerative disease, its effective display may be seen in 10–15 years. Wearable devices can help in such monitoring. Multiple sensors are shown in figure 6.11, and their combined input data are analyzed to understand the disease parameters are also depicted.

The aggregated feedback from multiple-devices is used to determine this neurodegenerative disease. Ambient noise and voice detection are done using microphones. Fine-motor skill detection is based on the swiping method from touch sensors. Eye movement, gaze and reflex action of pupillary nerves are registered from the camera. Activity information is gathered from devices like altimeters. Also used are barometers and weather sensory data. Heart-rate measurement and variability, saturation in oxygen level and such inputs are considered from photoplethysmography (PPG) and EEG and ECG sensors. Intertia measurement units (IMU) present data from sensors such as gyroscope, magnetometer, accelerometers

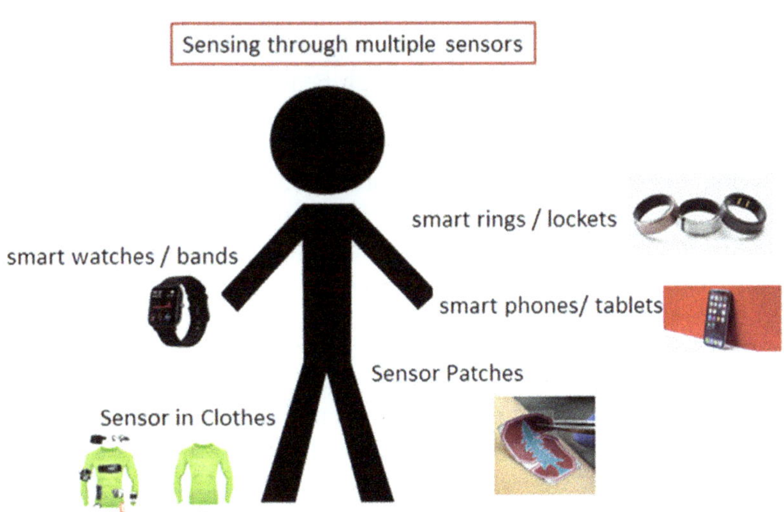


**Figure 6.11.** Digital biomarker sensors for analyzing Alzheimer's disease (courtesy: Kourtis *et al* [57]).

etc, to track activity and its structure. Other forms of information include temperature, moisture, location information, ambient visible/radiation levels etc. Muscular activity is signaled using an electromyography (EMG) sensor which is usually presented as a patch device. Other forms of patch device include galvanic skin response (GSR) to measure skin conductance and potential. Not only are these individual measures taken, but also the interactions in a social environment are monitored using proximity measuring devices as well as social network activity monitoring.

### 6.7.7 Monitoring Parkinson's disease

Parkinson's disease is a neurological movement disorder that leads to symptoms including tremors, rigidity, slowness of movements, etc. [71, 96, 97]. The effect of these limitations may be reduced by wearing accelerometers, inertial units, electromyography sensors and proper intake of medicines in a very timely manner. The positioning of such sensors can be done on various body parts for monitoring. Additional devices such as EEG may be used in conjunction.

### 6.7.8 Vision-related biosensors

3D vision is a new technology that is employed using stereovision concepts mainly for entertainment. The architecture of NeuroGlasses was described by Gong *et al* [80]. It is said to be a lightweight location-aware signal correction device. Such devices can be used to enhance our visual senses with virtual 3D viewing. Related virtual and augmented reality devices can help in medical and psychological betterment too [82, 84]. Examples include the emotional glove 'Emogle' that can be worn on the hand to feel sensations such as pain etc, generated through vibrations, thermal sensation, wind and electrical simulations from the glove [84]. Google glasses in conjunction with appropriate communication setup have been known to help in setting up a virtual operating theater too [94].

Wearable devices and biosensors have high beneficial applications in healthcare. However, there is also a related high privacy concern with respect to this form of digital data. On one hand, the analysis of data can help in understanding, predicting and monitoring various diseases, but the process of analysis requires high computation and big-data techniques which may be exploited by service providers and other stakeholders. Maintaining electronic healthcare records is also very crucial [81]. These, to a certain extent, can be accessed using mobile applications that tend to present a very easy form of usage. Another issue is maintaining the privacy and security of these records [92]. The security issues may be addressed by using recent techniques such as block-chaining, as shown by Chakraborty, Aich, and Kim [63]. In their proposed framework, the transaction and access management is performed using blockchain so that privacy is maintained. This framework can incorporate personal healthcare and external record management blockchains. Such a framework could also help in remote implementation of a complete healthcare system that can be accessed by all stakeholders.

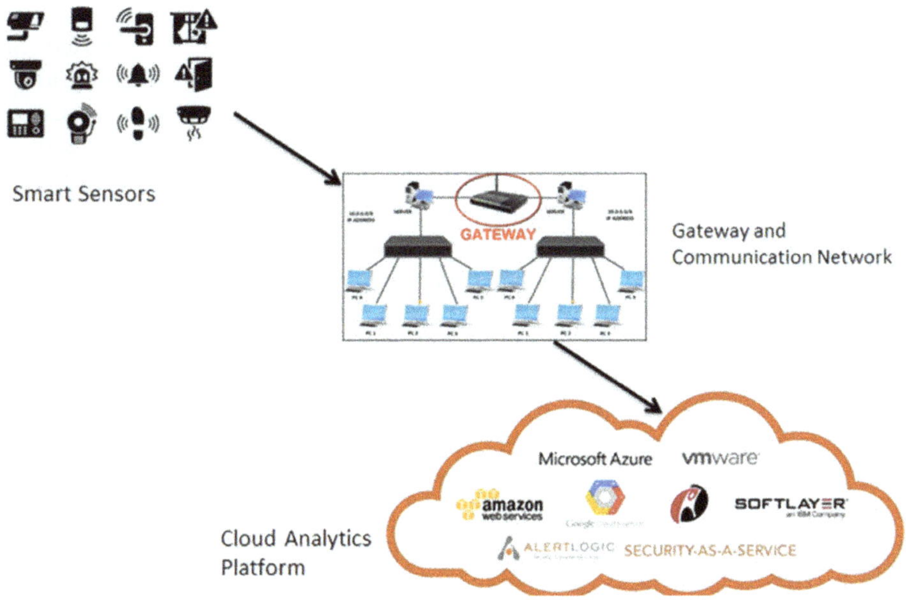


**Figure 6.12.** Overview of a typical IoT-based healthcare system that could incorporate wearable device networks.

A remote framework can also help stakeholders to share their data in a constrained manner with a particular person, such as a doctor/caregiver, or help in sending proper notification to proper authorities. Such a framework would incorporate concepts of an IoT-based healthcare system as shown by Alam *et al* [64] and Hiremath, Yang, and Mankodiya [91] depicted in figure 6.12. Such a system would have to encompass various types of sensors and sensor networks with different actuation capabilities, various data transfer rates and power utilization requirements. The application requirements such as high quality of service and reliability would have been given prominence. Such a system can be effectively used for understanding human activity [85] and as a personal health assistant too [86, 87]. The system would also have to incorporate existing and newer cellular and non-cellular technological requirements for easy upgradation.

Biosensor data is a huge management liability. These have high velocity, variability, and vivacity. Storage and retrieval require huge management capabilities. They must be compressed [68] too, have low-energy consumption, be Wi-fi or Bluetooth enabled for easy communication [69], be effective and include various design considerations [70] such as ubiquitous monitoring [72], using artificial intelligence-based computation [75, 76] or predictive analysis [77].

## 6.8 Conclusion

Smart devices have become today's essential technology upgrades in all our lives. This chapter discusses what smart systems are and how they are different from earlier devices. For example, large mobile phone use can be seen since 2000 but those phones are absolute due to change in technology. In the current scenario, even a few-months-old baby needs a smartphone due to features that are incorporated. Smart devices have properties like self-calibration, self-adaptability, self-healing and self-health assessment. Depending on the various applications, different sensors are interfaced with processors to build different smart sensors and their corresponding smart systems. In today's world every device needs to be smart to hit the market with success and to make more user-friendly devices.

Various devices, their parameters, their requirements and how they can be combined together in a system were discussed in this chapter. Devices such as ECG, Wearable ECG, EEG, and their usage as wearable devices were discussed in detail. Privacy issues and their solutions with the use of blockchaining framework were also put forth. IoT-based healthcare systems that can incorporate requirements from wearable sensors were also showcased.

Acknowledgements: we are grateful to the various authors whose works we have referred to in this literature.

## References

[1] Yurish S Y 2008 Self-adaptive intelligent sensors and systems: from theory to practical design *Int. Workshop on Robotic and Sensors Environments (Ottawa, ON)* pp x–xi

[2] Hunter G W, Stetter J R, Hesketh P and Liu C C 2010 Smart sensor systems *Electrochem. Soc. Interface* **19** 29–34

[3] Barcelo-Ordinas J M, Doudou M, Garcia-Vidal J and Badache N 2019 Self-calibration methods for uncontrolled environments in sensor networks: a reference survey *Ad Hoc Networks* **88** 142–59

[4] Rivera J, Carrillo M, Chacón M, Herrera G and Bojorquez G 2007 Self-calibration and optimal response in intelligent sensors design based on artificial neural networks *Sensors* **7** 1509–29

[5] Možek M, Vrtačnik D, Resnik D, Aljančič U, Penič S and Amon S 2008 Digital self-learning calibration system for smart sensors *Sens. Actuators* A **141** 101–8

[6] Pereira J D, Postolache O and Girao P S 2005 Adaptive self-calibration algorithm for smart sensors linearization *2005 IEEE Instrumentationand Measurement Technology Conf. Proc.* Vol. 1 pp 648–52

[7] Pertijs M, Meijer G, Pentijis M and Makinwa K 2014 Calibration and self-calibration of smart sensors *Smart Sensor Systems: Emerging Technologies and Applications* (New York: Wiley) pp 17–41

[8] Khalaschi E, Kalech M and Rokach L 2017 BG Negev Technologies and Applications Ltd *U.S. Patent* 9,728,014

[9] Elhoushi M, Georgy J, Noureldin A and Korenberg M J 2016 A survey on approaches of motion mode recognition using sensors *IEEE Trans. Intell. Trans. Syst.* **18** 1662–86

[10] Chandola V, Banerjee A and Kumar V 2009 Anomaly detection: a survey *ACM Computing Surveys (CSUR)* **41** 1–58

[11] Mehranbod N, Soroush M and Panjapornpon C 2005 A method of sensor fault detection and identification *J. Process Control* **15** 321–39

[12] Ellinas K, Tserepi A and Gogolides E 2017 Durable superhydrophobic and superamphiphobic polymeric surfaces and their applications: a review *Adv. Colloid Interface Sci.* **250** 132–57
[13] Huynh T P and Haick H 2016 Self-healing, fully functional, and multiparametric flexible sensing platform *Adv. Mater.* **28** 138–43
[14] Viter R and Iatsunskyi I 2019 Metal oxide nanostructures in sensing *Nanomaterials Design for Sensing Applications* pp 41–91 (Amsterdam: Elsevier)
[15] Huynh T P, Sonar P and Haick H 2017 Advanced materials for use in soft self-healing devices *Adv. Mater.* **29** 1604973
[16] Yasakau K A, Ferreira M G S, Zheludkevich M L, Terryn H, Mol J M C and Gonzalez-Garcia Y 2014 Novel and self-healing anticorrosion coatings using rare earth compounds *Rare Earth-Based Corrosion Inhibitors* (Cambridge: Woodhead Publishing) pp 233–66
[17] Hansen C J 2015 Microvascular-based self-healing materials *Recent Advances in Smart Self-Healing Polymers and Composites* (Cambridge: Woodhead Publishing) pp 129–57
[18] Han Y, Wu X, Zhang X and Lu C 2017 Self-healing, highly sensitive electronic sensors enabled by metal–ligand coordination and hierarchical structure design *ACS Appl. Mater. Interfaces* **9** 20106–14
[19] Yoon J H *et al* 2019 Extremely fast self-healable bio-based supramolecular polymer for wearable real-time sweat-monitoring sensor *ACS Appl. Mater. Interfaces* **11** 46165–75
[20] Makinwa K A, Pertijs M A, vd Meer J C and Huijsing J H 2007 Smart sensor design: the art of compensation and cancellation *ESSCIRC 2007–33rd European Solid-State Circuits Conf.* (Piscataway, NJ: IEEE) pp 76–82
[21] Yazdi N, Ayazi F and Najafi K 1998 Micromachined inertial sensors *Proc. IEEE* **86** 1640–59
[22] (https://sensing.honeywell.com/auto-zero-calibration-technique-pressure-sensors-technical-note.pdf) (Online resource accessed on 02-02-2020)
[23] Kouider M, Nadi M and Kourtiche D 2003 Sensors auto-calibration method-using programmable interface circuit front-end *Sensors* **3** 491–7
[24] Rudnitskaya A 2018 Calibration update and drift correction for electronic noses and tongues *Front. Chem.* **6** 433
[25] Rivera-Mejia J, Villafuerte-Arroyo J E, Vega-Pineda J and Sandoval-Rodriguez R 2015 Comparison of compensation algorithms for smart sensors with approach to real-time or dynamic applications *IEEE Sensors J.* **15** 7071–80
[26] Yurish S Y 2008 Self-adaptive smart sensors and sensor systems *Sens. Transducer J.* **94** 1–14
[27] Kirianaki N V, Yurish S Y, Shpak N O and Deynega V P 2002 *Data Acquisition and Signal Processing for Smart Sensors* (Chichester: Wiley) pp 72–8
[28] Hossein Motlagh N, Mohammadrezaei M, Hunt J and Zakeri B 2020 Internet of things (IoT) and the energy sector *Energies* **13** 494
[29] MSP430 Ultra-Low-Power Microcontrollers, Texas Instruments, USA, 2H, 2006 (http://ti.com/msp430)
[30] Lee E, Seo Y D and Kim Y G 2019 Self-adaptive framework based on MAPE loop for internet of things *Sensors* **19** 2996
[31] (https://techbriefs.com/component/content/article/tb/features/articles/33212) (online accessed on 12-01-2020)
[32] Sehrawat D and Gill N S 2019 Smart sensors: analysis of different types of IoT sensors *2019 3rd Int. Conf. on Trends in Electronics and Informatics (ICOEI)* (Piscataway, NJ: IEEE) pp 523–28

[33] Lin T, Rivano H and Le Mouël F 2017 A survey of smart parking solutions *IEEE Trans. Intell. Trans. Syst.* **18** 3229–53

[34] Chen Z, Zhou X, Wang X, Dong L and Qian Y 2017 Deployment of a smart structural health monitoring system for long-span arch bridges: a review and a case study *Sensors* **17** 2151

[35] Alías F and Alsina-Pagès R M 2019 Review of wireless acoustic sensor networks for environmental noise monitoring in smart cities *J. Sensors* **2019**

[36] Rizwan P, Suresh K and Babu M R 2016 Real-time smart traffic management system for smart cities by using internet of things and big data *2016 Int.Conf. on Emerging Technological Trends (ICETT)* (Piscataway, NJ: IEEE) pp 1–7

[37] Rohit G S, Chandra M B, Saha S and Das D 2018 Smart dual dustbin model for waste management in smart cities *2018 Third Int. Conf. for Convergence in Technology (I2CT)* (Piscataway, NJ: IEEE) pp 1–5

[38] Shahzad G, Yang H, Ahmad A W and Lee C 2016 Energy-efficient intelligent street lighting system using traffic-adaptive control *IEEE Sensors J.* **16** 5397–405

[39] Sarwar B, Bajwa I S, Jamil N, Ramzan S and Sarwar N 2019 An intelligent fire warning application using IoT and an adaptive neuro-fuzzy inference system *Sensors* **19** 3150

[40] Zhao Z, Wang J, Fu C, Liu Z, Liu D and Li B 2018 Design of a smart sensor network system for real-time air quality monitoring on green roof *J. Sensors* **2018** 1987931

[41] Aqib M, Mehmood R, Alzahrani A and Katib I 2020 A smart disaster management system for future cities using deep learning, GPUs, and in-memory computing *Smart Infrastructure and Applications* (Cham: Springer) pp 159–84

[42] Ntuli N and Abu-Mahfouz A 2016 A simple security architecture for smart water management system *Proc. Comput. Sci.* **83** 1164–9

[43] Mudumbe M J and Abu-Mahfouz A M 2015 Smart water meter system for user-centric consumption measurement *2015 IEEE 13th Int. Conf. on Industrial Informatics (INDIN)* (Piscataway, NJ: IEEE) pp 993–8

[44] Chowdhry D A, Hussain A, Rehman M Z U, Ahmad F, Ahmad A and Pervaiz M 2013 Smart security system for sensitive area using face recognition *2013 IEEE Conf. on Sustainable Utilization and Development in Engineering and Technology (CSUDET)* (Piscataway, NJ: IEEE) pp 11–4

[45] Hema N and Kant K 2019 Cost-effective smart irrigation controller using automatic weather stations *Int. J. Hydrol. Sci. Technol.* **9** 1–27

[46] Hema N 2018 AWS data driven smart irrigation system *Thesis* Jaypee Institute of Information Technology, Noida, India

[47] Hema N, Krishna K and Maringanti H B 2012 Site specific automated drip irrigation for palm trees using wireless sensor network *Infomation Technology, Automation and Precision Farming. Int. Conf. of Agricultural Engineering-CIGR-AgEng 2012: Agriculture and Engineering for a Healthier Life (Valencia, Spain, 8–12 July 2012)* (CIGR-EurAgEng)

[48] Hema N and Yadav J 2020 Secure home entry using Raspberry Pi with notification via telegram *6th Int. Conf. on Signal Processing and Communication (ICSC 2020)* (Piscataway, NJ: IEEE) pp 5–6

[49] (https://www2.deloitte.com/us/en/insights/focus/industry-4-0/smart-factory-connected-manufacturing.html) (online accessed on 17-02-2020)

[50] Dias D and Paulo Silva Cunha J 2018 Wearable health devices—vital sign monitoring, systems and technologies *Sensors* **18** 2414

[51] Yang G, Chen J, Jonsson F, Tenhunen H and Zheng L R 2012 A multi-parameter bio-electric ASIC sensor with integrated 2-wire data transmission protocol for wearable health-care system *2012 Design, Automation and Test in Europe Conf. and Exhibition (DATE)* (Piscataway, NJ: IEEE) pp 443–8
[52] Nambu I, Ebisawa M, Kogure M, Yano S, Hokari H and Wada Y 2013 Estimating the intended sound direction of the user: toward an auditory brain-computer interface using out-of-head sound localization *PLoS One* **8** e57174
[53] Yu Y, Zhou Z, Jiang J, Liu Y and Hu D 2015 Self-paced EEG-based brain-controlled car in real-world environment *2015 3rd Int. Conf. on Machinery, Materials and Information Technology Applications* (Atlantis Press)
[54] Ultimate guide to ECG (http://neurosky.com/biosensors/ecg-sensor/ultimate-guide-to-ecg/)
[55] Gao Y, Li H and Luo Y 2015 An empirical study of wearable technology acceptance in healthcare *Ind. Manage. Data Syst.* **115** 1704–23
[56] Menta A K, Subbiah I M and Subbiah V 2018 Bringing wearable devices into oncology practice: fitting smart technology in the clinic *Discov. Med.* **26** 261–70
[57] Kourtis L C, Regele O B, Wright J M and Jones G B 2019 Digital biomarkers for Alzheimer's disease: the mobile/wearable devices opportunity *NPJ Digital Med.* **2** 1–9
[58] Chung M, Fortunato G and Radacsi N 2019 Wearable flexible sweat sensors for healthcare monitoring: a review *J. R. Soc. Interface* **16** 20190217
[59] Epithelial tissue (https://cnx.org/contents/oWqVExrJ@3/Epithelial-Tissue#fig-ch04_02_01)
[60] Nag A, Mukhopadhyay S C and Kosel J 2017 Wearable flexible sensors: a review *IEEE Sensors J.* **17** 3949–60
[61] Hauke A, Simmers P, Ojha Y R, Cameron B D, Ballweg R, Zhang T, Twine N, Brothers M, Gomez E and Heikenfeld J 2018 Complete validation of a continuous and blood-correlated sweat biosensing device with integrated sweat stimulation *Lab Chip* **18** 3750–9
[62] Bazaev N, Zhilo N, Grinval'd V and Selishchev S 2018 Wearable dialysis: current state and perspectives *Wearable Technologies* (Rijeka: IntechOpen)
[63] Chakraborty S, Aich S and Kim H C 2019 A secure healthcare system design framework using blockchain technology *2019 21st Int. Conf. on Advanced Communication Technology (ICACT)* (Piscataway, NJ: IEEE) pp 260–4
[64] Alam M M, Malik H, Khan M I, Pardy T, Kuusik A and Le Moullec Y 2018 A survey on the roles of communication technologies in IoT-based personalized healthcare applications *IEEE Access* **6** 36611–31
[65] Meijer H A, Graafland M, Goslings J C and Schijven M P 2018 Systematic review on the effects of serious games and wearable technology used in rehabilitation of patients with traumatic bone and soft tissue injuries *Arch. Phys. Med. Rehab.* **99** 1890–9
[66] Lim C Y, Jang K J, Kim H W and Kim Y H 2013 A wearable healthcare system for cardiac signal monitoring using conductive textile electrodes *2013 35th Annual Int. Conf. of the IEEE Engineering in Medicine and Biology Society (EMBC)* (Piscataway, NJ: IEEE) pp 7500–3
[67] Butterworth I, Serallés J, Mendoza C S, Giancardo L and Daniel L 2015 A wearable physiological hydration monitoring wristband through multi-path non-contact dielectric spectroscopy in the microwave range *2015 IEEE MTT-S 2015 Int. Microwave Workshop Series on RF and Wireless Technologies for Biomedical and Healthcare Applications (IMWS-BIO)* (Piscataway, NJ: IEEE) pp 60–1

[68] Samie F, Bauer L and Henkel J 2015 An approximate compressor for wearable biomedical healthcare monitoring systems *2015 Int. Conf. on Hardware/Software Codesign and System Synthesis (CODES+ ISSS)* (Piscataway, NJ: IEEE) pp 133–42
[69] Zhang T, Lu J, Hu F and Hao Q 2014 Bluetooth low energy for wearable sensor-based healthcare systems *2014 IEEE Healthcare Innovation Conf. (HIC)* (Piscataway, NJ: IEEE) pp 251–4
[70] Cochrane K and Schiphorst T 2015 Developing design considerations for mobile and wearable technology m-Health applications that can support recovery in mental health disorders *2015 9th Int. Conf. on Pervasive Computing Technologies for Healthcare (PervasiveHealth)* (Piscataway, NJ: IEEE) pp 29–36
[71] Pasluosta C F, Gassner H, Winkler J, Klucken J and Eskofier B M 2015 An emerging era in the management of Parkinson's disease: wearable technologies and the internet of things *IEEE J. Biomed. Health Inform.* **19** 1873–81
[72] Hung K, Lee C C, Chan W M, Choy S O and Kwok P 2012 Development of a wearable system integrated with novel biomedical sensors for ubiquitous healthcare *2012 Annu. Int. Conf. of the IEEE Engineering in Medicine and Biology Society* (Piscataway, NJ: IEEE) pp 5802–5
[73] Khelil A, Shaikh F K, Sheikh A A, Felemban E and Bojan H 2014 Digiaid: a wearable health platform for automated self-tagging in emergency cases *2014 4th Int. Conf. on Wireless Mobile Communication and Healthcare-Transforming Healthcare Through Innovations in Mobile and Wireless Technologies (MOBIHEALTH)* (Piscataway, NJ: IEEE) pp 296–9
[74] Van Hoof C 2015 Frictionless wearable technology: the key to unleashing the power of wearable sensors for health and lyfestyle *2015 10th Int. Conf. on Design and Technology of Integrated Systems in Nanoscale Era (DTIS)* (Piscataway, NJ: IEEE)
[75] Sharma V, Som S and Khatri S K 2019 Future of wearable devices using IoT synergy in AI *2019 3rd Int. Conf. on Electronics, Communication and Aerospace Technology (ICECA)* (Piscataway, NJ: IEEE) pp 138–42
[76] Knickerbocker J, Budd R, Dang B, Chen Q, Colgan E, Hung L W and Narayanan R 2018 Heterogeneous integration technology demonstrations for future healthcare, IoT, and AI computing solutions *2018 IEEE 68th Electronic Components and Technology Conf. (ECTC)* (Piscataway, NJ: IEEE) pp 1519–28
[77] Dudakiya S, Galani H, Shaikh A, Thanki D, Late R A and Pawar S E 2016 Monitoring mobile patients using predictive analysis by data from wearable sensors *2016 Int. Conf. on Electrical, Electronics, and Optimization Techniques (ICEEOT)* (Piscataway, NJ: IEEE) pp 332–5
[78] Nagai S, Anzai D and Wang J 2017 Motion artefact removals for wearable ECG using stationary wavelet transform *Healthcare Technol. Lett.* **4** 138–41
[79] De D, Bharti P, Das S K and Chellappan S 2015 Multimodal wearable sensing for fine-grained activity recognition in healthcare *IEEE Internet Comput.* **19** 26–35
[80] Gong F, Xu W, Lee J Y, He L and Sarrafzadeh M 2011 Neuroglasses: a neural sensing healthcare system for 3-d vision technology *IEEE Trans. Inf. Technol. Biomed.* **16** 198–204
[81] Altenhoff B, Vaigneur H and Caine K 2015 One step forward, two steps back: the key to wearables in the field is the app *2015 9th Int. Conf. on Pervasive Computing Technologies for Healthcare (PervasiveHealth)* (Piscataway, NJ: IEEE) pp 241–4
[82] Kang J, Lee J and Jin S 2018 Personal sensory VR interface utilizing wearable technology *2018 Int. Conf. on Information and Communication Technology Convergence (ICTC)* (Piscataway, NJ: IEEE) pp 546–8

[83] Sultan N 2015 Reflective thoughts on the potential and challenges of wearable technology for healthcare provision and medical education *Int. J. Inf. Manage.* **35** 521–6

[84] Mohamedali F and Matoorian N 2016 Support dementia: using wearable assistive technology and analysing real-time data *2016 Int. Conf. on Interactive Technologies and Games (ITAG)* (Piscataway, NJ: IEEE) pp 50–4

[85] Subasi A, Radhwan M, Kurdi R and Khateeb K 2018 IoT based mobile healthcare system for human activity recognition *2018 15th Learning and Technology Conf. (L&T)* (Piscataway, NJ: IEEE) pp 29–34

[86] Sarfo F S and Ovbiagele B 2017 Mobile health for stroke: a promising concept for research and practice *Mhealth* **3** 4

[87] Rangel-Aldao R, Moreno J A and González-Taboada E Digital prevent: an evolving platform of preventative personal health 6–24

[88] Mamlin B W and Tierney W M 2016 The promise of information and communication technology in healthcare: extracting value from the chaos *Am. J. Med. Sci.* **351** 59–68

[89] Kale S, Mane S and Patil P 2017 Wearable biomedical parameter monitoring system: a review *2017 Int. Conf. of Electronics, Communication and Aerospace Technology (ICECA)* Vol. 1 (Piscataway, NJ: IEEE) pp 614–7

[90] Perego P, Andreoni G, Zanini R and Bellù R 2014 Wearable biosignal monitoring system for newborns *2014 4th Int. Conf. on Wireless Mobile Communication and Healthcare-Transforming Healthcare Through Innovations in Mobile and Wireless Technologies (MOBIHEALTH)* (Piscataway, NJ: IEEE) pp 271–4

[91] Hiremath S, Yang G and Mankodiya K 2014 Wearable internet of things: concept, architectural components and promises for person-centered healthcare *2014 4th Int. Conf. on Wireless Mobile Communication and Healthcare-Transforming Healthcare Through Innovations in Mobile and Wireless Technologies (MOBIHEALTH)* (Piscataway, NJ: IEEE) pp 304–7

[92] Lowens B, Motti V G and Caine K 2017 Wearable privacy: skeletons in the data closet *2017 IEEE Int. Conf. on Healthcare Informatics (ICHI)* (Piscataway, NJ: IEEE) pp 295–304

[93] Lewy H 2015 Wearable technologies–future challenges for implementation in healthcare services *Healthcare Technol. Lett.* **2** 2–5

[94] Datta N, MacQueen I T, Schroeder A D, Wilson J J, Espinoza J C, Wagner J P, Filipi C J and Chen D C 2015 Wearable technology for global surgical teleproctoring *J. Surg. Educ.* **72** 1290–5

[95] Lopez B D B, Aguirre J A A, Coronado D A R and Gonzalez P A 2018 Wearable technology model to control and monitor hypertension during pregnancy *2018 13th Iberian Conf. on Information Systems and Technologies (CISTI)* (Piscataway, NJ: IEEE) pp 1–6

[96] Lonini L, Dai A, Shawen N, Simuni T, Poon C, Shimanovich L, Daeschler M, Ghaffari R, Rogers J A and Jayaraman A 2018 Wearable sensors for Parkinson's disease: which data are worth collecting for training symptom detection models *NPJ Digital Med.* **1** 1–8

[97] Pardoel S, Kofman J, Nantel J and Lemaire E D 2019 Wearable-sensor-based detection and prediction of freezing of gait in Parkinson's disease: a review *Sensors* **19** 5141

**IOP** Publishing

Advances in Modern Sensors
Physics, design, simulation and applications
**G R Sinha**

# Chapter 7

## Smart stick for the visually impaired

**Chilakala Sudhamani**

In this world, everyone is interested in taking a trip around the world and wants to see each and every beautiful place. But there are some people in the world who are unable to find even their own belongings, referred to as blind or visually impaired. These people find difficulty in detecting the objects in front of them, and with walking inside and outside their homes. Sometimes it makes them feel bad and it can also be very dangerous for them to make a move. Therefore, they need the support of another to move inside or outside the home. However, it is not always possible for someone else to take care of them. Hence, to support the movement of blind people different devices have been developed with the advanced technologies in modern life. At the initial stage these people used a stick to identify the objects or obstacles in front of them. But with this stick, they are not able to identify distant objects and if any vehicle is coming towards him/her then it is very dangerous. So a smart stick was proposed, which helps in identifying obstacles in front of the user and also those some distance away. Many developments have been made to this smart stick in order to fulfill the requirements of visually impaired people.

Hence, in this chapter a solution is proposed with a smart stick to overcome major problems faced by the visually challenged. This stick includes many sensors like ultrasonic sensor, LDR densor, water sensor and GPS module. When any obstacle is detected in front of the user then the sensors will send an alert message to the user in the form of a beep sound or vibration. It has one more advantage: that it detects ups and downs like staircases, slopes on the road, and water depths etc. One main advanced feature is that the user can find the location of the stick by pressing the button of the GPS module. The smart stick is able to detect objects in front of the visually challenged and gives a suitable message by allowing him/her to move twice the normal speed because he/she feels safe. This stick is of low cost, light weight, with low power consumption, ability to fold and provides fast response.

doi:10.1088/978-0-7503-2707-7ch7  

## 7.1 Introduction

According to World Health Organization (WHO), globally at least 2.2 billion people have a vision impairment or blindness. Out of these people, at least 1 billion have a vision impairment that can be cured [1]. The vision impairment can be classified in two major types. They are distance vision impairment and near vision impairment. Distance vision impairment is further classified into four types based on visual acuity [2]:

- mild: if the presenting visual acuity is worse than 6/12, then a person has mild vision impairment;
- moderate: if the presenting visual acuity is worse than 6/18, then there is moderate vision impairment;
- severe: if the presenting visual acuity is worse than 6/60, then there is severe vision impairment;
- blind: if the presenting visual acuity is worse than 3/60, then one is said to be blind.
- if the visual acuity is more than 20/40 or 20/60, then it is said to be near vision impairment.

The major causes of vision impairment are uncorrelated refractive errors (43%), cataract (33%) and glaucoma (2%). In every day, blind people's lives are full of risks and they can't walk without the help of others. They cannot move independently and safely without any support. Therefore, to support the movement of the blind, different devices are developed based on the advanced technologies in modern life. To guide blind people around obstacles, stairs, uneven surfaces and holes an electronic device is introduced, known as an electronic travel aid (ETA) [3–6]. This device will help them to move independently anywhere they want safely without any danger and without depending on others.

In the early 1970s a laser cane was designed to identify the obstacles which are in the range of 1.5–3.5 m. This device uses three laser diodes as transmitters and three photo diodes as receivers [7]. A battery operated sonic torch was developed using an ultrasonic sensor [8]. This sensor will transmit the ultrasonic sound waves in front of the user and receive signals back which are reflected from the obstacle, if any obstacle is found in his/her path. The problem with this device is that it does not provide an accurate path. Therefore, to provide an accurate path as well as the position of objects, a device known as Mel dog was designed using an ultrasonic sensor and a laser sensor [9, 10]. But this device is heavy and large in size.

A white cane was designed to identify the slopes in the path of a blind person, i.e. steps, depths and geometric singularities [11]. A laser cane and smart cane were designed to alert them and to avoid dependency on others [12–14]. Additional features added to the white cane are voice and vibration alert outputs in order to provide quick alerts. With the use of this the smart cane can identify obstacles, alert the user and avoid accidents. But it cannot always be used because of its length; it is unusable crowded public places and where it is difficulet to identify objects.

Hence, to overcome the above problems and to identify objects more quickly and accurately, sensor inputs, i.e. laser, ultrasonic, infra red and water sensors and camera inputs are attached to the stick [15–18]. With these sensor inputs the person can identify the distance from the object, height and depth of the object, any water bodies in the path, speed of vehicles, direction of objects and light intensity. A device with an ultrasonic sensor as input and a speaker enabled voice headset is used as an output device to identify the obstacles in all directions (right, left, front and back) at a distance of 1 m [19]. One IR sensor and a set of ultrasonic sensors are used to identify obstacles and staircases at a distance of 4 m within 39 ms [20]. The output devices used in the blind stick are vibrator, buzzer alarm and voice enabled headset [21–23].

Hence, to overcome all these issues, blind people are using smart sticks to help find slope areas, steps, water sources and obstacles in front of the user. One of the major problems facing the user is identifying the location of the stick. So, in this chapter the major problems considered are:

- identifying the stick;
- identifying ups and downs or slopes in their path;
- identifying obstacles, holes and any water bodies in front of them.

Therefore, a new smart stick is proposed to overcome all these problems faced by a blind person.

## 7.2 Smart blind stick

The smart blind stick is used to detect obstacles like doors, walls, steps, trees etc, and also any slopes, water bodies in the path. Once it detects the obstacles, immediately a warning sound with vibration in the stick is sensed by the blind person to stop or change his/her path. To identify various obstacles in front of him/her various hardware components are connected to the microcontroller, like ultrasonic sensor, light dependent resistor (LDR), global positioning system (GPS), water sensor, flashing light and buzzer. The block diagram of a smart stick is shown in figure 7.1.

## 7.3 Hardware description

### 7.3.1 Arduino UNO

An arduino uno is actually a microcontroller based hardware device which is used to control different operations like communicating, controlling and operating different sensors, as shown in figure 7.2. This device consists of an ATMEGA 328 microcontroller, digital pins, analog pins, transmit and receive LEDs, reset button, ground pins, USB interface, and an external power adapter.

The primary task of this processor is to control the operations of other devices like sensors, buzzer, LDR etc, through a program code, which is stored in flash memory and a program data stored in data memory. Arduino Uno has 32 KB of flash memory for storing code, 2 KB of SRAM and 1 KB of EEPROM and operates with a clock speed of 16 MHz. The ATmega328/P is a low-power CMOS 8-bit

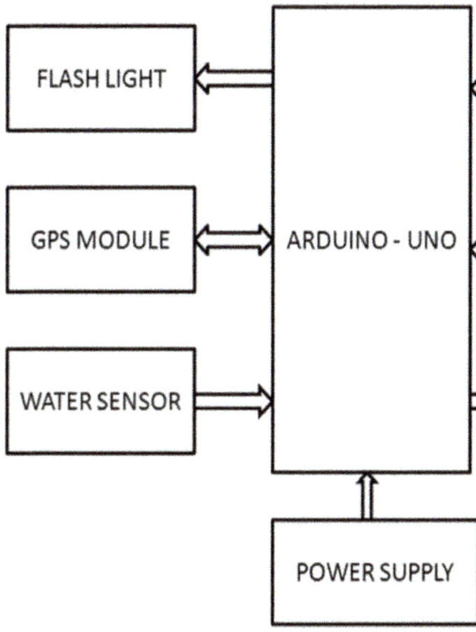


**Figure 7.1.** Block diagram of smart blind stick.

**Figure 7.2.** Arduino Uno microcontroller.

microcontroller based on the AVR enhanced RISC (reduced instruction set computer) architecture. It supports single register operations, logical operations, arithmetic operations and register operations. Most instructions are 16-bit but can support 32-bit instructions. The instructions are executed in single-level pipeline.

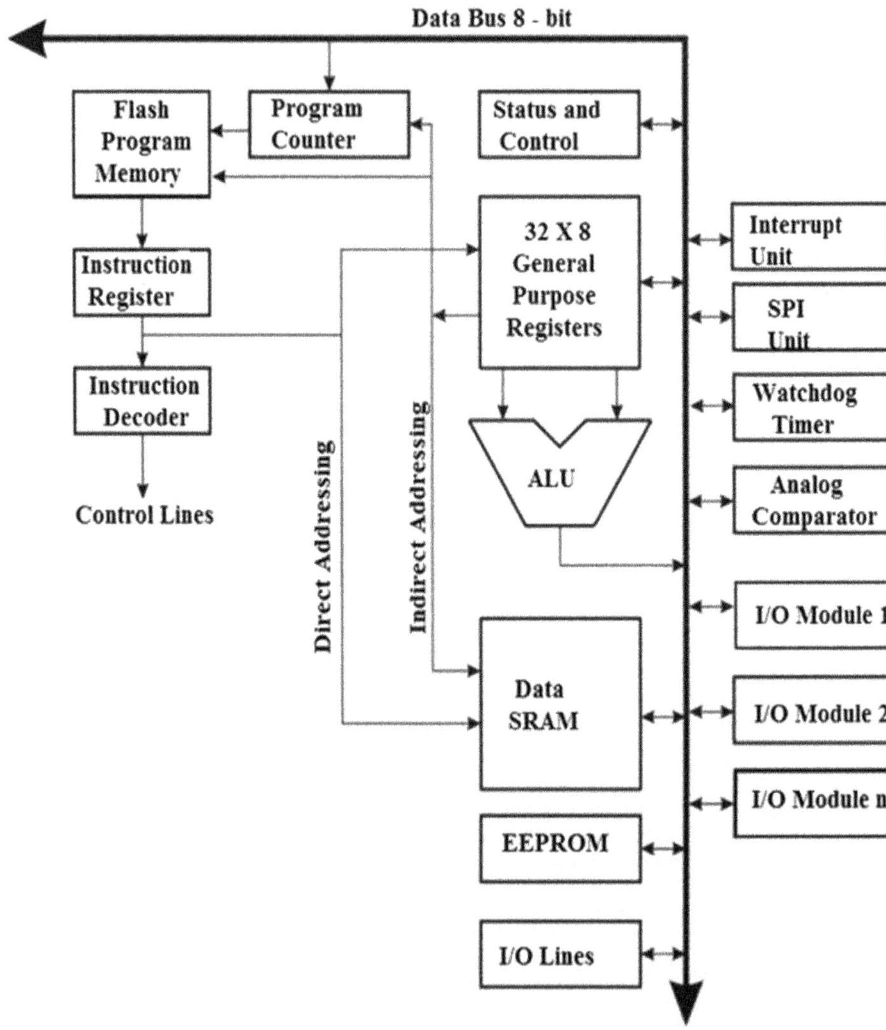


**Figure 7.3.** Arduino Uno architecture.

The architecture of Arduino Uno is shown in figure 7.3. This device receives the data from LDR, ultrasonic sensor, water sensor and GPS module. Based on the received data it sends control signals to the output devices like buzzer and vibrator in order to alert the user.

Arduino Uno consists of an ATMEGA 329P, a 28 pin microcontroller, which is shown in figure 7.4. It consists of 6 analog pins, 14 digital input and output pins, USB connector, power jack, automatic power switch, reset button, system clock and peripherals. Analog pins will acts like analog and digital converters. Sometimes analog pins can also be used as digital pins. The analog input powers are connected to the microcontroller using 6 analog pins and out of 14 digital input and output pins, 6 pins (pins 3, 5, 6, 9, 11) are used to connect PWM outputs, and pin 13 is used

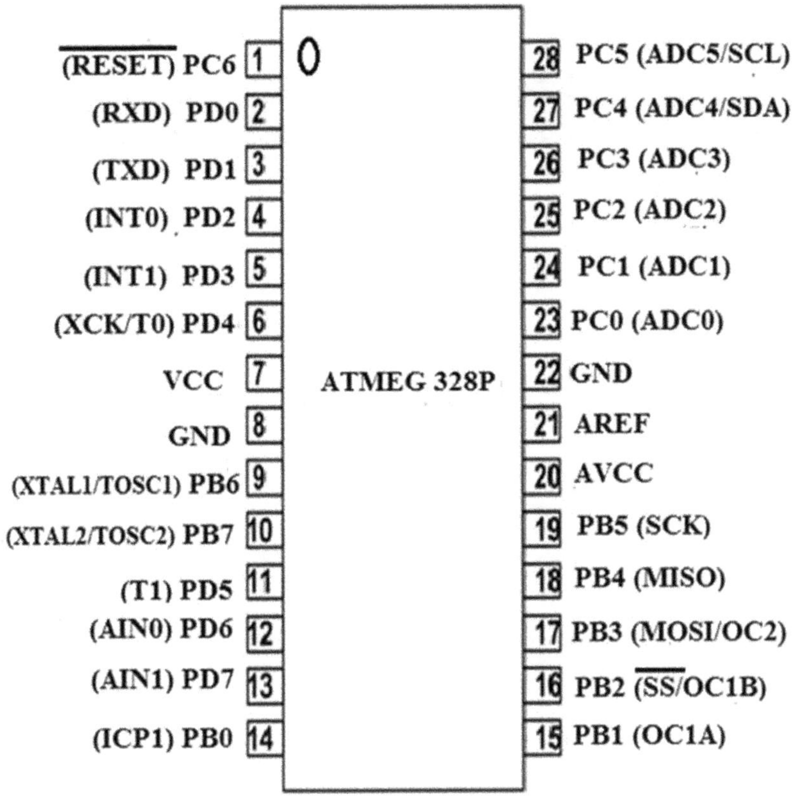


**Figure 7.4.** ATMEGA 328P pin diagram.

to connect LED. Pins 2 and pin 3 are used as receiver and transmitter for serial communications, respectively. Pins 4 and 5 are used to connect external interrupts. The power jack is used to connect power from external sources like a battery or an adaptor of 7–12 volts, or directly from a PC using USB. The reset button is used to reset the microcontroller when it is at low power.

*7.3.1.1 How to program Arduino Uno*

The Arduino development tool-chain includes the software tools required to write a program and it is very easy to program an Arduino. One can dump the program into the hardware device directly without the need of an expert and it can be done through a boot loader of 0.5 KB present on the Arduino kit. The Arduino software is open source; one can download and write the program easily using Arduino IDE in C++ language. After that the program is converted to machine language using AVR-GCC software and it is saved in the form of .hex file extension. The programming part includes: (a) declaration of variables; (b) initialing the variables; and (c) control code.

Once the program is done, save it and run for the output. If any errors are there check them in the error window; if not, verify the output on the console window. The

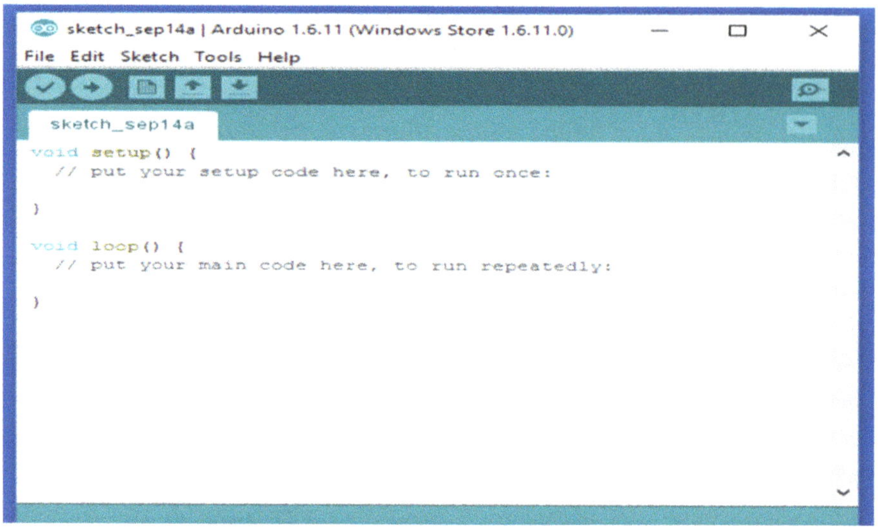


**Figure 7.5.** Arduino programming window.

Arduino tool consists of new, open, save, verify, upload, edit, sketch, tools and help etc, buttons on the toolbar. It contains many inbuilt functions which will change their colors automatically when typed. In programming, syntax plays a very important role to avoid any errors, i.e. opening and closing of parenthesis and brackets. A simple Arduino programming window is shown in figure 7.5.

To simulate an Arduino device, a free online Arduino simulator software tinker CAD is used. This tinker CAD software is very simple, easy to learn and not time consuming for simulation.

### 7.3.2 Ultrasonic sensor

Ultrasonic sensors are used to detect obstacles or objects in front through ultrasonic sound waves. The smart stick uses an HC-SR04 ultrasonic sensor to determine the distance of objects similarly to how bats identify them. An HC-SR04 ultrasonic sensor with transmitter and receiver modules is shown in figure 7.6. Whenever the user moves from one place to another, initially they need to switch on the power button on the smart stick to identify obstacles in his/her path and to continue their journey without any trouble. Whenever the device is switched on, immediately the ultrasonic sensor in the smart stick will emit sound waves towards his/her path. If any objects are found at a distance of 2 cm to 4 m, then the signals get reflected back from the objects towards the ultrasonic sensor which is present in the smart stick. The signals which are reflected back are received by the sensors. Based on those signals the device will calculate the distance of an object from the smart stick.

The time difference between transmitted and received signals will determine the distance between an object and the user, and it is calculated as

**Figure 7.6.** Ultrasonic sensor.

$$\text{Distance } (L) = \frac{T}{2} \times V \tag{7.1}$$

where $T$ is the time between transmission and reception of a sound wave, $V$ is the speed of the travelling wave. Using this distance the person can change his/her direction to continue to his/her journey without any problem. Ultrasonic sound waves are used in different applications for detecting different types of objects. Some of the characteristics of an ultrasonic sound wave are:

- It detects transparent objects—ultrasonic sound waves are reflected back from transparent objects like glass or liquid surfaces and are received by the same sensor. Therefore, it is used to detect transparent and non-transparent objects.
- It is resistant to moist and dirt—ultrasonic sound waves are not affected by moist and dust particles, i.e. detection of any obstacle is not affected by the moisture, sunlight, and dust particles in atmosphere.
- Detects complex shaped objects—complex objects like mesh or spring are also detected by ultrasonic sound waves.
- It is unable to detect soft materials like cloth.

The operation of an ultrasonic senor is affected by soft materials, for example cloth is difficult to detect, but it is not affected by sunlight or black material. Hence, with the use of an ultrasonic sensor any type of object or obstacle can be identified and thus the user is alerted.

*7.3.2.1 How does it work?*

The ultrasonic sensor radiates high frequency sound waves using transmitter pin. If any objects are found then the sound waves are reflected back, which are received by the transmitter echo pin. The working of an ultrasonic sensor is shown in figure 7.7. The distance between the user and an object in front of him/her is calculated based on the time difference between the transmitted and received signal.

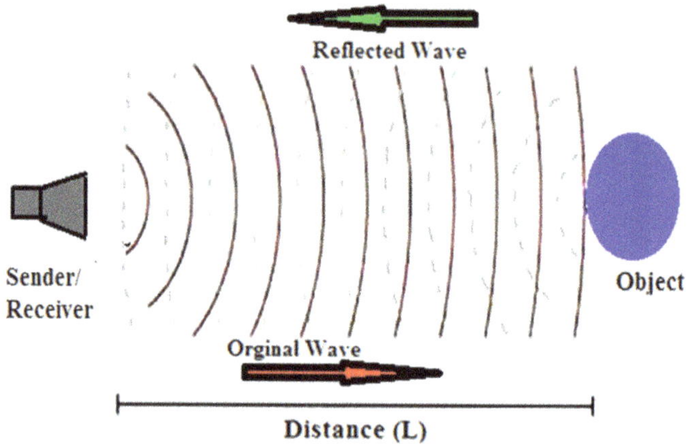


**Figure 7.7.** Working of an ultrasonic sensor.

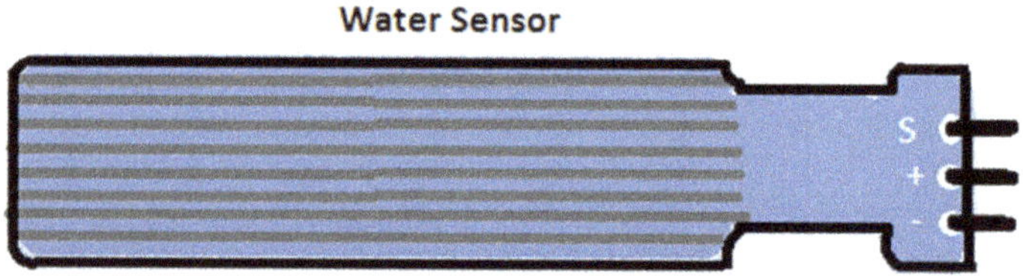


**Figure 7.8.** Water sensor.

### 7.3.3 Water sensor

This sensor is used to detect the presence of water or any other water bodies like drain leakages, spills and liquids in the direction of a user. It is very important to avoid danger because of water sources. The most commonly used water sensor is shown in figure 7.8. This device includes 1 MΩ resistor, electronic connector and several lines of connecting wires.

*7.3.3.1 How does it work?*

A water sensor is connected to an Arduino Uno for detecting the leakage and spill of water or liquid. It is mainly used to identify the presence or absence of water in front of the user. It will be connected to digital Pin 8 on the Arduino and to a vibrator output, which is to help the user identify the water source when the water sensor comes in contact with a source of water. The circuit diagram is shown in figure 7.9.

Applications:

- water sensors are used to measure the amount of gas left in the fuel tank, engine oil and cooling water in nuclear power plants and automobiles;
- to measure the water level in water tanks and storage tanks in industrial applications and also in home applications.

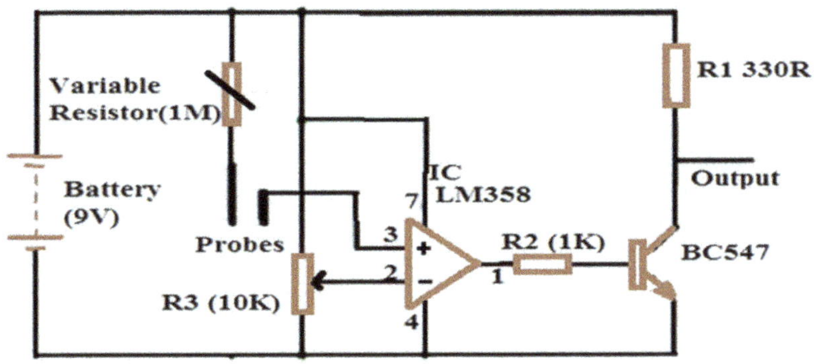


**Figure 7.9.** Circuit diagram of water sensor.

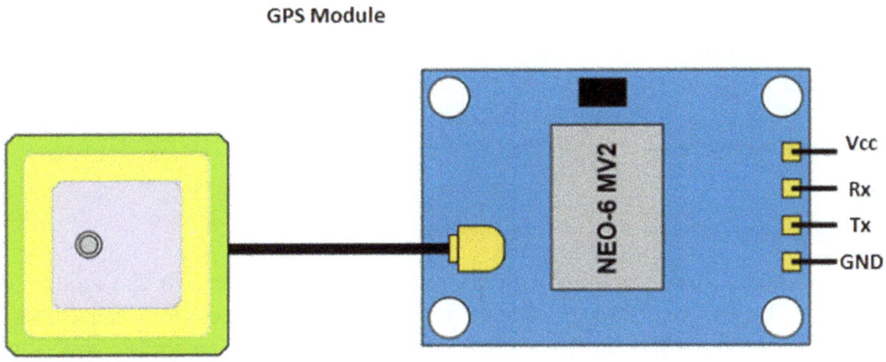


**Figure 7.10.** NEO-6 GPS module.

### 7.3.4 GPS module

A global positioning system is used to identify the location of a smart stick, which is an important task for a blind person. An NEO-6M GPS module is used to identify the location of a stick as shown in figure 7.10. This device includes a GPS receiver with an antenna of size 25 × 25 × 4 mm, which is used to receive the signals from the sensor which are transmitted by a blind person for identifying its position, anywhere which is near to him/her. Another device used is a battery backup, which is used to save data whenever the power accidentally shuts down. This device uses USART communication as receiver device to communicate with the Arduino Uno to identify the latitude and altitude of a device on the Earth.

### 7.3.5 LDR-light dependent resistor

Every person wants to identify whether it is night or day, or whether any vehicles are coming towards him/her with an intensity of light to avoid danger. Hence the smart stick is equipped with an LDR to determine the intensity of light. This device consists of an LDR or a photoresistor to know the resistance, i.e. the device will change its resistance with respective to incident electromagnetic waves which are falling on the device.

This device mainly works on the principle of photoconductivity, i.e. the conductivity of the material increases when it absorbs light from the outside. This is because the light energy present in the photons is absorbed by the electrons and jumps towards the conduction band. If more and more electrons are jumping from the valance band to the conduction band, then the current generated through this process will increase. In this process the conductivity increases and resistance decreases. Hence it is identified as day time. When there is no light energy, i.e. it is dark where the user is moving, then conductivity decreases and resistance increases, which is identified as night.

A simple and most commonly used LDR is shown in figure 7.11.

Applications of LDR.

Light dependent resistors are of low cost and very simple structure. Because of this, these devices are used in many applications such as:

(1) Used in cameras to identify the presence or absence of light;
(2) Used in light intensity meters, for counting packages on a conveyor belt, street lights, etc.

### 7.3.6 Alarm unit

The alarm unit is a device used to alert the user to obstacles. In the smart stick a buzzer and vibrator are used as output devices. These devices are used to alert the user to obstacles in his/her path, to avoid any dangers and to provide a safe journey. He/she can notice an alert through both buzzer and vibrator to avoid any problems, i.e. if he/she is moving on the road, they may not identify the buzzer sound because of noise or sounds coming from vehicles. In that case he/she may identify the obstacle from the vibrator alert. The function and working of buzzer and vibrator is explained below in detail.

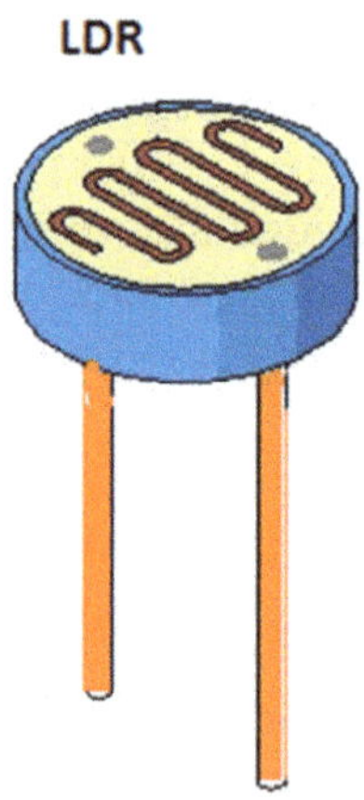


**Figure 7.11.** Light dependent resistor.

*7.3.6.1 Buzzer*

The buzzer acts like a transducer, i.e. it converts electrical energy into mechanical energy, which is shown in figure 7.12. The buzzer input is electrical or an oscillating signal and output is in the form of sound waves whose frequency range is 20 Hz–20 KHz, which is the audible frequency range of humans. The major function of the buzzer in this smart stick is to warn the user about the obstacle by producing sound waves.

*7.3.6.2 Vibrator*

A vibrator is another type of output device and it consists of a motor which is improperly balanced. A vibrator is shown in figure 7.13.

It is observed from the figure that the off-centered weight is attached to the motor's rotational shaft. The amount of vibration depends on the amount of weight attached to the motor. These motors are also found in cell phones, gaming controllers and pagers.

## 7.4 Results

A smart walking stick with microcontroller, ultrasonic sensor, LDR, GPS module, water sensor and buzzer is shown in figure 7.14. A water sensor is fixed at the end of a smart stick to identify any water bodies; an ultrasonic senor is fixed above the water sensor to identify obstacles in front of the user and other sensors are placed above the ultrasonic sensor. At first, the outputs of each sensor are verified individually by connecting it to the microcontroller and then all are connected together. The proposed smart walking stick with required components is shown below.

**Figure 7.12.** Buzzer.

**Figure 7.13.** Vibration motor.

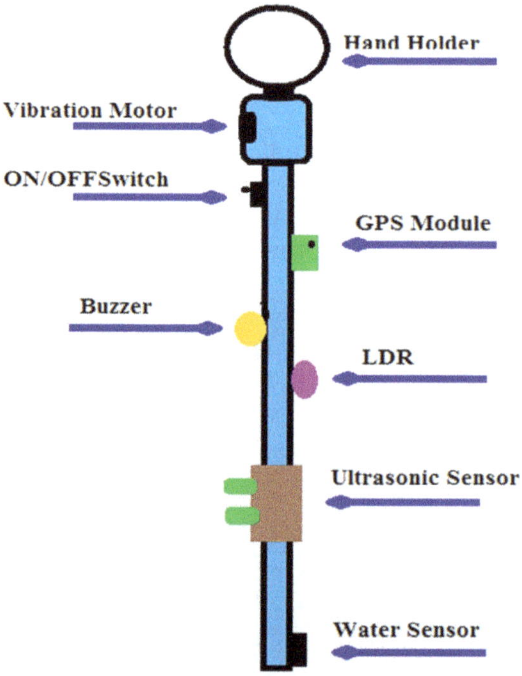


**Figure 7.14.** Proposed walking stick.

### 7.4.1 Ultrasonic sensor

An ultrasonic sensor is used to detect obstacles at a given range from the device. This device will send ultrasonic signals towards a desired direction and if any obstacle is identified then the signals are reflected back towards the same device. Once the device calculates the distance between the person and obstacle then it will be announced by the buzzer. In this manner the obstacles are identified by the person. To do this the trigger pin of the ultrasonic sensor is connected to pin 9, echo pin is connected to pin 10 and buzzer is connected to pin 2 of Arduino, as shown in figure 7.15.

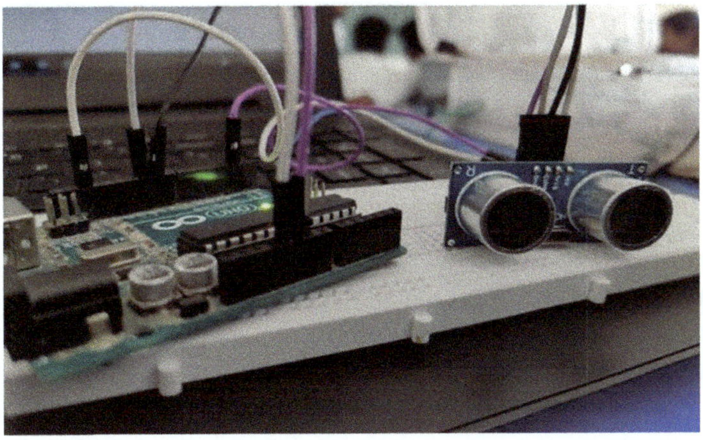

**Figure 7.15.** Ultrasonic sensor connected to Arduino Uno.

Once connections are made, switch on the power supply, then the distance of an object will be displayed on the serial monitor and this is announced by the buzzer. By using an ultrasonic sensor the person will know about obstacles in his/her path. This helps the user to reach his destination without any danger. An ultrasonic sensor can identify any kind of object within the range of 2–400 cm. The distance of an object using an ultrasonic sensor is shown in the serial monitor in figure 7.16.

### 7.4.2 Detection of water by water sensor

The water sensor is connected to the microcontroller, i.e. the Arduino Uno to verify it is working. The analog input pin A0 of the Arduino Uno is connected to the source pin, $V_{CC}$ is connected to the positive pin and GND is connected to the negative pin of the water sensor, the buzzer is connected to the output port which is shown in figure 7.17.

Depending on the depth of water, output is displayed on the serial monitor. The buzzer which is connected to the output port alerts the user to change his/her direction. The level or depth of the water is measured by conducting an experiment and those values are displayed on the serial monitor, and shown in figure 7.18. The same is identified by the user through a buzzer sound.

### 7.4.3 Detection of light by using LDR

NPN transistor, LED, resistors and power supply are used to detect the intensity of light. The transistor base terminal is connected to one end of the LDR and resistor, the other end of the LDR is connected to an emitter junction, the resistor's other end is connected to the negative terminal of the LED and the same is connected to the transistor's collector terminal. Power supply is given at the LED's positive terminal and the emitter, which is shown in figure 7.19.

The output of this circuit is connected to the analog pin of the Arduino Uno. Whenever light falls on the LDR, the resistance of the LDR will decrease and voltage value will increase. This voltage is connected to the analog pin of the

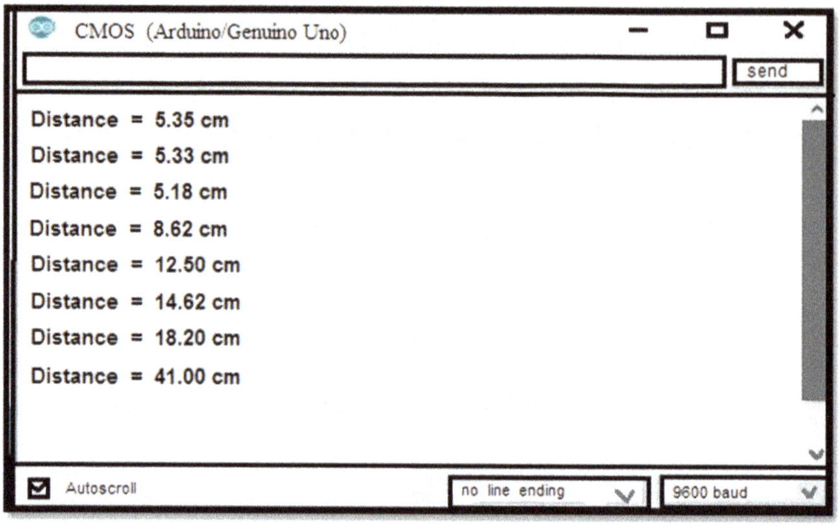


**Figure 7.16.** Ultrasonic sensor readings.

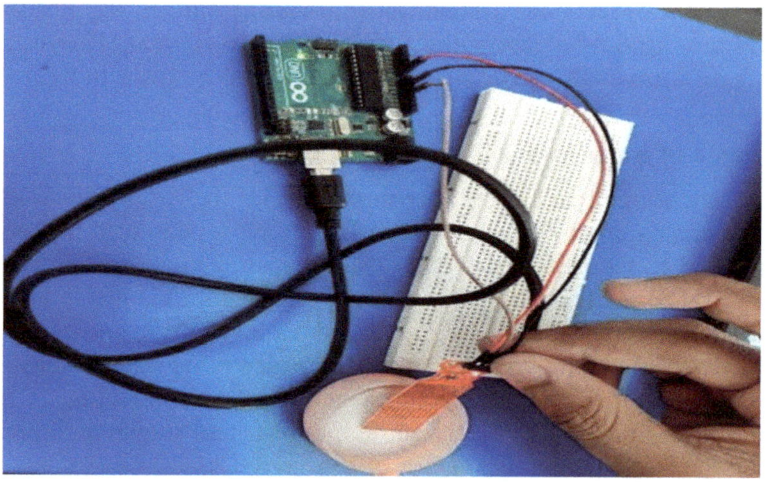

**Figure 7.17.** Water sensor working.

Arduino. As the intensity of light decreases, resistance of LDR increases and at the same time voltage at the analog pin decreases, which will be considered as night.

### 7.4.4 Location of the stick

To find the location of a smart stick, a GPS module is used and this GPS module consists of transmitter, receiver, *V*cc and GND. The transmitter and receiver are connected to pins 3 and 4 of the Arduino and *V*cc to 5 v, respectively. The latitudes and longitudes are observed in the serial monitor, shown in figure 7.20.

Water Level : 25mm to 30mm
Water Level : 20mm to 25mm
Water Level : 15mm to 20mm
Water Level : 15mm to 20mm
Water Level : 15mm to 25mm
Water Level : 10mm to 15mm
Water Level : 10mm to 15mm
Water Level : 5mm to 10mm

**Figure 7.18.** Serial monitor reading for water sensor.

**Figure 7.19.** LDR working.

Now all these sensors are connected to the microcontroller and placed in a smart stick. This stick will help the user to identify any types of obstacles, water bodies, light intensity and location of the stick. With the help of this stick any blind person can move safely without needing the assistance of other people.

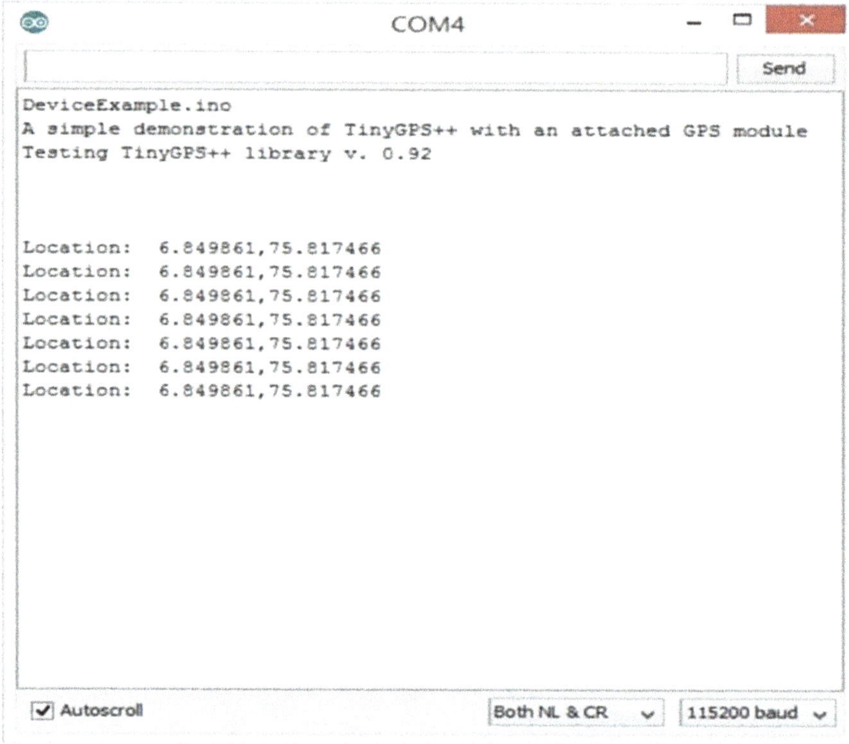


**Figure 7.20.** Latitudes and longitudes of location.

## 7.5 Conclusion

Blind people or visually impaired people have difficulty detecting objects in front of them whenever they are walking inside and outside the house, which makes it dangerous. Therefore, to support the movement of blind people, different devices have been developed with advanced technologies of modern life. In this chapter we considered a smart blind stick, which helps in identifying the obstacles which are in front of the user. A solution is identified with a smart stick for helping the user, which includes different sensors like an ultrasonic sensor, LDR, water sensor, and GPS module. When any obstacle is detected in front of the person then the sensors will send an alert message to the user in the form of a beep sound or vibration. It has one more advantage: that it detects up and down, staircases, and slopes on the road etc. One main advanced feature is that the user can find the location of the stick by pressing the button through a GPS module. The stick is capable of detecting objects in front of the user and gives a suitable message empowering him/her to move twice his normal speed because she/he feels safe. The smart stick is of low cost, gives fast response, consumes low power, is lightweight and also able to fold. With this stick, any person can move and do his/her work without the need of others, and it is very safe.

These smart sticks can in future be replaced by ultrasonic smart glasses to identify obstacles, which would be easy to carry and small in size. In future, the visually impaired could ride a bike if the ultrasonic sensor is replaced by fast detecting sensors.

## References


[1] World Health Organization 2014 *Visual impairment and blindness*, Fact sheet N '282'

[2] National Disability Policy *A Progress Report – October 2014*. National Council on Disability

[3] Ulrich I and Borenstein J 2001 The guide cane-Applying mobile robot technologies to assist the visually impaired *IEEE Trans. Syst. Man Cybern. A: Syst. Humans* **31** 131–6

[4] Terlau T, Penrod W M and Sonar K 2008 *Curriculum Handbook* Available from: http://aph.org/manuals/ksonar.pdf

[5] Whitney L 2009 *Smart Cane to Help Blind Navigate* Available from: https://www.cnet.com/news/smart-cane-to-help-blind-navigate/

[6] Hans du Buf J M, Barroso J, Rodrigues J M F, Paredes H, Farrajota M, Fernandes H, Jos J, Teixeira V and Saleiro M 2011 The smart vision navigation prototype for blind users *Int. J. Digital Content Technol. Appl.* **5** 351–61

[7] Borenstein J and Ulrich I 1997 The guide cane—a computerized travel aid for the active guidance of blind pedestrians *IEEE Int. Conf. Rob. Autom.* vol. 34–35 pp 97–100

[8] Debnath N, Hailani Z A, Jamaludin S and Ir. Dr. Syed Abdul Kader Aljunid 2001 An electronically guided walking stick for the blind *23rd Annual Int. Conf. of the IEEE Engineering in Medicine and Biology Society (Istanbul, Turkey)*

[9] Tachi S, Tanie K, Komoriya K and Abe M 1985 Electrocutaneous communication in a guide dog robot (Meldog) *IEEE Trans. Biomed. Eng.* **32** 461–9

[10] Kang S J, Ho Y, Kim and In H Moon 2001 Development of an intelligent guide-stick for the blind *IEEE Int. Conf. on Robotics & Automation Seoul (Korea)*

[11] Yuan D and Manduchi R 2005 Dynamic environment exploration using a virtual white cane *Proc. IEEE Comput. Soc. Conf. Comput. Vision Pattern Recog.* vol.1 pp 243–9

[12] Wahab M H A *et al* 2011 Smart cane: assistive cane for visually-impaired people *Int. J. Comput. Sci.* **8** 1–4

[13] Tahat A A 2009 A wireless ranging system for the blind long-cane utilizing a smart-phone *IEEE Proc. of the 10th Int. Conf. on Telecommunications (Zagreb, Croatia) vol. 8 pp 111–7*

[14] Bolgiano D and Meeks E Jr 1967 A laser cane for the blind *IEEE J. Quantum Electron.* **3** 268

[15] Shoval S, Borenstein J and Koren Y 1998 Auditory guidance with the navbelt – a computerized travel aid for the blind *IEEE Trans. Syst. Man Cybern.* **28** 459–67

[16] Magatani K, Sawa K and Yanashima K 2001 Development of the navigation system for the visually impaired by using optical beacons *Proc. 23rd Annual Int. Conf. IEEE Engineering in Medicine and Biology Society* pp 1488–90

[17] Sainarayanan G, Nagarajan R and Yaacob S 2007 Fuzzy image processing scheme for autonomous navigation of human blind *Appl. Soft Comput. J.* **7** 257–64

[18] Meijer P B L 1992 An experimental system for auditory image representations *IEEE Trans. Biomed. Eng.* **vol. 39** 112–21

[19] Olakanmi O O 2014 A multidimensional walking aid for visually impaired using ultrasonic sensors network with voice guidance *Int. J. Intel. Syst. Appl.* **08** 53–9

[20] Nada A, Mashelly S, Fakhr M A and Seddik A F 2015 Effective fast response smart stick for blind people *Second Int. Conf. on Advances in Bio-Informatics and Environmental Engineering (Italy)* pp 1–7

[21] Sathya D, Nithyaroopa S, Betty P, Santhoshni G, Sabharinath S and Ahanaa M J 2018 Smart walking stick for blind person *Int. J. Pure Appl. Math.* **118** 4531–6

[22] Anwarl A and Aljahdali S 2017 A smart stick for assisting blind people *IOSR J. Comput. Eng.* **19** 86–90

[23] Bourbakis N G and Kavraki D 2001 An intelligent assistant for navigation of visually impaired people *Proc. of the 2001 IEEE 2nd Int. Symp. on Bioinformatics and Bioengineering Conf.* pp 230–5

**IOP** Publishing

Advances in Modern Sensors
Physics, design, simulation and applications
**G R Sinha**

# Chapter 8

## Smart and wearable sensors

**Vandana Khare and K Rangacharya**

Sensors are key factors in any application and convert physical variables into electrical signals. This is not enough in the current era and they need to evolve into something more sophisticated to play a technically and economically appropriate role. A smart sensor can accomplish this by taking inputs from the physical environment and using built-in calculation resources to perform predefined functions based on the detected inputs. Wearable human interaction devices are technologies with various applications including improvement of human comfort and safety, and monitoring different health conditions. The discussion supports the architectures of smart sensors, blocks of smart sensor circuits, smart sensors as wearables and examples of wearable technology.

### 8.1 Introduction

A smart sensor is a device that takes physical or mechanical inputs from the environment, converts these inputs into electrical signals and ultimately calibrates itself because it is an intelligent device [1].

It consists of analog/digital converters and a communication interface with some intelligent control through inbuilt compute resources executed in the processor [2–4]. Hence, a smart sensor is a combination of 'transducer, actuator, and multiple sensors'. Let's take a look at each of these components in detail.

A sensor always performs environmental measurements such as those of temperature, light, sound, magnetic, etc, and converts them into electrical signals. Physical stimuli are measured with maximum and minimum ranges as the output signals are bounded. The sensitivity of the sensor may differ from the specified value [5, 6] and it is always reported as the sensitivity error. These sensors have a lot of limitations in practice.

Actuators convert certain types of stored energy into motion or action. This stored energy is in the form of liquid pressure, force, compressed air in a cylinder, etc [7–9]. A heater is one of the best examples of actuators used in sensors and can be used to control the temperature in automated molding equipment. Another practical example of an actuator is MEMS (micro-electro-mechanical-systems) in which MEMS thermal actuators are used [10].

Collectively, sensors and actuators are called 'transducers'. Transducers convert one form of signal energy into another form and this phenomenon is applicable in electrochemical, electromechanical and electromagnetic applications [11–14]. Accelerometers are the best examples of transducers and are mostly used for electromechanical applications.

As stated above, a 'smart sensor' is a collection of multiple sensors, transducers and actuators. Smart sensors are generally used for analysis, localization, communication and mobilization to facilitate rapid development [15–17].

## 8.2 Features of smart sensors

1. Automatic range selection and calibration of data.
2. Self-memory element to store local variable values.
3. Self-tuning control.
4. Communication initiation through a serial bus.
5. Higher accuracy and better reliability [18–21].

## 8.3 Evaluation of smart sensors

1. First-generation sensors contain some electronic chips along with MEMS. Silicon chips are also combined occasionally.
2. Second-generation sensors are purely analog amplification systems and analog-to-digital conversion is directly available on the chips.

### 8.3.1 Third-generation

**Third-generation** smart sensors are a hybrid combination of a transducer, signal conditioning devices and a computing device for the linearization of digital intelligence data. Figure 8.1 shows the architecture of third-generation smart sensors. The engaged part from second to third generation is a transducer along with integrated signal conditioning components. Integrated IC output is available at the analog stage so data transfers from ADC are converted into digital form completely.

In figure 8.1, the data is received by the transducer, processed by the signal conditioning unit, presented to the ADC unit where it is converted into digital form and ultimately stored in ROM or RAM according to requirement. The computing device is used to calculate the values of temperature and humidity and display them on the host computer. The drawback of a third-generation sensor is that the ADC along with the processor is externally connected to ROM and RAM [21–24].

### 8.3.2 Fourth-generation

**Fourth-generation** sensors provide two-way communication, self-testing, and inbuilt memory cells. This resonates with the definition of a smart sensor as it can act as an

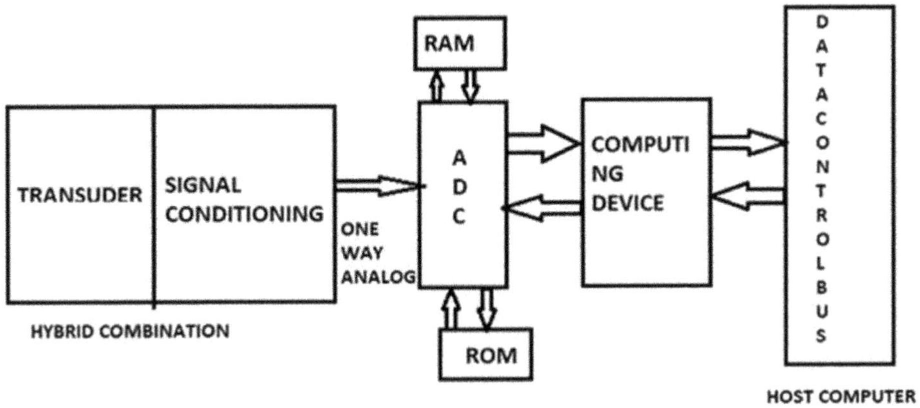


**Figure 8.1.** Architecture of third-generation smart sensor.

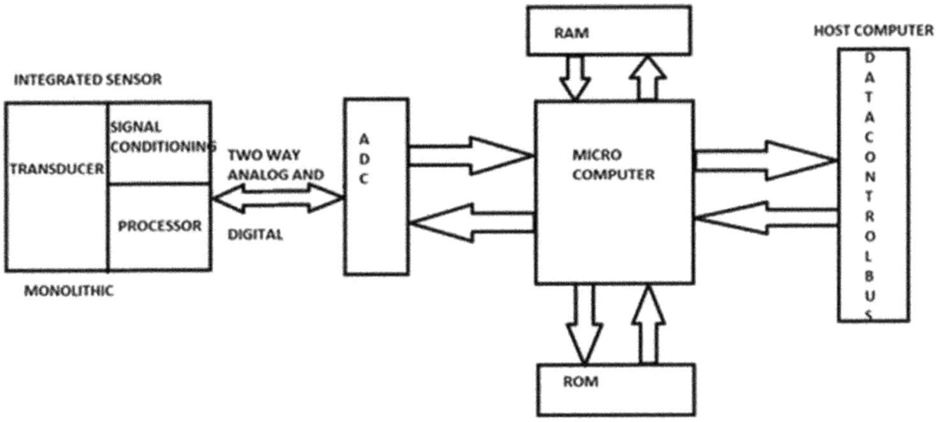


**Figure 8.2.** Fourth-generation smart sensor architecture.

intelligent device with data calibrations, without the need for external hardware. The processor used in the sensor can take charge of the decision making capability in the sensing system. Figure 8.2 gives the architecture of fourth-generation sensors. Note that the optimized design contains the integration of three components—transducer, signal conditioning component, and processor.

In figure 8.2, the processor can send and receive signals from the ADC. Here another element, i.e., the micro-computer, is used to calibrate the signal and calculate the values which are then stored in memory and directly given to the data controller bus.

### 8.3.3 Fifth-generation

Fifth-generation smart sensors are more advanced as they provide bidirectional communication, as can be seen in figure 8.3. Inbuilt memory cells are called 'Field Programmable Read-Only Memory (PROM)'. The sensor can also perform digital

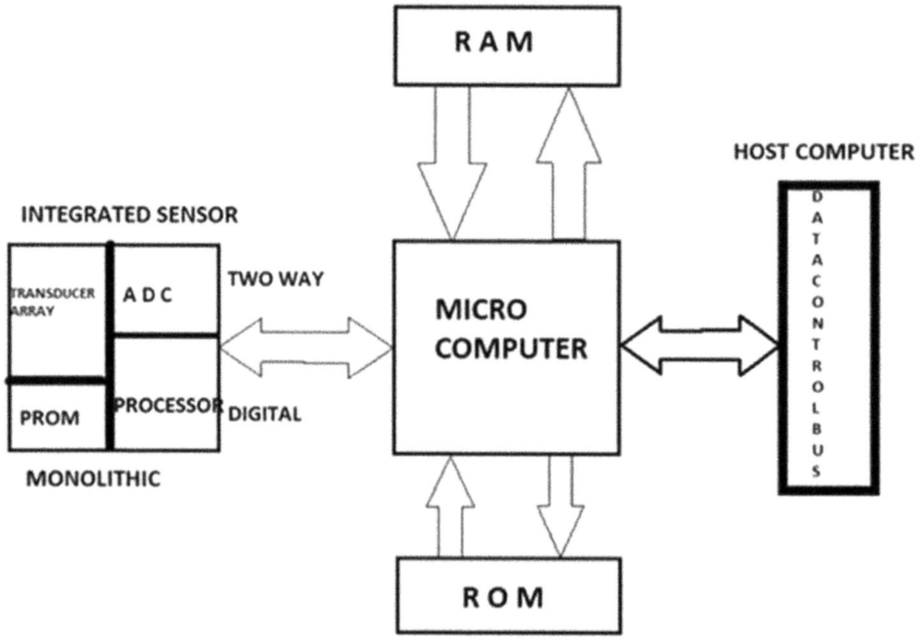


**Figure 8.3.** Fifth-generation smart sensor architecture.

signal processing tasks. These sensors are also called VLSI sensors because most of the sensing part is of the integrated type.

In fourth-generation sensors, the process up to digital conversion is integrated into one chip whereas, in fifth-generation sensors, all the components are fixed on one chip. The digital form is directly available from the integrated sensor and calibrated automatically. These values are stored by the micro-computer and sent through the data controller bus for display [23–26].

## 8.4 Design of a smart sensor

While designing a smart sensor, three aspects need to be considered:

1. data acquisition;
2. data transfer;
3. data processing.

### 8.4.1 Data acquisition

This is a sampling technique used to convert measurements of analog signals from the physical environment into digital numeric values that can be manipulated by the micro-computer. Examples of analog signals in real-time aspects include temperature, pressure, force, etc. Data acquisition systems are usually designed by software programs. There are many open-source software packages for collecting data from different hardware components [27–29].

A channel in a data acquisition system fundamentally represents an input source that is usually considered as a sensor which wishes to measure, digitize and perform further processing. On the basis of channel, data acquisition systems are classified into two methods:

1. multi-channel data acquisition systems;
2. single-channel data acquisition systems.

- A multi-channel acquisition system has multiple input sources and every source is connected with a different channel. The channels may be multiplexed with a single ADC or independent with their own complete systems.
- In a single-channel acquisition system, all conversions are attributed to that particular channel only. As is obvious, the cost of DAQs having independent channels is usually on the higher side [29, 30].

There are different ways to set up sensor systems, among which setup parameters control and measuring systems are the most considerable. At present, trends for automatic control and measurement systems use two traditional methods which are discussed here briefly:

- methods using time-division channeling, based on multiplexing sensors, i.e. based on time-shared data acquisition from each sensor;
- methods using space-division channeling, based on simultaneous data acquisition from all the sensors.

### 8.4.2 Data transfer

Data transfer refers to data communication systems that may be digital-to-analog or analog-to-digital in nature. The sensor acquires data in an analog form that needs to be converted into digital form for processing. Data transfer units include parallel and serial communication components. Smart sensor networking is mainly based on the wireless network for connecting to a system. Wireless connectivity can be achieved by a serial communication system.

Digital data transmission from a digital output smart sensor to a remote central computer demands additional measures for error protection of transmitted data. This can be achieved by means of protective coding using various redundant codes [31]. Cyclic codes are most widely used for this purpose. According to this, the common coding algorithm can be presented as fxi(Txi)→Nparallel→Nserial→Ncyclic code. The cyclic (n, n info) code is the protective code, each code combination of which is expressed as a polynomial having a degree not exceeding $(n - 1)$ and divided on the generating multinomial (polynomial) $P(x)$ of $nc = n - ni$ degree, where $nc$ is the number of check bits, $ni$ is the number of information bits and $n$ is the code block length. The checking by the cyclic redundant code presents a method similar to the method of the control sum calculation, differing in that the cyclic redundant code has, as a rule, a two-byte length and is calculated based on the polynomial of the divider. The cyclic code combination represents a 16-bit remainder of the division. Using such a method, at each moment of time, a block of certain size is transmitted. Any transferred bit influences the cyclic code, therefore such an error check is the most effective among similar methods.

### 8.4.3 Data processing

Data processing is always involved in processing information. The various processes involved are as follows:

1. Validation: the process of ensuring that the data is correct and relevant.
2. Sorting: the process of sorting a package of data and arranging it in a sequence.
3. Summarization: the process of compression of data is called summarization.
4. Aggregation: this process is involved in the concatenation of multiple inputs of data into a package.
5. Analysis: the process of organization or collection of data to get the required results or to calibrate results in various conditions. This also includes verification of data by intelligence.
6. Classification: division of input data into subcategories depending on the given algorithm. This process is involved in robotic intelligent sensor data processing.

## 8.5 Consequences

### 8.5.1 Advantages of smart sensor

1. A smart sensor works for conditioning and controlling of signals at reduced load for a centralized overall computing system.
2. Direct digital control provides high accuracy which is not achievable by analog data operation or central process computing.
3. Individual sensor controlling provides monitoring and controlling of more than one variable.
4. Collection of sensors in a network reduces the number of modules in the central network.
5. Smart sensor usage gives high performance. This is because independent usage of sensors needs an individual transducer network which increases the load of central computing. This load is engaged by the smart sensor or by using a collection of sensors with a combination of processing units.
6. Architectures based on multi-drop sensor network systems allow reduction in the number of interconnecting cables.
7. Smart sensors are highly reliable due to reduction in number of interconnecting cables and inclusion of a built-in test feature. The built-in test feature adds high-level confidence in the measurement channel.
8. These systems are scalable and flexible with expansion or reduction in measurement channels using IBIM (Intelligence Bus Interface Module).

### 8.5.2 Disadvantages

1. Designing of smart sensors requires predefined embedded functions.

2. The cost of design increases slightly due to the embedding of different items into one architecture.

## 8.6 General applications

i. Optical sensors: an optical sensor measures the physical quantity of light and also senses reflected light. Most of the applications of optical sensors are as smart sensors in cameras, measurement of liquid level, pressure, temperature, displacement, vibrations, force, radiation, pH-value, and heart-beat measurements. How an optical sensor works as a smart sensor can be easily expressed by showing the system of heart-beat monitoring. This is a robust application in which light-emitting lasers are used to measure heart rate using light. An LED shines through the skin and reflects back from the blood absorption moment. The fluctuations of light level are sensed by the light detector. The fluctuations of current rate can be taken as input and the system of embedded functions calibrates its levels and gives the heart rate as output, this has been shown in a block diagram in figure 8.4.
ii. Temperature sensors: a temperature sensor consists of BJT or MOSFET circuits as smart sensing elements and outputs duty-cycle-modulation (DCM) signals. During this time, BJT arrays generate three voltages which are converted into currents.These currents are amplified, added and subtracted. This operation finally results in currents but these currents are temperature coefficients and are almost linearly related to temperature. This performance architecture is built by three blocks: the BJT array block, modulation circuit block and IC (microcontroller) as shown in figure 8.5.
iii. Integrated multi-core sensor network: a multicore sensor network consists of numerous arrays. Each array of sensors works to generate a signal with respect to the analog type. Figure 8.6 shows the network architecture. In this core network, different types of sensor arrays club together as a single network. The core outputs may be dependent and this dependency can be controlled by the IC or microcontroller. The overall output signal is transferred from the core network to the data controller network where the data control signal is modulated, added or subtracted before being converted to digital form. The received formatted data can be accessed and controlled by the IC according to the request. Multicore sensor networks

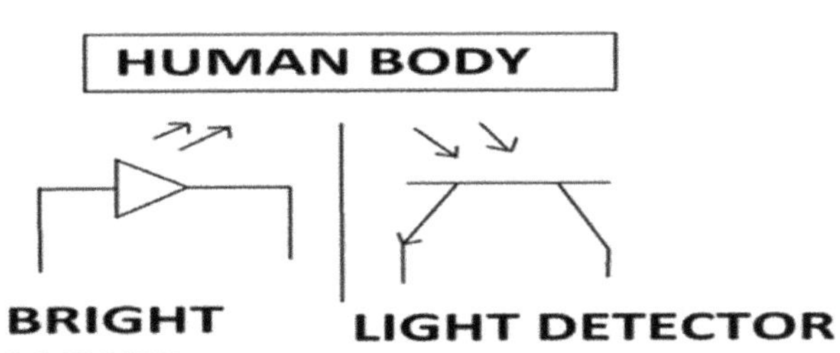


**Figure 8.4.** Basic block of heart-beat measuring system included in the smart sensor network.

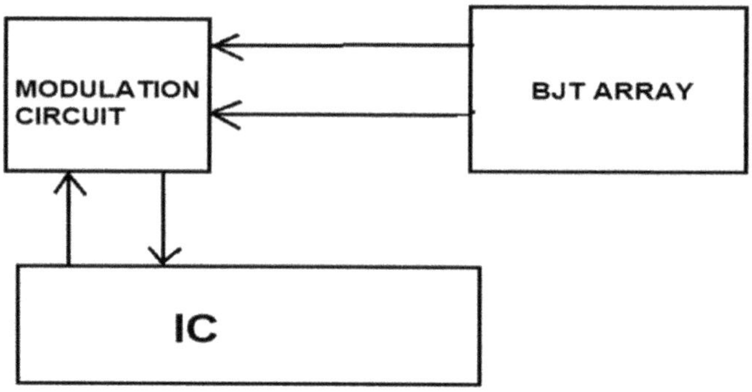


**Figure 8.5.** Blocks of smart temperature sensor.

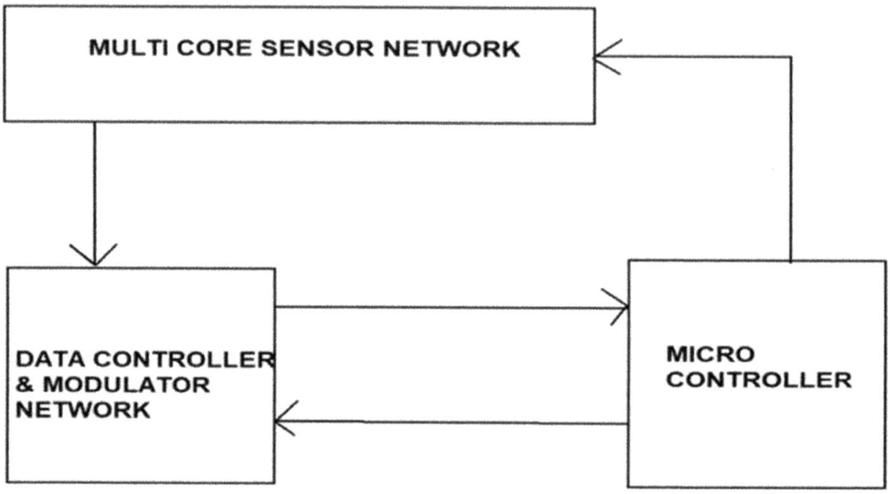


**Figure 8.6.** Multi-core sensor architecture.

have many features including self-calibration, self-authentication of controller request, power optimization and non-dependency of previous inputs as well as outputs during operation.

## 8.7 Wearable sensors

In recent years, several changes in technology have taken place to accommodate comfortable living. These changes are mostly related to hands-on electronic gadgets, health monitoring devices, security-based devices, and smart wearable devices.

### 8.7.1 Need for wearable sensors

1. Efficient health monitoring systems are needed to address health issues that are very difficult to monitor manually. Hence, there is an urgent need to monitor a human body by an automatic system. Health monitoring signals

include heart rate, ECG, photoplethysmography (PPG), blood pressure, galvanic skin response, body temperature, etc.
2. Electronic gadgets are heavy but their need is definitely felt. Wearable devices including Bluetooth microphones, wearable mobiles, wrist bands etc, are hence needed.
3. Technology is changing very rapidly from featured mobile technology to smartphone technology, from low-level programming languages to scripting, from machine learning to deep learning, etc. Hence, electronic gadgets need to be replaced by wearable ones to keep up with this change.

### 8.7.2 Smart sensor as a wearable sensor

Health monitoring in wearable forms needs to use sensors for monitoring but the problem is that they require processor circuits to read sensor values while in operation. The smart sensor provides self-calibration and wireless connectivity to the host network. A sensor with self-calibration and an inbuilt monitoring system, connected to a host network by wired or wireless interfaces, is called a smart sensor network. If this network is active and is appended with a battery, it transforms into a comfortable device and is called a wearable smart sensor.

## 8.8 Wearable sensor devices

1. Wristwatches: measure movement, pulse, and sleep.
2. Electronic T-shirts: a system for ubiquitous health and activity monitoring which consists of T-Shirts with integrated wireless sensor nodes, hosts and base stations.
3. BP monitor using PPG: applications for live BP monitor using PPG, so it is comfortable to wear.
4. Rings: these are of two types: one measures pulse and the other handles voice and GPS related bugs.
5. Employee IDs: in this era of sensors, inbuilt sensors in the ID card are either far-field communication (FFC) or near-field communication (NFC) networks.

### 8.8.1 Wristwatches architecture and performance

In the wristwatch architecture shown in figure 8.7, in which a microcontroller is connected with a sensor node. The sensor node acts as a smart network inbuilt with a PPG sensor, a pedometer or an activity system. This sensor node monitors a person $24 \times 7$ and calibrates values that are then transferred to the controller.

Sensor nodes consist of ADCs, modulators, and amplifiers along with the sensor array. The node acts as a smart network and the output from the node is transferred to a microcontroller.

The network system consists of a Bluetooth network with accessible security systems, which are required for mobile connection. Through Bluetooth connectivity, we can access sensor data from a mobile phone using pre-developed apps.

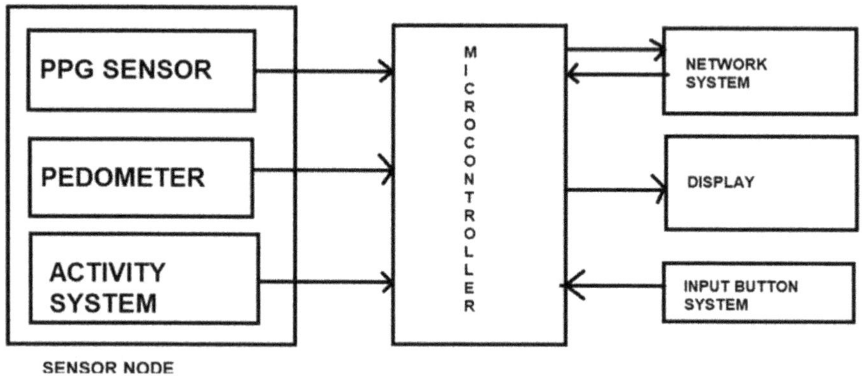


**Figure 8.7.** Architecture of wristwatch smart network.

**PPG:** Photoplethysmography is a simple optical technique used to detect volumetric changes in blood during peripheral circulation. PPG produces low-intensity infrared light (IR). When this light travels to biological tissues, it is absorbed mainly by the blood. When the intensity of light changes, the flow of blood also changes and the resultant blood flow through vessels is directly proportional to the voltage signal which is received by the PPG. Hence, small changes in the volume of blood flow can be detected by this system.

**PEDOMETER:** a pedometer is a device used to track movement of a person. It considers shakes of hands and legs followed by distance measurement using GPS. Finally, the distance is calculated by multiplying the number of steps with the step length and calibration is performed by the sensor node network.

### 8.8.2 Electronic T-Shirt architecture and working principle

Electronic T-Shirts consist of a large number of active nodes to monitor different types of health measurements such as ECG and blood pressure. In this system, the sensor board is architected as a multi-hop network. This multi-hop-network consists of an accelerometer, an A/D converter, wireless transceivers with three nodes and another node with electrodes, an AC-coupled instrumentation amplifier, a low/high pass filter and a final gain stage.

In figure 8.8, Node 1 is the accelerometer, Node 2 is the A/D converter, Node 3 represents the wireless transceivers and Node 4 is a special network for modulation. Figure 8.9 shows the block diagram of Node 4.

**Operation:** Figure 8.9 shows the architectural blocks of a smart ECG monitoring system. It consists of a three-axis accelerometer sensor that records electrical activity of the heart versus time. Whenever the heartbeat increases, the developed voltage also increases. The accelerometer reads these values and transfers them to Node 4. The voltage is taken from the electrical potential between various points on the body. Node 4 modulates the voltage readings processed by the microcontroller through the wireless network and the resulting values are hosted through a wireless communication network. This simple operation results in the working of this smart wearable T-Shirt.

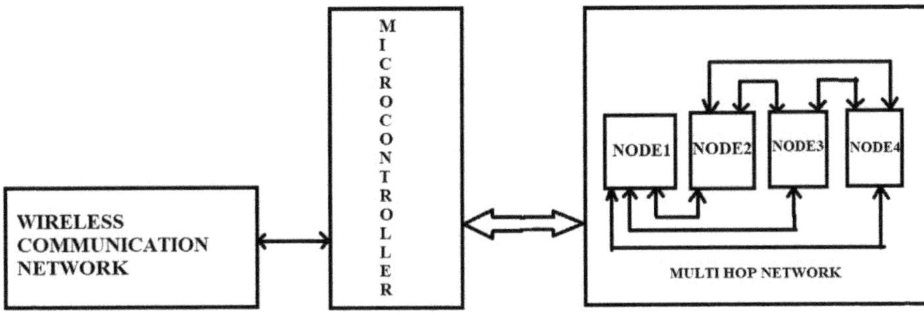


**Figure 8.8.** Architecture of electronic T-Shirt system.

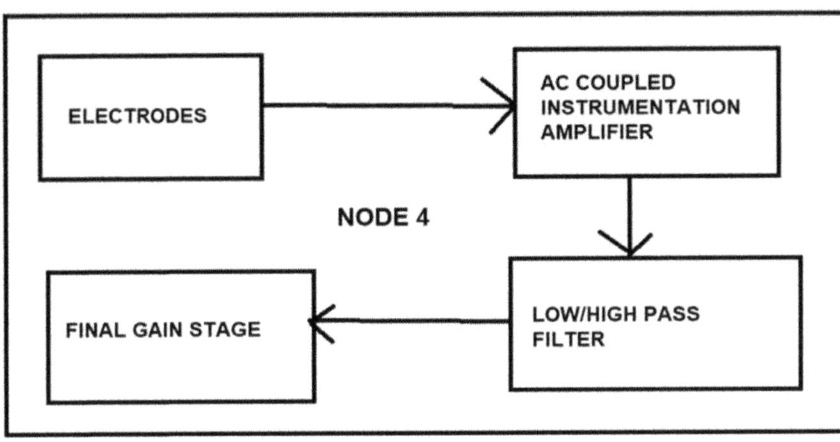


**Figure 8.9.** Block diagram of Node 4.

### 8.8.3 BP monitoring using PPG

BP (blood pressure) is a common test in the medical area and a direct indicator for hypertension. Due to its higher importance among all risk factors of cardiovascular diseases, it is the most responsible parameter to monitor in people affected by hypertension.

Traditionally, BP can be tested by a sphygmomanometer consisting of an inflatable cuff, a measuring unit and a mechanism for inflation. BP measurement can also be efficiently done by photoplethysmography (PPG) wearable devices discussed above.

The model estimates BP by using PPG alone employing certain signal processing techniques and deep learning. The application can be studied from two aspects: first based on its hardware architecture and second focusing on the software model.

Figure 8.10 shows hardware architecture of wearable devices consisting of one processing unit supported by traditional machine learning as well as deep learning concepts. The PPG values are transmitted to the processing unit, among which the

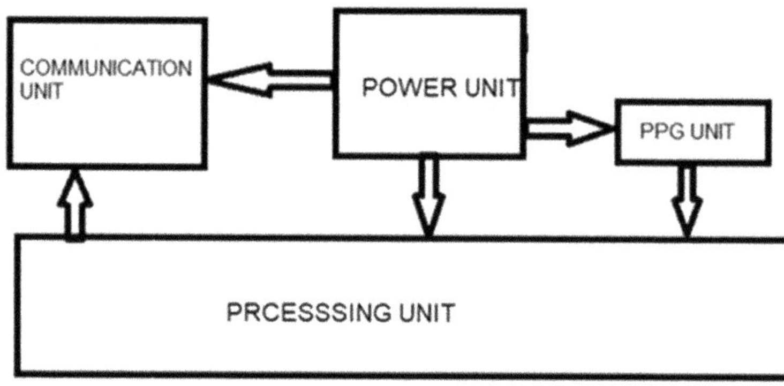


**Figure 8.10.** BP estimation only with PPG hardware architecture.

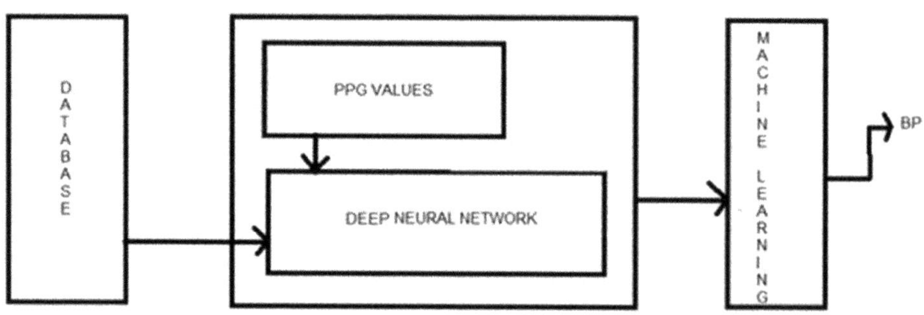


**Figure 8.11.** BP estimation software model.

raw data will be taken by the software model from which the blood pressure results are outputted. The value can be read using the communication unit which can also be connected with display or server etc.

In the process of BP calibration, the most important role is that of the software model, which uses concepts in DATABASES, deep neural network (DNN) and machine learning. Figure 8.11 shows the BP estimation software model.

DATABASE: this is the collection of researched pre-defined PPG signals, and may contain a large number of signals. To decrease the processing time of calibration, a subset of the raw data can be chosen by using some regression techniques unwanted frequencies can be removed using Butterworth filters etc.

DNN and machine learning: these concepts are useful in removing shorter waves and unbounded waves. The longer waves are transmitted to a DNN which then learns some features. These features are fed into some well-known deep learning models like the Leave-One-Subject-Out experiment to get systole blood pressure and diastole blood pressure.

## 8.9 Conclusion

Smart sensors always work effectively with proper calibration. Sometimes, these sensors are also called active sensors. For increasing reliability of electronic devices, sensors are replaced with smart sensors which are very useful in daily life. Smart wearables are intelligent devices that always work with a specified principle.

Earlier technologies like first-generation smart sensors consist of electronic chips. Second-generation sensors consist of analog-to-digital converters on the chip. Third-generation sensors consist of the transducer and signal conditioning on the chip. Fourth-generation sensors modify and add extra features on the chip. The fifth-generation sensors are an instance of advanced technology with all components on a single chip. Nowadays, wearable sensors are mostly used in smart watches, wearable rings, and wearable t-shirts. Smart watches and wearable rings are also used for heartbeat monitoring.

## References


[1] Sehrawat D and Gill N S 2019 Smart sensors: Analysis of different types of IoT sensors *2019 3rd Int. Conf. Trends in Electronics and Informatics (ICOEI)* doi:10.1109/icoei.2019.8862778

[2] Wakita A, Tanji M, Kitada S, Shibutani M, Uchiyama H and Inakage M (n.d.) *A coordination model for wearable fashion. Ninth IEEE Int. Symp. on Wearable Computers (ISWC'05)*. doi:10.1109/iswc.2005.2

[3] Jagtap N, Wadgaonkar J and Bhole K 2016 *Smart wrist watch. 2016 IEEE Students' Conf. Electrical, Electronics and Computer Science (SCEECS)*. doi:10.1109/sceecs.2016.7509273

[4] Schmalzel J, Figueroa F, Morris J, Mandayam S and Polikar R 2005 An architecture for intelligent systems based on smart sensors *IEEE Trans. Instrum. Meas.* **54** 1612–6

[5] Hogenbirk E J, Verhoeven H-J and Huijsing J H 1995 An integrated smart sensor for flow and temperature with I2C bus interface: FTS2 *Proc. Int. Sym. Circuits Syst.* pp 2225–8

[6] Chifor B C, Bica I, Patriciu V V and Pop F 2018 A security authorization scheme for smart home internet of things devices *Future Gener. Comput. Syst.* **86** 740–9

[7] Kwaaitaal T 1993 The fundamentals of sensors *Sens. Actuators* A **39** 103–10

[8] Nesa N and Banerjee I 2017 IoT-based sensor data fusion for occupancy sensing using Dempster–Shafer evidence theory for smart buildings *IEEE Internet Things J.* **4** 1563–70

[9] Nanotechnology - Enabled Sensing 2009 *Report on the National Nanotechnology Initiative Workshop, Arlington, Virginia, May 5–7*

[10] Hunter G W, Stetter J R, Hesketh P J and Liu C-C *Smart Sensor Systems.* Available online at https://electrochem.org/dl/interface/wtr/wtr10/wtr10_p029-034.pdf

[11] Fujita H 1996 Future of actuators and microsystems *Sens. Actuators* A **56** 105–11

[12] Fujita H 1998 Microactuators and micromachines *Proc. IEEE* **86** 1721–32

[13] Schomburg W K, Ahrens R, Bacher W, Goll C, Meinzer S and Quinte A 1998 AMANDA—low-cost production of microfluidic devices *Sens. Actuators* A **70** 153–8

[14] Heikenfeld J, Jajack A, Rogers J, Gutruf P, Tian L, Pan T and Kim J 2018 Wearable sensors: Modalities, challenges, and prospects *Lab Chip* **18** 217–48

[15] Slapničar G, Mlakar N and Luštrek M 2019 Blood pressure estimation from photoplethysmogram using a spectro-temporal deep neural network *Sensors* **19** 3420

[16] Allen J 2007 Photoplethysmography and its application in clinical physiological measurement *Physiol. Meas.* **28** R1–39

[17] Geddes L A, Voelz M H, Babbs C F, Bourland J D and Tacker W A 1981 Pulse transit time as an indicator of arterial blood pressure *Psychophysiology* **18** 71–4

[18] Yang W, Yang K, Jiang H, Wang Z, Lin Q and Jia W 2014 Fetal heart rate monitoring system with mobile internet *2014 IEEE Int. Symp. Circuits and Systems (ISCAS) (Australia)*
[19] Mukherjee S, Ghosh A and Sarkar S K 2019 Arduino based wireless heart-rate monitoring system with automatic SOS message and/or call facility using SIM900A GSM Module *2019 Int. Conf. Vision Towards Emerging Trends in Communication and Networking (ViTECoN)* doi:10.1109/vitecon.2019.8899504
[20] Dutta A, Banerjee A, Bose A, Auddy A, Rana T K and Bhattacharyya S 2017 Heart tracer – The route to your heart *2017 8th Annual Industrial Automation and Electromechanical Engineering Conf (IEMECON)* doi:10.1109/iemecon.2017.8079555
[21] Patil O R, Gao Y, Li B and Jin Z 2017 CamBP. Proc. 2017 ACM Int. Joint Conf. Pervasive and Ubiquitous Computing and Proc *2017 ACM Int. Symp. Wearable Computers on - UbiComp '17* doi:10.1145/3123024.3124428
[22] Slapničar G, Mlakar N and Luštrek M 2019 Blood pressure estimation from photo-plethysmogram using a spectro-temporal deep neural network *Sensors* **19** 3420
[23] Yousef Q, Reaz M B I and Ali M A M 2012 The analysis of PPG morphology: Investigating the effects of aging on arterial compliance *Measurement Sci. Rev.* **12** 266–71
[24] Loukogeorgakis S, Dawson R, Phillips N, Martyn C N and Greenwald S E 2002 Validation of a device to measure arterial pulse wave velocity by a photoplethysmographic method *Physiol. Meas.* **23** 581–96
[25] Xing X and Sun M 2016 Optical blood pressure estimation with photoplethysmography and FFT-based neural networks *Biomed. Opt. Express* **7** 3007
[26] Agarwal R, Bills J E, Hecht T J W and Light R P 2010 Role of home blood pressure monitoring in overcoming therapeutic inertia and improving hypertension control: A systematic review and meta-analysis *Hypertension* **57** 29–38
[27] Kurylyak Y, Lamonaca F and Grimaldi D A neural network-based method for continuous blood pressure estimation from a PPG signal *Proc. 2013 IEEE Int. Instrumentation and Measurement Technology Conf. (I2MTC)* (*Minneapolis)* 6–9 May 2013) pp 280–3 doi:10.1109/I2MTC.2013.6555424
[28] Gesche H, Grosskurth D, Küchler G and Patzak A 2011 Continuous blood pressure measurement by using the pulse transit time: Comparison to a cuff-based method *Eur. J. Appl. Physiol.* **112** 309–15
[29] Alian A A and Shelley K H 2014 Photoplethysmography *Best Pract. Res. Clin. Anaesthesiol.* **28** 395–406
[30] Wieringa F P, Mastik F and van der Steen A F W 2005 Contactless multiple wavelength photoplethysmographic imaging: A first step toward 'SpO2 camera' technology *Ann. Biomed. Eng.* **33** 1034–41
[31] Kirianaki N, Yurish S, Shpak N and Deynega V 2002 *Data Acquisition and Signal Processing for Smart Sensors* (New York: Wiley)

# Chapter 9

## Cognitive and biosensors: an overview

**Samrudhi Mohdiwale, Mridu Sahu, G R Sinha and Silvia Liberata Ullo**

Cognitive sensors and biosensors are a most widely accepted technology, applicable in a variety of areas, alone or in combination. This chapter aims to provide a comprehensive literature on these sensors, recent trends of research and research challenges, that can be explored in the near future. Machine learning and security concerns uplift the search for literature in this study. A variety of applications are also explored to provide a wider understanding on the area of cognitive sensors and biosensors.

### 9.1 Introduction and background

The sensor is a device or a module that can sense the stimuli generated by physical or environmental activity. In this technological era, we are surrounded by millions of sensors monitoring our health and environment. To know the heart rate, for instance, step counters, built in mobile sensors, are used nowadays. To become aware of weather and temperature changes, specific sensors are deployed and used for sensing the surrounding environment. Environment monitoring has a broader perspective, which includes besides the weather and temperature monitoring, crop and agriculture monitoring, traffic monitoring, dam monitoring for flood issues and many more [1–4]. Deployment of sensors, data gathering and pre-processing, handling of big data, energy consumption, applicability are just some issues addressed while working on sensors, and which imply the presence of a telecommunications network.

Wireless sensor networks (WSNs) are used in a variety of applications, ranging from area monitoring, agriculture and industrial monitoring, health monitoring, threat detection, urban mobility and traffic monitoring, and smart grid monitoring [5–7, 121–127]. The areas covered by WSNs are very vast and increasing day by day, and up to now, they have been using the industrial, scientific, medical (ISM) band of transmission, which is unlicensed and free for communication. Due to the increase in the volume of data generated, and area covered by WSNs, the ISM band is suffering

doi:10.1088/978-0-7503-2707-7ch9

from its limited transmission band, which increases the traffic congestion in the network and the probability of data loss [8].

To deal with these issues, cognitive sensors technology has emerged, replicating what happened in more complex systems such as radars, where the introduction of cognitive aspects has greatly improved their technical capabilities [128].

Biosensors represent a further evolution of cognitive sensors, as a result of the technological revolution where the semiconductor and chip technology, providing a large and systematic production, has helped in realizing receptors from the body, able to transduce the biological data and provide the outputs to the connected devices. Based on the transducer type, biosensors are classified into electrochemical, electronics, optical, piezoelectric, gravimetric and pyroelectric biosensors [9, 10]. Figure 9.1 represents a pictorial view of cognitive sensors and biosensors applications.

In this chapter, details on cognitive sensors and biosensors are presented along with their research challenges and application perspective.

## 9.2 Cognitive sensors

The term cognitive means involvement of intellectual ability such as thinking, memory and reasoning [11]. Sensors which use these cognitive abilities are termed as cognitive sensors. For many WSNs and other networks, which work on a licensed spectrum of a particular band, this latter is heavily underutilized. Only 5%–10% of the bandwidth is utilized properly and the rest is unused [8, 12]. To deal with this issue, cognitive sensor networks have evolved, and turned into cognitive radio sensor networks (CRSNs), where the cognitive sensors utilize the intelligent scheme of sensing and exploiting the share of unused spectrum, by introducing the cost limited

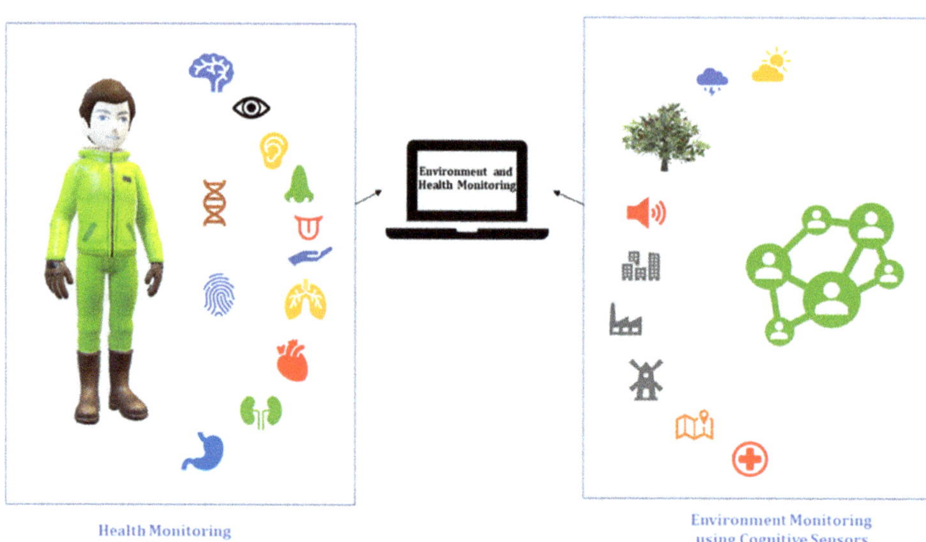


**Figure 9.1.** Cognitive and biosensors: a pictorial view.

to the spectrum utilized for the specific time and location. Cognitive sensors have functionalities of spectrum sensing, decision making, sharing and mobility management for secondary users [8, 13]. These additional functionalities allow cognitive sensors to sense the unused spectrum and also make the decision for the best portion of it. They also allow avoiding collisions among multiple secondary users, asking for the same spectrum. Moreover, when primary users want back the spectrum taken by secondary users, CRSNs manage the mobility of the spectrum in such situations, by working on the spectrum assignment with priority.

### 9.2.1 Research challenges

The cognitive sensor network is a growing area of research with multiple research opportunities. In this section, different challenges obtained from different papers found in the literature are presented, highlighting opportunities for new researches in the area.

- *Increased power consumption and sensor lifetime*: The cognitive capability in sensors introduces high power requirement and increased resources which ultimately increase the overall costs. The network should be designed such that it reduces the power consumption and improves the utility of the model [14, 15]. However, there are many researches ongoing in this direction using optimal node selection, multichannel cooperative sensing etc [16, 17], and therefore new ideas and solutions for the problem will maintain its gravity.
- *Spectrum allocation*: The spectrum can be requested by multiple secondary users simultaneously, hence there should be an efficient strategy for fair spectrum allocation with a prioritization scheme [18].
- *Interference management*: The cognitive sensor network faces one of the major challenges of interference, caused by cognitive nodes when the spectrum is occupied by primary nodes or other secondary nodes. Interference increases the power consumption [19]. The nodes must be designed such that they minimize interference and maximize spectrum utility.

### 9.2.2 Application of cognitive sensors

In CRSN, the quality of services refers to the reliability, communication delay, latency and resource allocation or distribution. Although different routing protocols exist for the quality-of-service measurement, they still need to be explored for the different applications [20, 21, 111–117, 129–131]. Table 9.1 shows the application of cognitive sensors along with the main challenges and methods that have been investigated.

### 9.2.3 Cognitive sensors and machine learning

Machine learning is a widely used concept nowadays which is gaining its popularity in almost every application area [22]. Cognitive sensors and networking are also one of them. In this section we will deal with some problems of cognitive sensor networks, where machine learning is applicable.

**Table 9.1.** Applications of cognitive sensor: methods and challenges.

| Application area | Challenge | Method used |
|---|---|---|
| Industrial safety control and monitoring systems | Resource allocation, QoS, energy management, security | Spectrum handoff strategy, energy harvesting, artificial intelligence (AI) Techniques, joint resource optimization |
| Healthcare | Efficient data transmission and monitoring, personalized resource service, high latency, time and computation | Simulated annealing, edge cognitive computing, AI |
| Smart grid applications: power system monitoring, dynamic pricing, fraud detection, load and power management | Electromagnetic interference, multi-path effects, equipment noise, latency, throughput | Cross layer framework, scheduling, bio inspired optimization methods, dynamic channel selection |
| Home monitoring | Real time monitoring | RF based module, distributed framework |
| Military applications and surveillance | Security and location | RFID/CRSN |
| Disaster management | Accuracy and early detection | Dynamic channel selection, convolutional neural networks |
| Environment monitoring | Efficient structural health monitoring | Dynamic cognitive sensors and artificial neural networks |

- *Load balancing*: As energy consumption issue is discussed in the previous sections, increase in network traffic leads to more energy consumption. To reduce the energy consumption, traffic should be organized such that the resource utilization is optimized. To attain the desired quality of service (QoS), load balancing using machine learning came into the picture. This has been achieved using reinforcement learning which detects the traffic in the paths and diverts it to alternative paths to reduce the load and corresponding energy consumption [23].
- *Channel selection*: The channel selection is an important issue in cognitive sensor networks, where channels should be selected by considering channel quality and communication matrices. To obtain the suitable channel, machine learning algorithms are employed in [24–27].
- *Modulation scheme identification*: Higher order statistics, spectral and phase features are explored in the literature to demonstrate the choice of the modulation scheme. Machine learning methods need to be explored, to get better modulation schemes and to improve the network lifetime [28].

- *Wireless network parameter identification*: The usage pattern of the consumers can help to predict the network requirements, in terms of their location and data services. Machine learning techniques, such as support vector machine, K-nearest neighbors, for instance, are utilized in research to find the best classifiers suitable for this problem [28].
- *State prediction*: The *status* of primary users, active or inactive, can be found using machine learning models, able to retrieve features from different sources and calculate the state of the users [28].

Deep learning techniques are also used in this area for primary user classification and related applications [29, 30]. Apart from supervised techniques of machine learning, unsupervised methods also put a tight command on cognitive sensors to improve the QoS and to solve many problems related to networking [31]. Unsupervised methods, such as clustering for example, are used for optimal offloading, efficient network formation, reduced interference and dimensionality reduction. Reinforcement learning is also used in this area for energy harvesting, spectrum access and sensing [28].

### 9.2.4 Cognitive sensors and security threats

Security is one of the major concerns nowadays in every area. In the field of cognitive sensors and networking, security and vulnerabilities are mainly raised in cognitive capabilities and reconfigurability of the network. This section deals with attacks and security threats in the cognitive sensor network. Figure 9.2 represents the major threats faced by the cognitive sensor system.

The details on major security threats are as follows:

- *Primary user attack*: In this type of attack, cognitive users (CUs), represented by secondary users, receive a false signal related to a primary user. As per the principles of CRSNs, whenever there is a request received by a primary user, CUs need to leave the spectrum and search for other availability. This degrades the performance of the network [32]. Obviously, the nodes generating these incumbent signals are of the greedy or malicious type.
- *Spectrum sensing falsification attack*: The transmitted signal suffers with signal loss due to fading and multi-path transmission. To solve this issue, collaborative spectrum sensing techniques are used, where all the observations are sent to a fusion center (FC) in charge of taking the final decision. When the FC receives false transmissions from the nodes, it may take a wrong decision, based on false transmission. In this case, a spectrum sensing falsification attack is in progress or has occurred [32–34].
- *MAC layer and IEEE 802.22 threats*: MAC and physical layers show the closest coupling among them, and try to avoid interference by collaboration. Common control channels (CCCs) enable exchange of control information. MAC spoofing, congestion and jamming attacks are the threats faced by the CCCs [32, 35].

**Figure 9.2.** Security threats in a cognitive sensor network.

- *Cross layer attack*: Attacks targeted on multiple layers of the IEEE 802.22 model are called cross layer attacks. In this type of attack, malicious users aim to acquire the whole bandwidth. To solve this issue, the transmission control protocol (TCP) layer can be designed for secure operations, to aware the CCCs [32, 36].
- *Software defined radio security*: Another challenging domain in cognitive networks is security related to hardware and software protection. A software based threat involves installation of bugs or viruses into the system, by causing authenticity and data integrity-based problems. Hardware based threats cause increase in transmission power, and utilization of resources in excess, for instance. Details of these types of attack are also provided in [32].

## 9.3 Biosensors

Biosensors are receiver-transducer systems that can sense the biological activities and provide data related to it. The systematic workflow of biosensors is presented in figure 9.3.

Analytes are the cells, antibodies, enzymes or any other biological materials that are of interest for specific biosensors. These analytes are sensed by the receptors. The latter are very sensitive parts connected to the analytes that make the transducers become activated. Each transducer converts the input of the receptors into the signal of interest. The transducer output can be electrical, heat, magnetic, optical etc, as per the choice of biosensor and its application. Based on transducer output, the biosensors can be categorized as shown in figure 9.4. The study of these categories has been provided by [37].

As a matter of fact, the technological era is a breakthrough in the area of biotechnology and medicine. The biosensors are designed in such a smart way so that they can be wearable and assistive by using certain advanced techniques, like Internet of Things (IoT), machine learning and other biotechnological advances. Smart healthcare provides solutions for disease prevention, monitoring, diagnosis, treatment and management. IoT, the cloud, artificial intelligence and microelectronics are just some areas helping in smart healthcare management [38].

### 9.3.1 Research challenges

Biosensors are used in various applications ranging from healthcare to industry. Cognitive capability of sensors builds a model that fuses the capability of sensing and decision making as per cognition. Apart from the challenges faced by cognitive sensors, discussed previously, in this section, research challenges faced by smart sensors applicable in different fields are discussed.

- *Quality cost tradeoff*: Smart sensor technology is applicable in various fields, however, it faces the challenges of providing quality data with affordable

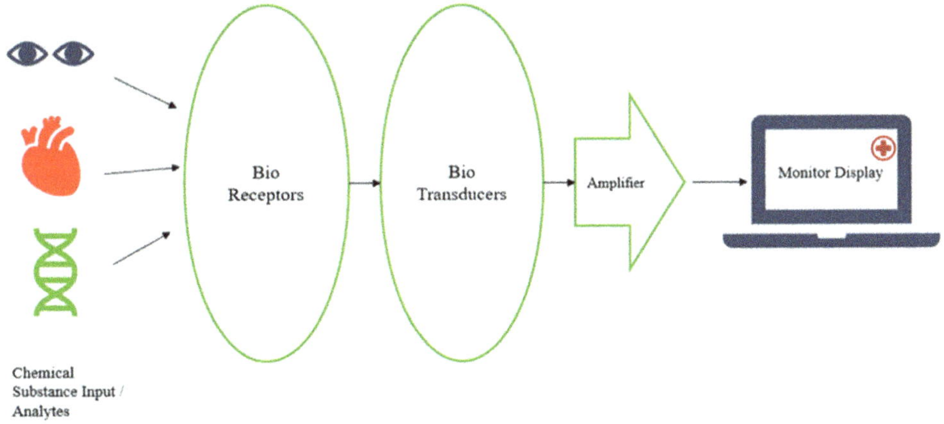


**Figure 9.3.** Working of biosensors.

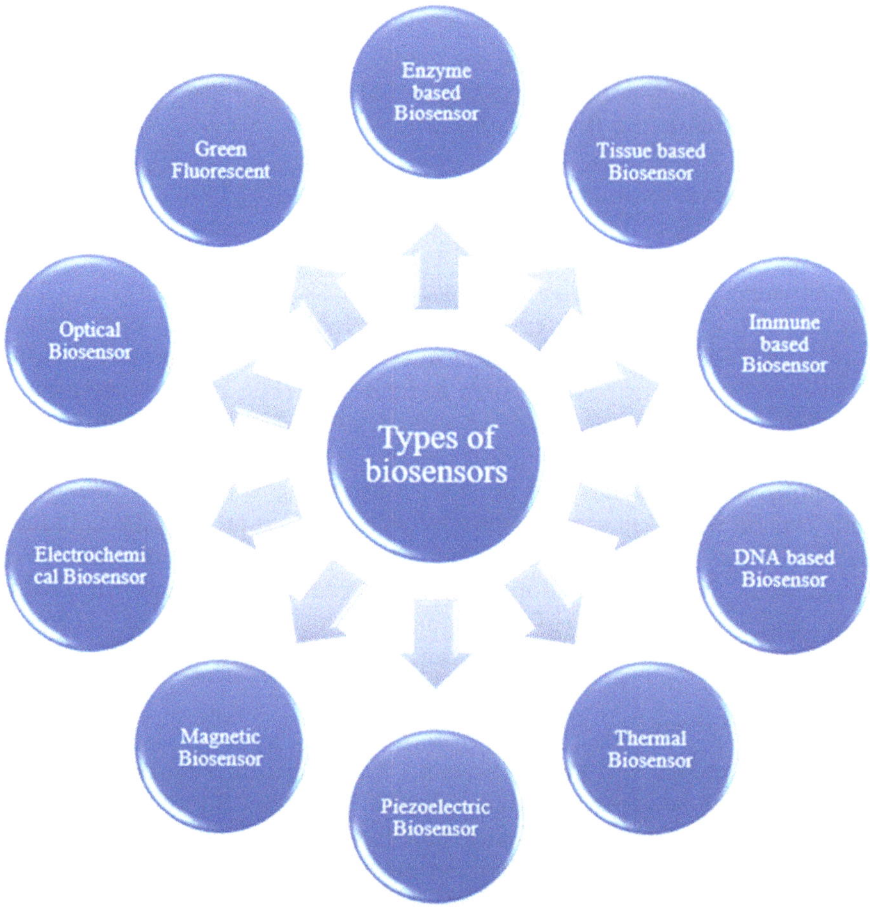


**Figure 9.4.** Types of biosensors.

pricing. As cost increases, quality of sensed signal gets improved [38, 39]. For example, dry EEG (electroencephalography) electrodes are flexible and cheap, but sensing quality of brain activity is lower than wet EEG electrodes having higher cost [40, 41].

- *Data fusion*: Multiple sensors involved in any acquisition process lead to a problem of data fusion, and to compatibility issues. The presence of heterogeneous sensors causes in fact incompatibility among data or extracted features, that need to be processed carefully [39]. This challenge needs to be explored in its various dimensions to avoid any integration and computational problems.
- *Data integrity*: The recorded data from biosensors, and any smart or wearable sensor, must ensure data integrity. But there is no standard procedure that has been followed to store data, and this has increased the complexity of analysis and decision making based on multiple data of a similar type [38].

This challenge, standardization of data storage, should be addressed to improve data integrity.

- *Communication, privacy and security issues*: Smart sensors allocated for agriculture and industrial applications, healthcare and monitoring, are connected via cognitive sensor networks or IoT based applications, which require communication in different stages of data transmission and analysis. The applicability and choice of the best suitable technology for specific applications need to be explored [38]. Yet, the acquired data from individuals, and from living or nonliving things, are subjected to privacy issues. Channel connections imply privacy losses in different situations, such as under hackers attack, to give an example, by causing serious security issues. To maintain the privacy and security, the designing of unassailable systems and accurate security measures need to be found.
- *Big data challenges*: Sensors are generating data with high velocity, volume and variety with the intention of detailed in-depth analysis and prediction. The diverse and large data need to be managed properly for analysis. The Big Data management faces challenges related to the expertise level reached in the specific area, and to the operation and research related issues, like those identified in the healthcare sector, to give an example [42]. The area of expertise and technological advances in one field must be explored and the experts from another area must have to learn and understand the new concepts. In addition, some considerations are needed as regarding the higher resources for big data sharing and management.
- *User friendliness*: The technology suffers the challenges of compatibility for different platforms with high technical expertise. The smart sensors are still not evolved to their peak, but nevertheless a common platform seems to be the solution where the data can be stored and made accessible by any layman for analysis. Therefore, the designed model needs to be user friendly for every perspective and to allow general purpose access [38].

### 9.3.2 Application of biosensors

Biosensors alone and with cognitive capability are applicable in a variety of areas. Details on different applications are presented in this section, and also shown in figure 9.5. Different applications along with protocols of transmission and processing methods, such as machine learning or any other similar techniques, have been discussed in the subsequent subsections.

The details on some major applications are as follows:

- *Healthcare and drug production*: Healthcare is a vast area and biosensors are used in multiple areas of healthcare. Some of the important area are discussed in the table 9.2. Biosensors are primarily used for disease and disorder detection although, now they are used in recent areas such as advertisement, waste management, Forensics, assisted living and biometrics, as well.

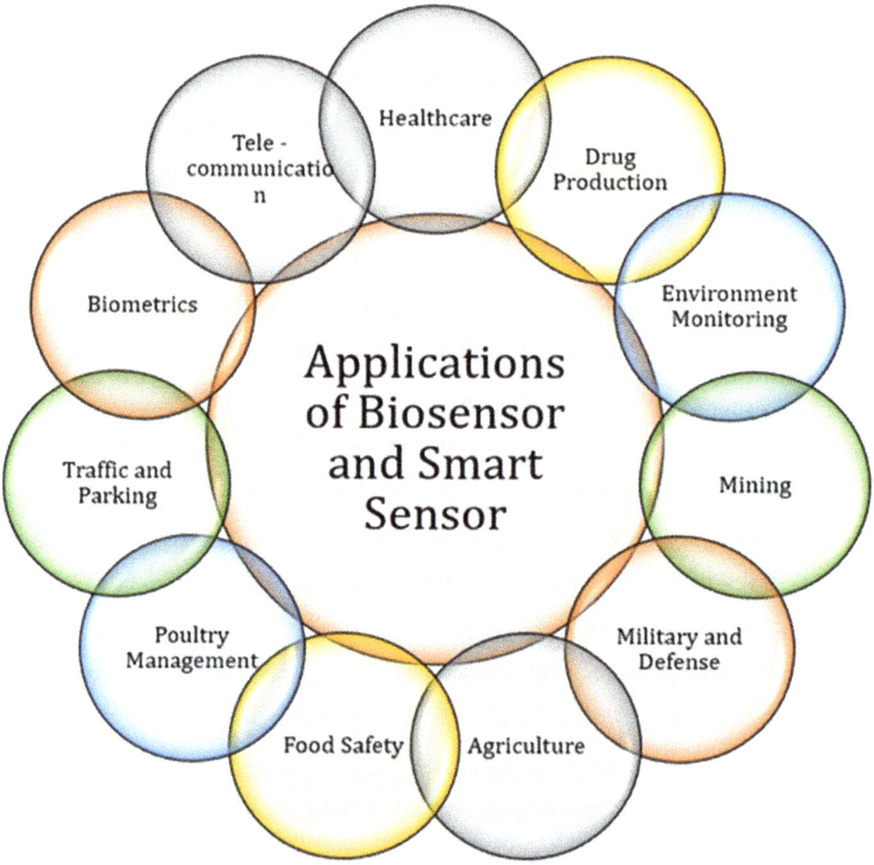


**Figure 9.5.** Applications of biosensors and smart sensors.

- *Environment, agriculture, food monitoring*: Other most popular areas where biosensors and WSN are used are environmental, agriculture and food monitoring. The sensors placed in this area must be robust and reliable for real-time applications. In this section, water quality monitoring, air quality monitoring, soil health monitoring, agriculture monitoring, fruit health monitoring, food quality and state monitoring, aquatic culture monitoring, poultry monitoring, using different technologies are presented as shown in table 9.3.

- *Military, defense, mining application*: Militant activity monitoring, enemy movement detection and identification are the key motivational requirements of research in this area. Security is a very critical concern nowadays along with the physical fitness of the military. Keeping all that in mind, biosensors are used in different military areas, such as warfare agent detection, location and tracking, psychological monitoring, military security with CRSNs. In this section, application of biosensors in the area of mining is also presented in table 9.4.

**Table 9.2.** Biosensors in healthcare and drug production.

| Application | Types of sensor | Positioning | Transmission network | Processing method (machine learning or any other) |
|---|---|---|---|---|
| Disease/disorder diagnosis and prediction [43–45] | Smart biosensors | Respective body part | Bluetooth, zigbee, wi-fi etc | Clinical, machine learning, IoT |
| Assisted living [53–55] | Cognitive or smart sensors | House or device/ wearable | WSN, CRSN/bluetooth, zigbee, wi-fi | Machine learning, IoT |
| Biometrics and security [56–58] | Wearable sensors/smart sensors | Device/house/monitoring field | CRSN/WSN | Machine learning, cloud, security protocols, blockchain |
| Advertisement [46] | Wearable + cognitive sensor | Body/monitoring screen | CRSN/bluetooth | Data science and artificial intelligence |
| Healthcare waste management [47–49] | Cognitive sensor/IoT/ wearable sensor | Building | CRSN/RFID | IoT, machine learning |
| Forensics and drug discovery [50–52] | Biosensors | Testing element | Bluetooth, zigbee, wi-fi | IoT |

**Table 9.3.** Biosensors in environment, agriculture, food monitoring.

| Application | Types of sensor | Positioning | Transmission network | Processing method |
|---|---|---|---|---|
| Water monitoring [59–61] | Biosensors and cognitive sensors | Water bodies | Acoustic waves, optical fibres, wi-fi, Zigbee | WSN, machine learning, cognitive ocean network |
| Air monitoring [62–65] | Environmental biosensors | Indoor/outdoor buildings | Http API, zigbee, optical fibre | WSN, machine learning, IoT |
| Soil monitoring [66–68] | Environmental biosensors/cognitive sensors | Field area | RFID, LoRa, wi-fi | WSN, machine learning, IoT |
| Agriculture monitoring [69–72] | Environmental biosensors/cognitive sensors | Field area | Wi-fi, zigbee, bluetooth etc | WSN, machine learning, IoT |
| Fruit monitoring [73–75] | Biosensors chip | Cold storage/field | Zigbee | WSN, machine learning, IoT |
| Food monitoring [76–78] | Heterogeneous sensors, electrochemical sensor | Food storage/crop field | Wi-fi, zigbee, bluetooth etc | WSN, satellite imaging, machine learning, IoT |
| Aquatic culture monitoring [79–81] | Biosensors chip | Aquatic crop area | Optical fiber, zigbee, HC-12 trans-receiver | WSN, biomass monitoring |
| Poultry monitoring [82–84] | Optical biosensors, environmental sensors, acoustic sensors | Poultry farm | RFID, wi-fi | API, machine learning, IoT, WSN, imaging |

**Table 9.4.** Biosensors in military, defense and mining application.

| Application | Types of sensor | Positioning | Transmission network | Processing method |
|---|---|---|---|---|
| Military biological warfare agent detection [85] | Biosensors and smart sensors | Handy devices | N/A | Manual or machine learning |
| Tracking and localization [86–88] | Biosensors and smart sensors, cognitive sensors | Body | CRSN | WSN |
| Psychological monitoring [89] | Biosensors, EEG, fMRI | Body | Bluetooth | Machine learning |
| Military security [90, 91] | Cognitive sensors | Field area | CRSN | Security protocols, AI |
| Mining [118–120] | Biosensors, smart sensors | Field area | Bluetooth, zigbee, wi-fi | WSN, machine learning |

**Table 9.5.** Biosensors in smart cities.

| Application | Types of sensor | Positioning | Transmission network | Processing method (machine learning or any other) |
|---|---|---|---|---|
| Parking [92–94] | Optical biosensors | Parking area and road traffic area | RFID, zigbee | IoT, WSN, machine learning |
| Drainage [95–97] | Environmental biosensors | Surface area | Wi-fi, zigbee | IoT, WSN, machine learning |
| Traffic management [92, 98, 99] | Optical biosensors, CMOS sensor | Road side | Wi-fi, zigbee | IoT, WSN, machine learning |
| Healthcare [100, 101] | Heterogeneous biosensors | GPS based or any other | Wi-fi, zigbee | Cloud computing, IoT, ICT |
| Environment monitoring [102–104] | Environmental biosensors | Different area | Wi-fi, RFID | Cloud computing, IoT, WSN, machine learning |

- *Smart cities*: Smart cities are equipped with smart sensors that have the capability of monitoring and providing schematic solutions. In smart cities, parking, drainage, traffic, healthcare and environment monitoring are some of the major concerns that must be taken into consideration for smart

**Table 9.6.** Biosensors in telecommunication and industry.

| Application | Types of sensor | Positioning | Transmission network | Processing method (machine learning or any other) |
|---|---|---|---|---|
| Telecommunication [105–107] | Biosensors and cognitive sensors | Antenna | Wi-Fi, ISM or band specific | WSN, CRSN, machine learning, optimization |
| Industry monitoring [108–110] | Bio sensors, smart sensors | Industrial area | LoRaWAN, wi-fi | IoT, WSN, cloud |

monitoring. In this section details of sensors in smart cities have been provided in table 9.5.

- *Telecommunication and industry*: Telecommunication is the less explored area for biosensor-based technology, and efficient energy management of the telecommunication networks has been explored by few researchers. IoT and machine learning based paradigms still need to be explored in the future for telecommunication and smart sensor-based applications. Details on sensor applications in this area are presented in table 9.6.

Focusing on the industrial area, biosensors and smart sensors are used in many plants and industries for monitoring and security purposes. Although their use is delicate and critical in some areas, they play an important role in monitoring.

## 9.4 Conclusion

This chapter has given an overview of cognitive sensors and biosensors, by highlighting how cognitive sensors have the capability of working with logic and understanding the situation better than other sensors usually employed in WSNs. Their property has been utilized in cognitive sensor networks for spectrum allocation in different paid spectrum bands, that can be used for optimizing the available resources. Different challenges and applications of cognitive sensors have also been discussed. In the second part of the chapter, an extensive overview of biosensors and smart sensor technology has been provided. It can be seen that biosensors are used in almost every area, and they are proving their capability in different challenging applications. However, it is still necessary for a deeper exploration of the right sensors for a high efficiency and low cost scenario. From this chapter, readers can search for related applications and start their own research in any specific direction.

## Acknowledgment

The authors would like to thank National Institute of Technology Raipur for providing infrastructure to conduct the research.

## References

[1] Addabbo P, Focareta M, Marcuccio S, Votto C and Ullo S L 2016 Contribution of Sentinel-2 data for applications in vegetation monitoring *Acta IMEKO* **5** 44–54

[2] Addabbo P, Focareta M, Marcuccio S, Votto C and Ullo S L 2016 Land cover classification and monitoring through multisensor image and data combination *2016 IEEE Int. Geoscience and Remote Sens. Symp. (IGARSS)* (Piscataway, NJ: IEEE) 902–5

[3] Addabbo P, Angrisano A, Bernardi M L, Gagliarde G, Mennella A, Nisi M and Ullo S 2017 A UAV infrared measurement approach for defect detection in photovoltaic plants *2017 IEEE Int. Metrol. AeroSpace (MetroAeroSpace)* (Piscataway, NJ: IEEE) 345–50

[4] Ullo S L, Addabbo P, Di Martire D, Sica S, Fiscante N, Cicala L and Angelino C V 2019 Application of dinsar technique to high coherence Sentinel-1 images for dam monitoring and result validation through *in situ* measurements *IEEE J. Select. Topic Appl. Earth Observ. Remote Sens.* **12,** 875–90

[5] Ali A, Ming Y, Chakraborty S and Iram S 2017 A comprehensive survey on real-time applications of WSN *Future Internet* **9** 77

[6] Kassim M R M and Harun A N 2016 Applications of WSN in agricultural environment monitoring systems *2016 Int. Conf. Info. Commun. Technol. Converg. (ICTC)* (Piscataway, NJ: IEEE) 344–9

[7] Murillo A F, Pe~na M and Martnez D 2012 Applications of WSN in health and agriculture *2012 IEEE Colombian Commun. Conf. (COLCOM)* (Piscataway, NJ: IEEE) 1–6

[8] Ahmad A, Ahmad S, Rehmani M H and Hassan N U 2015 A survey on radio resource allocation in cognitive radio sensor networks *IEEE Commun. Surv. Tutor.* **17,** 888–917

[9] Sawant S 2017 Development of biosensors from biopolymer composites *Biopolymer Composites in Electronics* (Amsterdam: Elsevier) pp 353–83

[10] Pejcic B, De Marco R and Parkinson G 2006 The role of biosensors in the detection of emerging infectious diseases *Analyst* **131** 1079–90

[11] (https://merriam-webster.com/dictionary/cognitive).

[12] Zhao Q and Sadler B M 2007 A survey of dynamic spectrum access *IEEE Sign. Proc. Mag.* **24** 79–89

[13] Haykin S 2005 Cognitive radio: brain-empowered wireless communications *IEEE J. Sel. Areas Commun.* **23** 201–20

[14] Gür G and Alagöz F 2011 Green wireless communications via cognitive dimension: an overview *IEEE Netw.* **25** 50–6

[15] Huang X, Han T and Ansari N 2015 On green-energy-powered cognitive radio networks *IEEE Commun. Surv. Tutor.* **17** 827–42

[16] Kakalou I and Psannis K E 2018 Sustainable and efficient data collection in cognitive radio sensor networks *IEEE Trans. Sustain. Comput.* **4** 29–38

[17] Najimi M, Ebrahimzadeh A, Andargoli S M H and Fallahi A 2014 Lifetime maximization in cognitive sensor networks based on the node selection *IEEE Sensor J.* **14** 2376–83

[18] Byun S-S, Balasingham I and Liang X 2008 Dynamic spectrum allocation in wireless cognitive sensor networks: improving fairness and energy efficiency *2008 IEEE 68th Vehi. Technol. Conf.* (Piscataway, NJ: IEEE) 1–5
[19] Bukhari S H R, Rehmani M H and Siraj S 2015 A survey of channel bonding for wireless networks and guidelines of channel bonding for futuristic cognitive radio sensor networks *IEEE Commun. Surv. Tutor.* **18** 924–48
[20] Sun W, Yuan X, Wang J, Han D and Zhang C 2010 Quality of service networking for Smart Grid distribution monitoring *2010 First IEEE Int. Conf. Smart Grid Commun.* (Piscataway, NJ: IEEE) 373–8
[21] Shah G A, Gungor V C and Akan O B 2013 A cross-layer qos-aware communication framework in cognitive radio sensor networks for Smart Grid applications *IEEE Trans. Ind. Inf.* **9** 1477–85
[22] Bkassiny M, Li Y and Jayaweera S K 2012 A survey on Machine Learning techniques in cognitive radios *IEEE Commun. Surv. Tutor.* **15** 1136–59
[23] Kumar K A, Avinash J and Poornima G 2018 Qos aware load balancing in cognitive wireless sensor networks using machine learning concepts *2018 3rd IEEE Int. Conf. Recent Trends Electron., Info. Commun. Technol. (RTEICT)* (Piscataway, NJ: IEEE) 2377–81
[24] Silva V F, Macedo D F and Leoni J L 2014 Spectrum decision in wireless sensor networks employing machine learning *2014 Brazil. Symp. Comput. Netw. Distr. Syst.* (Piscataway, NJ: IEEE) 386–93
[25] Politanskyi R and Klymash M 2019 Application of artificial intelligence in cognitive radio for planning distribution of frequency channels *2019 3rd Int. Conf. Adv. Info. Commun. Technol. (AICT)* (Piscataway, NJ: IEEE) 390–4
[26] Tanwongvarl C and Chantaraskul S 2015 Performance comparison of learning techniques for intelligent channel assignment in cognitive wireless sensor networks *2015 Seventh Int. Conf. Ubiquitous and Future Networks* (Piscataway, NJ: IEEE) 503–7
[27] Ma X, Ning S, Liu X, Kuang H and Hong Y 2018 Cooperative spectrum sensing using extreme learning machine for cognitive radio networks with multiple primary users *2018 IEEE 3rd Adv. Info. Technol., Electron. Auto. Contr. Conf. (IAEAC)* (Piscataway, NJ: IEEE) 536–40
[28] Zhou X, Sun M, Li G Y and Juang B-H F 2018 Intelligent wireless communications enabled by cognitive radio and machine learning *China Commun.* **15** 16–48
[29] Cui Y, Jun Jing X, Sun S, Wang X, Cheng D and Huang H 2015 Deep learning based primary user classification in cognitive radios *2015 15th Int. Symp. Commun. Info. Technol. (ISCIT)* (Piscataway, NJ: IEEE) 165–8
[30] Zheng S, Chen S, Qi P, Zhou H and Yang X 2020 Spectrum sensing based on deep learning classification for cognitive radios *China Commun.* **17** 138–48
[31] Smailagic A 2004 Unsupervised machine learning and cognitive systems in learning user state for contextaware computing *2004 Int. Conf. Mach. Learn. App. 2004. Proc.* (Piscataway, NJ: IEEE) 1–2
[32] Fragkiadakis A G, Tragos E Z and Askoxylakis I G 2012 A survey on security threats and detection techniques in cognitive radio networks *IEEE Commun. Surv. Tutor.* **15** 428–45
[33] Chen Y 2012 Collaborative spectrum sensing in the presence of secondary user interferences for lognormal shadowing *Wireless Commun. Mob. Comput.* **12** 463–72
[34] Li C, Raghunathan A and Jha N K 2009 An architecture for secure software defined radio *2009 Design, Auto. Test Europe Conf. Exhibition* (Piscataway, NJ: IEEE) 448–53

[35] Sharma R K and Rawat D B 2014 Advances on security threats and countermeasures for cognitive radio networks: a survey *IEEE Commun. Surv. Tutor.* **17** 1023–43
[36] Wang W, Sun Y, Li H and Han Z 2010 Cross-layer attack and defense in cognitive radio networks *2010 IEEE Global Telecommun. Conf. GLOBECOM 2010* (Piscataway, NJ: IEEE) 1–6
[37] Mehrotra P 2016 Biosensors and their applications-a review *J. Oral Biol. Craniofac. Res.* **6** 153–9
[38] Tian S, Yang W, Le Grange J M, Wang P, Huang W and Ye Z 2019 Smart healthcare: making medical care more intelligent *Global Health J.* **3** 62–5
[39] Wang Y, Cang S and Yu H 2019 A survey on wearable sensor modality centred human activity recognition in health care *Expert Syst. Appl.* **137** 167–90
[40] Gargiulo G, Calvo R A, Bifulco P, Cesarelli M, Jin C, Mohamed A and van Schaik A 2010 A new EEG recording system for passive dry electrodes *Clin. Neurophysiol.* **121** 686–93
[41] Mathewson K E, Harrison T J and Kizuk S A 2017 High and dry? Comparing active dry EEG electrodes to active and passive wet electrodes *Psychophysiology* **54** 74–82
[42] Chen P-T, Lin C-L and Wu W-N 2020 Big data management in healthcare: adoption challenges and implications *Int. J. Inf. Manage.* **53** 102078
[43] Tuli S, Basumatary N, Gill S S, Kahani M, Arya R C, Wander G S and Buyya R 2020 Healthfog: an ensemble deep learning based smart healthcare system for automatic diagnosis of heart diseases in integrated IoT and fog computing environments *Future Gener. Comput. Syst.* **104** 187–200
[44] Mani N, Singh A and Nimmagadda S L 2020 An IoT guided healthcare monitoring system for managing realtime notifications by fog computing services *Procedia Comput. Sci.* **167** 850–9
[45] Singh R P, Javaid M, Haleem A and Suman R 2020 Internet of things (IoT) applications to fight against covid-19 pandemic *Diab. Metab. Syndr. Clin. Res. Rev.* **14** 521–4
[46] Chylinski M, Heller J, Hilken T, Keeling D I, Mahr D and de Ruyter K 2020 Augmented reality marketing: a technology-enabled approach to situated customer experience *Austr. Market. J. (AMJ)* in press
[47] Dubey S, Singh P, Yadav P and Singh K K 2020 Household waste management system using IoT and machine learning *Procedia Comput. Sci.* **167** 1950–9
[48] Gaynor M and Waterman J 2016 Design framework for sensors and RFID tags with healthcare applications *Health Policy Technol.* **5** 357–69
[49] Zhanga L, Hua J, Guob C and Xuc H Dynamic power optimization for secondary wearable biosensors in e-healthcare leveraging cognitive WBSNs with imperfect spectrum sensing. *Future Gen. Comput. Syst.* **112** 67–92
[50] Sousa E H S, de Franca Lopes L G, Gonzalez G and Gilles-Gonzalez M-A 2017 Drug discovery targeting heme-based sensors and their coupled activities *J. Inorg. Biochem.* **167** 12–20
[51] Ferreira P C, Atade V N, Chagas C L S, Angnes L, Coltro W K T, Paix~ao T R L C and de Araujo W R 2019 Wearable electrochemical sensors for forensic and clinical applications *TrAC, Trends Anal. Chem.* **119** 115622
[52] Servida F and Casey E 2019 IoT Forensic challenges and opportunities for digital traces *Digit. Invest.* **28** S22–9
[53] Syed L, Jabeen S, Manimala S and Alsaeedi A 2019 Smart healthcare framework for ambient assisted living using IoMT and big data analytics techniques *Future Gener. Comput. Syst.* **101** 136–51

[54] Debauche O, Mahmoudi S, Manneback P and Assila A 2019 Fog IoT for health: a new architecture for patients and elderly monitoring *Procedia Comput. Sci.* **160** 289–97
[55] Yacchirema D, de Puga J S, Palau C and Esteve M 2018 Fall detection system for elderly people using IoT and big data *Procedia Comput. Sci.* **130** 603–10
[56] Tanwar S, Parekh K and Evans R 2020 Blockchain-based electronic healthcare record system for healthcare 4.0 applications *J. Info. Secur. App.* **50** 102–407
[57] Wang Z, Luo N and Zhou P 2020 Guardhealth: blockchain empowered secure data management and graph convolutional network enabled anomaly detection in smart healthcare *J. Parallel Distrib. Comput.* **142** 1–12
[58] Karakaya N, Alptekin G I and İncel Ö D 2019 Using behavioral biometric sensors of mobile phones for user authentication *Procedia Comput. Sci.* **159** 475–84
[59] Wang C Y *et al* 2012 Photon stimulated ozone sensor based on indium oxide nanoparticles II: ozone monitoring in humidity and water environments *Sens. Actuators* **B 164** 37–42
[60] Wang D, Liang P, Jiang Y, Liu P, Miao B, Hao W and Huang X 2018 Open external circuit for microbial fuel cell sensor to monitor the nitrate in aquatic environment *Biosens. Bioelectron.* **111** 97–101
[61] Lu H, Wang D, Li Y, Li J, Li X, Kim H, Serikawa S and Humar I 2019 Conet: a cognitive ocean network *IEEE Wirel. Commun.* **26** 90–6
[62] Loy-Benitez J, Heo S and Yoo C 2020 Soft sensor validation for monitoring and resilient control of sequential subway indoor air quality through memory-gated recurrent neural networks-based autoencoders *Control Eng. Pract.* **97** 104330
[63] Salman N, Kemp A H, Khan A and Noakes C 2019 Real time wireless sensor network (WSN) based indoor air quality monitoring system *IFAC-Papers OnLine* **52** 324–7
[64] Li J, Lu X and Wang W 2020 Leak monitoring and localization in Baghouse filtration system using a distributed optical fiber dynamic air pressure sensor *Opt. Fiber Technol.* **57** 102–218
[65] Kaivonen S and Ngai E C-H 2020 Real-time air pollution monitoring with sensors on city bus *Digit. Commun. Netw.* **6** 23–30
[66] Deng F, Zuo P, Wen K and Wu X 2020 Novel soil environment monitoring system based on RFID sensor and lora *Comput. Electron. Agric.* **169** 105169
[67] Chen C, Yan X, Ma Y, Yu S, Xu Q, Grantz D A, Lammers P S, Wang Z, Sun Y and Cheng Q 2019 Monitoring near-surface soil water content using an innovative perforated cylinder coaxial dielectric sensor *J. Hydrol.* **573** 746–54
[68] Majone B, Viani F, Filippi E, Bellin A, Massa A, Toller G, Robol F and Salucci M 2013 Wireless sensor network deployment for monitoring soil moisture dynamics at the field scale *Procedia Environ. Sci.* **19** 426–35
[69] Khattab A, Habib S E, Ismail H, Zayan S, Fahmy Y and Khairy M M 2019 An IoT-based cognitive monitoring system for early plant disease forecast *Comput. Electron. Agric.* **166** 105028
[70] Lavanya G, Rani C and Ganeshkumar P 2019 An automated low cost IoT based fertilizer intimation system for smart agriculture *Sustain. Comput. Info. Syst.* in press
[71] Dong X, Vuran M C and Irmak S 2013 Autonomous precision agriculture through integration of wireless underground sensor networks with center pivot irrigation systems *Ad Hoc Netw.* **11** 1975–87
[72] Kundu M, Krishnan P, Kotnala R and Sumana G 2019 Recent developments in biosensors to combat agricultural challenges and their future prospects *Trends Food Sci. Technol.* **88** 157–78

[73] Junxiang G and Jingtao X 2011 Fruit cold storage environment monitoring system based on wireless sensor network *Procedia Eng.* **15** 3466–70

[74] Kang H and Chen C 2020 Fast implementation of real-time fruit detection in apple orchards using deep learning *Comput. Electron. Agric.* **168** 105–8

[75] Nturambirwe J F I and Opara U L 2020 Machine learning applications to non-destructive defect detection in horticultural products *Biosyst. Eng.* **189** 60–83

[76] Kumaravel S, Balamurugan T, Jia S-H, Lin H-Y and Huang S-T 2020 Ratiometric electrochemical molecular switch for sensing hypochlorous acid: applicable in food analysis and real-time *in situ* monitoring *Anal. Chim. Acta.* **1106** 168–75

[77] Gao T, Tian Y, Zhu Z and Sun D-W 2020 Modelling, responses and applications of time-temperature indicators (ttis) in monitoring fresh food quality *Trends Food Sci. Technol.* **99** 311–22

[78] Vij A, Vijendra S, Jain A, Bajaj S, Bassi A and Sharma A 2020 IoT and machine learning approaches for automation of farm irrigation system *Procedia Comput. Sci.* **167** 1250–7

[79] Madsen J D 1993 Biomass techniques for monitoring and assessing control of aquatic vegetation *Lake Reserv. Manage.* **7** 141–54

[80] Vishwakarma V, Gurav A, Patel H and Sahasrabudhe S 2018 Acqua culture monitoring system *2018 Int. Conf. Smart City Emerg. Technol. (ICSCET)* (Piscataway, NJ: IEEE) 1–4

[81] Qi L, Han Y, Zhang X, Xing S and Fu Z *et al* 2012 Real time monitoring system for aquatic cold-chain logistics based on WSN *Chinese Soc. Agric. Machin.* **43** 134–40

[82] Ben Sassi N, Averos X and Estevez I 2016 Technology and poultry welfare *Animals* **6** 62

[83] Raj A A G and Jayanthi J G 2018 IoT-based real-time poultry monitoring and health status identification *2018 11th Int. Symp. Mechatron. App. (ISMA)* (Piscataway, NJ: IEEE) 1–7

[84] Ellis D I, Broadhurst D, Kell D B, Rowland J J and Goodacre R 2002 Rapid and quantitative detection of the microbial spoilage of meat by fourier transform infrared spectroscopy and machine learning *Appl. Environ. Microbiol.* **68** 2822–8

[85] Pohanka M, Skla P and Kroe M 2007 Biosensors for biological warfare agent detection *Def. Sci. J.* **57** 185

[86] Durisic M P, Tafa Z, Dimic G and Milutinovic V 2012 A survey of military applications of wireless sensor networks *2012 Mediterranean Conf. Embedded Comput. (MECO)* (Piscataway, NJ: IEEE) 196–9

[87] Ball S, Ferguso A and Rondeau T W 2005 Consumer applications of cognitive radio defined networks *First IEEE Int. Symp. New Front. Dynam. Spectr. Access Netw.* DySPAN 2005 (Piscataway, NJ: IEEE) 518–25

[88] Hinman R D 2006 Application of cognitive radio technology to legacy military waveforms in a jtrs (joint tactical radio system) radio *MILCOM 2006–2006 IEEE Military Commun. Conf.* (Piscataway, NJ: IEEE) 1–5

[89] Karstoft K-I, Statnikov A, Andersen S B, Madsen T and Galatzer-Levy I R 2015 Early identification of posttraumatic stress following military deployment: application of machine learning methods to a prospective study of Danish soldiers *J. Effect. Disorder* **184** 170–5

[90] Tyugu E 2011 Artificial intelligence in cyber defense *2011 3rd Int. Conf. Cyber Con-ict* (Piscataway, NJ: IEEE) 1–11

[91] Mody A N, Reddy R, Kiernan T and Brown T X 2009 Security in cognitive radio networks: an example using the commercial IEEE 802.22 standard *MILCOM 2009–2009 IEEE Military Commun. Conf.* (Piscataway, NJ: IEEE) 1–7

[92] Chinrungrueng J, Sunantachaikul U and Triamlumlerd S 2007 Smart parking: an application of optical wireless sensor network *2007 Int. Symp. App. Internet Workshops* (Piscataway, NJ: IEEE) 66

[93] Pala Z and Inanc N 2007 Smart parking applications using RFID technology *2007 1st Annu. RFID Eurasia* (IEEE) 1–3

[94] Wang H and He W 2011 A reservation-based smart parking system *2011 IEEE Conf. Comput. Commun. Workshops (INFOCOM WKSHPS)* (Piscataway, NJ: IEEE) 690–5

[95] Velladurai V, Saravanan M, Vigneshbabu R, Karthikeyan P and Dhlipkumar A 2017 Human safety system in drainage, unused well and garbage alerting system for smart city *2017 Int. Conf. on I-SMAC (IoT in Social, Mobile, Analytics and Cloud)(I-SMAC)* (Piscataway, NJ: IEEE) 6–9

[96] Malik H, Kandler N, Alam M M, Annus I, Le Moullec Y and Kuusik A 2018 Evaluation of low power wide area network technologies for smart urban drainage systems *2018 IEEE Int. Conf. Environ. Eng. (EE)* (Piscataway, NJ: IEEE) 1–5

[97] Chan K-H, Cheang C-S and Choi W-W 2014 Zigbee wireless sensor network for surface drainage monitoring and food prediction *2014 Int. Symp. Antennas Propagation Conf. Proc.* (Piscataway, NJ: IEEE) 391–2

[98] Balid W, Tash H and Refai H H 2016 Versatile real-time traffic monitoring system using wireless smart sensors networks *2016 IEEE Wireless Commun. Netw. Conf.* (Piscataway, NJ: IEEE) 1–6

[99] Ghazal B, ElKhatib K, Chahine K and Kherfan M 2016 Smart traffic light control system *2016 Third Int. Conf. Electric., Electron., Comput. Eng. App. (EECEA)* (Piscataway, NJ: IEEE) 140–5

[100] Solanas A *et al* 2014 Smart health: a context-aware health paradigm within smart cities *IEEE Commun. Mag.* **52** 74–81

[101] Pramanik M I, Lau R Y, Demirkan H and Azad M A K 2017 Smart health: big data enabled health paradigm within smart cities *Expert Syst. Appl.* **87** 370–83

[102] Siregar B, Nasution A B A and Fahmi F 2016 Integrated pollution monitoring system for smart city *2016 Int. Conf. ICT Smart Society (ICISS)* (Piscataway, NJ: IEEE) 49–52

[103] Shah J and Mishra B 2016 IoT enabled environmental monitoring system for smart cities *2016 Int. Conf. IoT App. (IOTA)* (Piscataway, NJ: IEEE) 383–8

[104] Jamil M S, Jamil M A, Mazhar A, Ikram A, Ahmed A and Munawar U 2015 Smart environment monitoring system by employing wireless sensor networks on vehicles for pollution free smart cities *Procedia Eng.* **107** 480–4

[105] Sorrentino M, Bruno M, Triro A and Rizzo G 2019 An innovative energy efficiency metric for data analytics and diagnostics in telecommunication applications *Appl. Energ.* **242** 1539–48

[106] Wakamatsu S, Watanabe S, Ishii T, Koyama M, Aizawa H and Kurosawa S 2007 Oscillation frequency stability of qcm-biosensor in liquid based on electronic circuit technology for telecommunication *IEEE Int. Freq. Contr. Symp. Joint 21st European Freq. Time Forum* (Piscataway, NJ: IEEE) 2007 16–9

[107] Ahmed A A G-E 2019 Benefits and challenges of internet of things for telecommunication networks *Telecommunication Network Trends Development* (Rijeka: IntechOpen)

[108] Arslanova F, Kamenev E and Odintsov E 2019 Integration of microplasma sensor ces into oil and gas industry and mining industry *Topical Issues of Rational Use of Natural Resources 2019, Volume 1: Proc. XV Int. Forum-Contest of Students and Young*

*Researchers under the Auspices of UNESCO (St. Petersburg Mining University, Russia, 13–17 May 2019)* (Boca Raton, FL: CRC Press) 386

[109] Ferreira P, Miranda R N, Cruz P M and Mendonca H S 2019 Multi-protocol lorawan/wifi sensor node performance assessment for industry 4.0 energy monitoring *2019 IEEE-APS Topical Conf. Antennas Propagation in Wireless Commun. (APWC)* (Piscataway, NJ: IEEE) 403–7

[110] Karthick S, Muralidharan S, Lee H-S, Kwon S-J and Saraswathy V 2019 Reliability and long-term evaluation of go-mno2 nano material as a newer corrosion monitoring sensor for reinforced concrete structures *Cem. Concr. Compos.* **100** 74–84

[111] Oyewobi S S and Hancke G P 2017 A survey of cognitive radio handoff schemes, challenges and issues for industrial wireless sensor networks (CR-IWSN) *J. Netw. Comput. Appl.* **97** 140–56

[112] Kumar M, Arun R, Vimala and Aravind Britto K R 2019 A cognitive technology based healthcare monitoring system and medical data transmission *Measurement* **146** 322–32

[113] Wan S, Gu Z and Ni Q 2020 Cognitive computing and wireless communications on the edge for healthcare service robots *Comput. Commun.* **149** 99–106

[114] Fadel E *et al* 2017 Spectrum-aware bio-inspired routing in cognitive radio sensor networks for smart grid applications *Comput. Commun.* **101** 106–20

[115] Hilal A R *et al* 2018 A distributed sensor management for large-scale IoT indoor acoustic surveillance *Future Gener. Comput. Syst.* **86** 1170–84

[116] Muhammad K, Ahmad J and Baik S W 2018 Early fire detection using convolutional neural networks during surveillance for effective disaster management *Neurocomputing* **288** 30–42

[117] Bayrakdar M E 2020 Enhancing sensor network sustainability with Fuzzy logic based node placement approach for agricultural monitoring *Comput. Electron. Agric.* **174** 105461

[118] Schnitzler N, Ross P-S and Gloaguen E 2019 Using machine learning to estimate a key missing geochemical variable in mining exploration: application of the random forest algorithm to multi-sensor core logging data *J. Geochem. Explor.* **205** 106344

[119] Zrelli A and Ezzedine T 2018 Design of optical and wireless sensors for underground mining monitoring system *Optik* **170** 376–83

[120] He H *et al* 2018 Self-powered smelling electronic-skin based on the piezo-gas-sensor matrix for real-time monitoring the mining environment *Sens. Actuators* B **267** 392–402

[121] Ullo S *et al* 2018 Application of wireless sensor networks to environmental monitoring for sustainable mobility *IEEE Int. Conf. Environ. Eng. (EE)* (Piscataway, NJ: IEEE)

[122] Gallo M *et al* 2018 A flexible mobility system based on CHIP architectures: the NETCHIP research project *2018 IEEE Int. Conf. Environ. Electric. Eng. 2018 IEEE Industrial and Commercial Power Syst. Europe (EEEIC/I&CPS Europe)* (Piscataway, NJ: IEEE)

[123] Carminati M *et al* 2019 Prospects of distributed wireless sensor networks for urban environmental monitoring *IEEE Aerosp. Electron. Syst. Mag.* **34** 44–52

[124] Di Bisceglie M, Ullo S L and Vaccaro A 2012 The role of cooperative information spreading paradigms for smart grid monitoring *2012 16th IEEE Mediterranean Electrotechn. Conf.* (Piscataway, NJ: IEEE)

[125] Ullo S L and Vaccaro A 2012 Gossip sensor networks for power quality monitoring in smart grids *Int. Conf. Renew. Energ. Power Qual. ICREPQ*

[126] Ullo S, Vaccaro A and Velotto G 2010 The role of pervasive and cooperative sensor networks in smart grids communication *MELECON 2010–2010 15th IEEE Mediterranean Electrotechn. Conf.* (Piscataway, NJ: IEEE)

[127] Ullo S L and Vaccaro A 2017 Decentralized protocols for power quality monitoring in pervasive networks-case study: the smart grids *Power Quality* (Cambridge: Cambridge Scholars Publishing) ch 8

[128] Ullo S L 2018 Evolution of cognitive radars toward intelligent systems architectures *2018 5th IEEE Int. Workshop Metrol. AeroSpace (MetroAeroSpace)* (Piscataway, NJ: IEEE)

[129] Ullo S L *et al* 2019 Landslide geohazard assessment with convolutional neural networks using sentinel-2 imagery data *IGARSS 2019–2019 IEEE Int. Geosci. Remote Sens. Symp.* (Piscataway, NJ: IEEE)

[130] Frattolillo F, Landolfi F and Ullo S L 2008 Energy consumption estimation in hybrid sensor networks running assisted navigation algorithms *IGARSS 2008–2008 IEEE Int. Geosci. Remote Sens. Symp.* Vol. 3 (Piscataway, NJ: IEEE)

[131] Frattolillo F, Quarantiello N and Ullo S L 2007 Implementing assisted navigation in hybrid sensor networks *2007 IEEE Int. Geosci. Remote Sens. Symp.* (Piscataway, NJ: IEEE)

# Chapter 10

## Sensor technologies combined with AI helping in smart transport systems as driverless cars

**Chan Myae Kyaw Win, Myat Thu Kha, May Thin Khaing, Chan Myae Khin, G R Sinha, Phyu Phyu Khaing and Divya Prakash Shrivastava**

This chapter presents a study of the role of sensors and sensor technologies in a smart transport system as a revolution combining artificial intelligence (AI) and smart sensors, especially in driverless cars. We discuss an interesting evolution of driverless cars that originated at the beginning of the twentieth century and the modern trends of self-driving cars which are ongoing in the twenty-first century. The fundamental concepts of advanced sensors and developments of driverless car technologies are also studied. The emphasis is also on studying industries and companies which are working on driverless cars and their technologies with some impact analysis. We have investigated how these technologies helped in creating a revolution in modern transportation systems in our daily lives. Nevertheless, there are some challenges for manufacturers and users, which are reported in the chapter. Finally, the future developments of autonomous cars and their impact on society and apprehensions are also discussed.

### 10.1 History of driverless cars using smart sensors

Self-driving cars, also called as driverless cars, are a brilliant technological outcome of sensor technology [1–3] and AI [4]. In this section, we discuss the history of driverless cars; their theories are divided into two main categories, namely automated and autonomous cars. The sensors used in such cars are not ordinary sensors but smart sensors that can operate with the help of suitable machine learning methods and AI techniques. The technologies are replacing the human and the rapid changes in the technologies are helping to have self-controlled cars or driverless cars. The driverless option in self-driving cars is facilitated by machine learning [5] and modern sensors. The sensors used in the cars are smart sensors equipped with AI and other similar soft computing tools. While there has been much progress in the area of

doi:10.1088/978-0-7503-2707-7ch10  

automated vehicles, their operation still has challenges to face and therefore the cars are not running on roads as alternative public transport.

In earlier times, probably the cars were not conceptualized because AI and the modern sensor technologies were not there. The driverless car initially started as a radio-controlled car [6], which was also called American Wonder [6] in 1925. It was equipped with radio control Co. [6] as radio equipment, and was developed by a US Army electrical engineer. In August 1925, American Wonder drove in New York streets, passing through the heavy traffic, and the car had a transmitting antenna which was operated by a second car behind by sending out radio impulse signals. The antenna in the front car was used to receive the signals to operate the inner electrical motors for the movements. Unfortunately, the demonstration was not very successful and the car crashed into the other vehicle carrying the photographers. In 1930, the World's Fair showcased radio-controlled electric cars that could move with electromagnetic fields supplied by circuits set in highway roads. In 1950, General Motors demonstrated its Firebird that had an electronic guiding system.

In 1960 and the decade after, tests were held for driverless cars working with cables. A driverless car interacting with magnetic cables which were embedded in the road was tested by the United Kingdom's transport and road research laboratory. Using cables, the cars operated successfully without any problems related to speed, weather, atmospheric conditions, and they worked well. The next initiative was a driverless car powered and controlled by buried cables and the relay of computer messages was tested, but due to cabling issues the cars were not a good choice.

In figure 10.1, the history of cars [7] with technological advancements can be seen. In 1925, it started with the radio-controlled option and now the cars are operating as taxis in a few developed nations. The developments in the figure are self-explanatory and the operation of cars in completely driverless mode became possible just because of the sensors and AI.

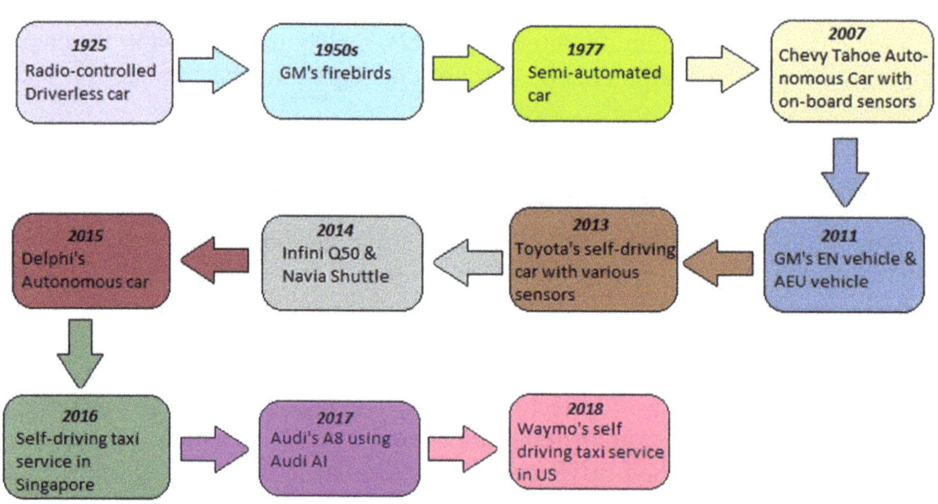


**Figure 10.1.** History of technological advances and self-driving cars.

Smart sensors [8], wireless sensors [9] and LIDARs [2, 10–12] have become the most important components in the self-driving car system due to the fact that these sensors provide real time and accurate data which is interpreted by suitable machine learning methods [5, 13–15] and AI [4, 16, 17].

## 10.2 Automation levels

As the progress of self-driving cars continued to move ahead, the level of automation [1, 5, 18] kept on enhancing. This is possibly due to fusion of sensor technology with modern machine learning tools, which makes the operation of cars powerful and efficient. A typical flow of automation levels has been depicted in figure 10.2 with features and controls available in all levels of automation. We can see that the very first level, Level 0 [19] has fully manual control; and Level 1 [20] has improvement due to an automated braking system. Acceleration and steering controls could be facilitated in Level 2 [20] but this level of automation still needs human intervention. Safety related functions were managed by cars using some automation and still requiring manual operation in Level 3 [21]. Level 4 [22, 23] becomes empowered with almost full automation capabilities but human intervention in adaptive operation of various controls is still there. Finally, Level 5 [22] becomes a fully automated version that can operate in all possible conditions.

Advances in automation levels have progressively enabled the capabilities and flexibilities of driverless cars, and the role of sensors plays the most significant part here in this achievement. The sensors included basic types of sensors in the beginning that were of bulky type mostly operated in mechanical or electro-mechanical ways; then optical sensors and acoustic sensors were used; and finally cognitive sensors [20], smart sensors [24], and wireless sensors [9] are being used in all levels of full automation today. Safety, accident-free drives and seamless driving are realized just because of smart sensors which have tremendous vision abilities that can guide the cars to avoid any type of accident and crash. The smart sensors and internet of things (IoT) [25] have made the driving seamless and accident-free in addition to providing security, which is related to the network and the intruder who can penetrate into the security and hack the operation of cars.

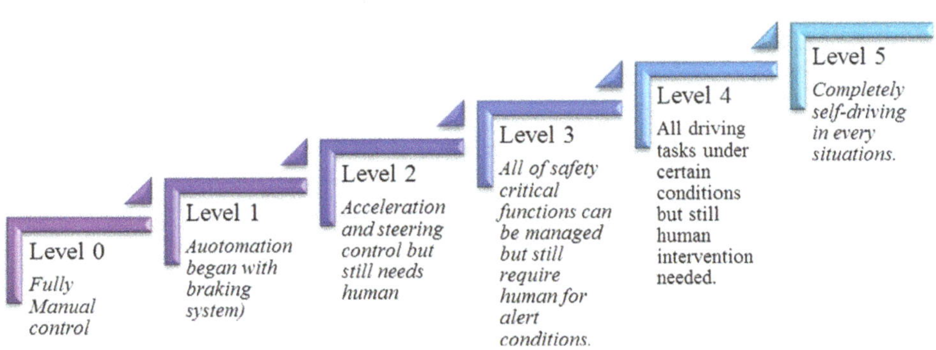


**Figure 10.2.** Levels of automation.

Nowadays, fully autonomous cars are being developed and the attempt is being made to make them popular for public transport as soon as possible across the world. The engineers, software professionals and researchers are trying to get the absolute trust in the invention of driverless car by wise design of the requirements of autonomous cars. The various versions of driverless car have been upgraded using various sensors like global positioning system (GPS) [26], radar, ultrasonic transducer, light detection and ranging (LiDAR) [2, 10, 11, 27] etc. In implementation of the self-driving car, the automation levels of driving were actually defined by the Society of Automotive Engineers (SAE) [22, 23, 28].

## 10.3 Sensors and other technologies used by manufacturing companies

Depending on different levels of automation, car manufacturers design the various driverless cars with several functionalities and capabilities. The literature suggests numerous companies which are working on the invention and manufacturing of driverless cars in different ways. The list of companies includes Apple, Huawei, Toyota, GM, Ford [19, 22, 29] etc, and major companies can be seen in figure 10.3.

With the development of new technologies, smart invention of driverless cars has been growing. The important facts derived from the expectation of people are convenient transport, reduction of traffic jams and safety matters, despite the well-known fact that the technologies are not fully developed enough to guarantee one hundred percent safety. But the possibility is enhanced with the help of smart and cognitive sensors, wireless technology, prediction techniques, classification algorithms and AI. Technologies required in designing and building autonomous vehicles include AI, software and various types of modern sensors that can help seamless driving with no crashes. Among the technologies, AI plays the most important role in implementation of cars, since AI helps better navigation and guiding of self-driving cars. However, testing driverless cars on the public roads is still a challenging task and these challenges are overcome by the companies working

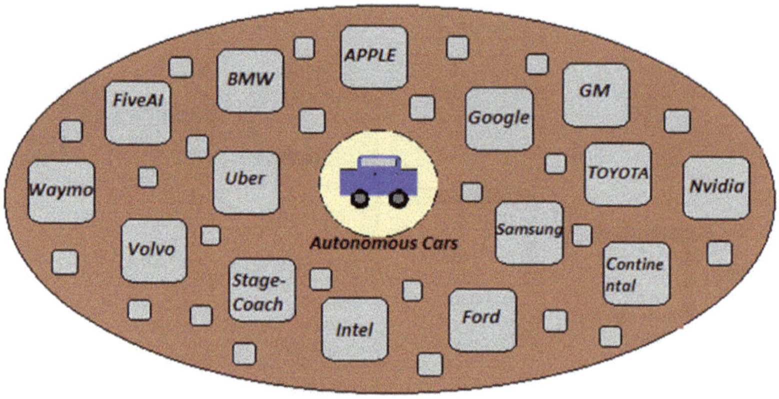


**Figure 10.3.** Manufacturing companies.

on driverless cars such as Google, Uber, Tesla [22, 30–34] and so on. An ordinary car works with a number of functions like a display system, auto-mirror system, braking system, fuel system, gear system, cooling engine system and other electrical systems. Driverless cars must have an increased number of functionalities for the purpose of smart sensing, data capture, data interpretation, guiding, mapping and controlling various operations of the cars. This necessitates new and modern technologies such as AI, smart sensors, image processing techniques, LIDAR and machine learning algorithms.

Nowadays, many car manufacturers are developing and testing driverless cars, and few of them are Google, Waymo, Ford, Tesla, Audi, BMW, GM, Apple, Volkswagen and Volvo [22, 30, 32, 34, 35]. Waymo is a self-driving car of Alphabet developed in collaboration with Intel [36] to control the processing power in the self-driving technology. Autonomous cars utilize a sequence of tasks, that include data from sensors, image processing, machine learning and neural networks to get the autonomous system. The neural network recognizes the data and sends to machine learning algorithms. The data are from cameras and other sensors in detecting traffic, light, trees and street signs.

For instance, Google's Waymo project is a nearly fully autonomous car with some possibility of human intervention to override the car if necessary. This is not self-driving in the fullest sense but it can control itself and operate in all conditions with little human intervention. Major features of the cars by Google are free steering, adaptive cruise control and lane-centering steering. In all these salient features, cars become helpless without using smart and intelligent sensors. In fact, Google was the first company to design and build a car with no steering wheel, accelerator or brake pedal. In 2014, Google unveiled the prototype of the driverless car having two seats for passengers and a monitor for displaying weather information, the current speed and a small countdown animation to launch. This can be requested for pick up at the user location with the destination set just by using a smartphone and any suitable mobile App. Once the journey is finished, the screen displays a message to remind the passengers of their personal belongings. Google did not stop there, but continued its efforts to create a new and upgraded prototype of the self-driving car with more automation controls, functions, and of course more intelligent sensors [15].

Self-driving cars are still in testing modes in most of the countries where the cars are being tested for public transport purposes. Google's driverless car Waymo uses a range of sensors and other technologies for understanding the environment and directing to whatever destination the rider wishes to reach. LiDAR provides a remote sensing method to observe the environment. Figure 10.4 shows all technologies used in the car and we can see that LiDAR and sensors are major components among the technologies and requirements in the design process. A typical diagram of a model of driverless car is shown in figure 10.5, which also highlights cameras, sensors, GPS sensors, LiDAR sensors, RADAR sensors and orientation sensors [26]. Figure 10.6 depicts the functionalities of RADAR and LiDAR sensors.

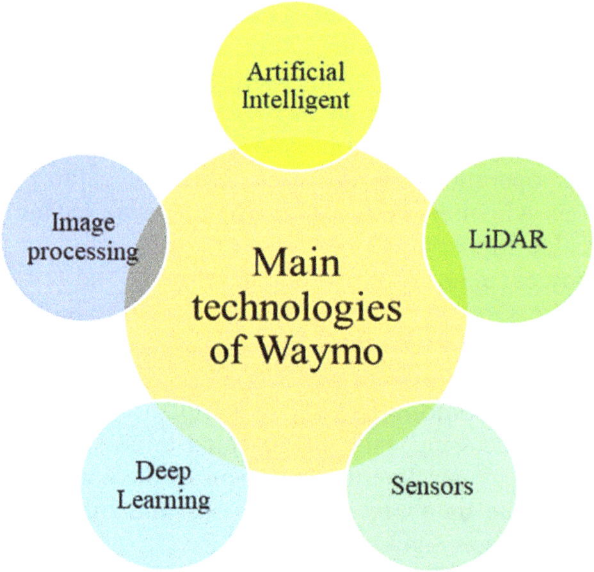


**Figure 10.4.** Technologies used in Waymo.

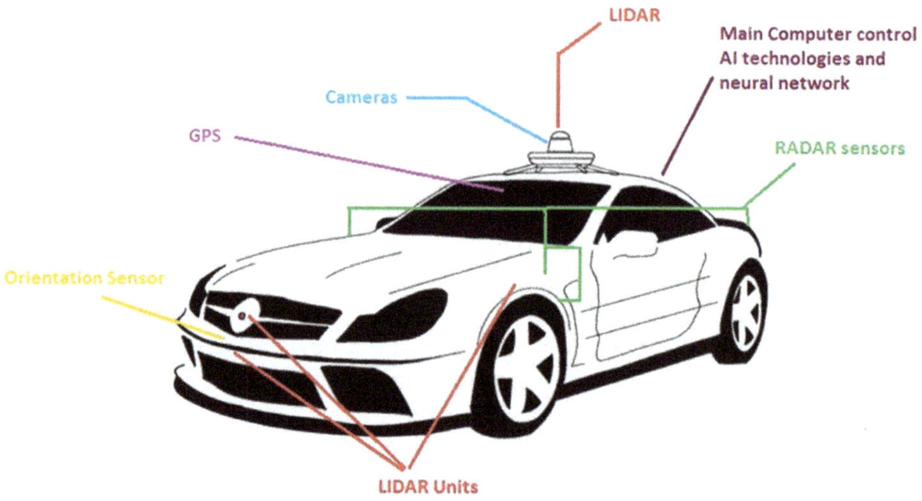


**Figure 10.5.** A typical diagram of a driverless car.

Besides Google, there are other companies also involved in putting their efforts into developing driverless technologies. Tesla Motors has also presented its autonomous car whose prototype is a little similar to that of Google, and it has an 'autopilot' function which requires a driver to perform the whole process and the driver is responsible for the process. Ford [22] is involved in upgrading the technologies for autonomous cars and working to create autonomous cars which are able to scan the road signs to get information about a permitted speed limit. If

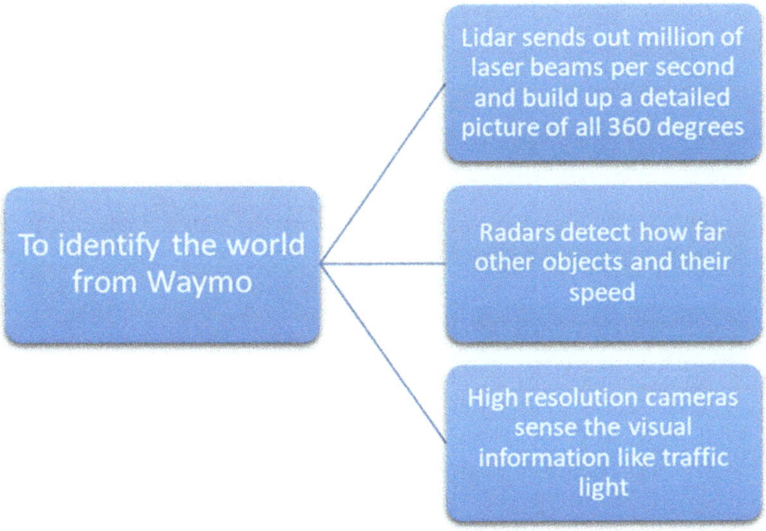


**Figure 10.6.** RADAR and LiDAR functions.

the speed becomes high, then this affects the braking system and limits the amount of fuel. Nissan, Audi, Bosch [30, 32, 33] and other car manufactures are in the list for testing and developing autonomous cars.

## 10.4 Design components

This section discusses the components and tools used in the design and development of self-driving cars. Actually, the brainchild of the autonomous concept began with an autonomous urban challenge held in 2003, and now the driverless car system plays an important role in the driverless assistance system. The driverless system includes:

- global path planning;
- local path planning.

Global path planning[12, 37] provides a valid path which contains driving orientation and destinations, whereas local path planning [38] is used to avoid static or dynamic obstacles by ensuring the car is driven in the center of the lane. Path planning is a critical problem for autonomous cars and determines the success of the operation of cars. The path planning techniques are based on a hierarchical hybrid algorithm which helps autonomous driving in a complex urban environment. Local path planning generates a path that can be followed by the car practically from a start point to a short-term destination and avoids obstacles on the road. One such planning mechanism is shown in figure 10.7. We can see in the figure that both the planning methods are connected in a certain manner so that the car is driven to the final destination without any obstacles. The GPS device involves GPS sensors and a GIS (global information system) that contains all the information about goals which, with the help of a suitable algorithm (Dijkstra's algorithm as an example),

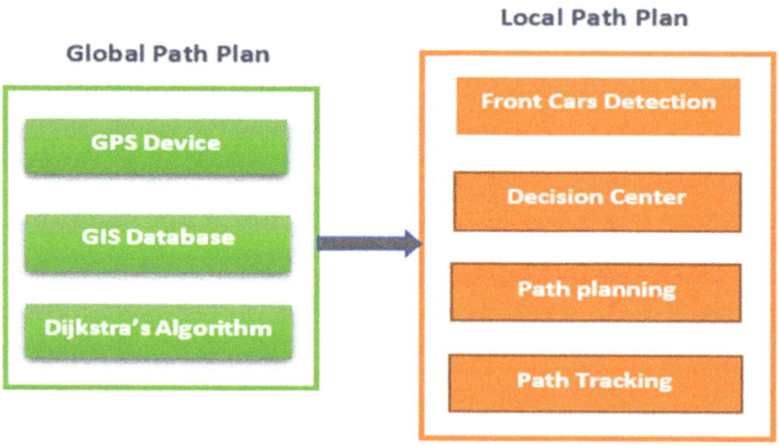


**Figure 10.7.** Path planning system used in driverless cars.

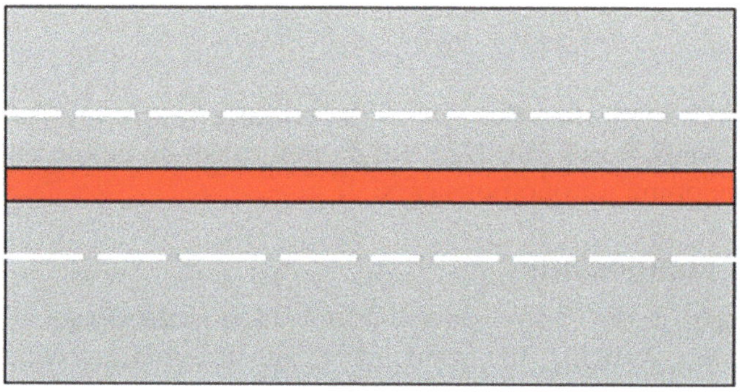

**Figure 10.8.** GPS data lane (GPS lane is in the center, dark).

takes the car to the local path planning system. The local path planning system includes a front car detection system which further involves a number of sensors, like vision based sensors and cameras. Path planning, tracking or navigation is made with the help of a decision center associated with the local path planning system.

When a car is running on the road, the planning system assists in navigation with the help of a controller who can navigate to the destination. This requires some accurate GPS data using a data line (figure 10.8) to the route, and hence a high accuracy GIS map is required, as shown in figure 10.9. The path navigation becomes more accurate when data captured and analyzed is intelligent with the help of smart GPS sensors and machine learning algorithms.

In implementation of the Chevy Tahoe Autonomous Car [7] that was chosen in the 2007 DARPA challenge in the urban area, a number of sensors were used on-board. The steering, throttle and brake management are controlled with a remote system and the sensors. The major types of sensors used in this project are:

- Velodyne HDL-64 lIDAR (HDL)x1;
- Applanix POS-LV 220/420 GPS/IMU (APLX) x1;

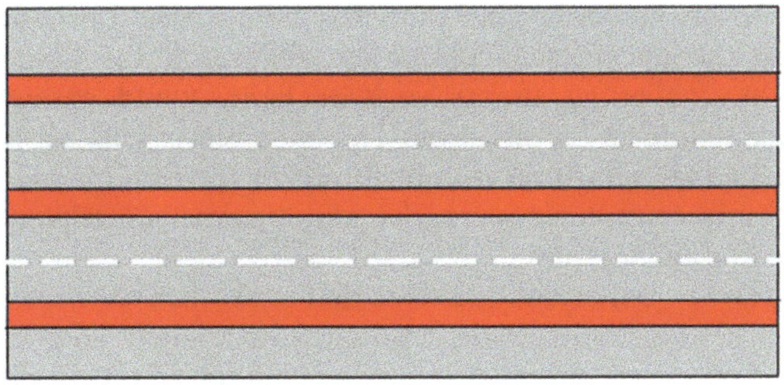

**Figure 10.9.** Highly accurate GPS path.

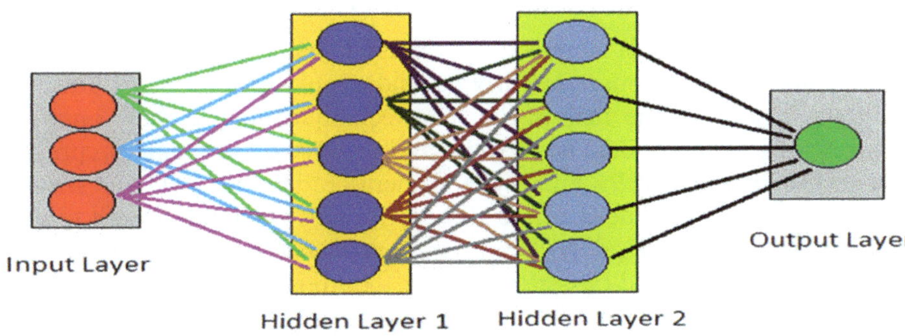


**Figure 10.10.** A deep neural network.

- SICK LMS 291-S05/S14 lIDAR (LMS) x6;
- Point grey firefly (PGF) x2;
- IBEO Alasca XT LIDAR (XT) x2;
- Continental ARS 300 Radar (ARS) x5;
- Continental ISF 172 lIDAR (ISF) x2.

Machine learning is the most important tool or method for navigation and path planning. There are numerous methods for it reported in the literature. In driverless cars, deep learning [13, 39–43] is employed which overcomes all drawbacks of ordinary neural network based methods and gives extremely high computational capability and throughput. The deep neural network can handle a large number of hidden layers due to pooling and convolution properties. Figure 10.10 shows a typical architecture of a deep neural network, having only two hidden layers, but the number of hidden layers may be hundreds also.

## 10.5 Sensor technology

As discussed in the previous section, sensor technologies are the most important technologies that empower the concept and implementation of driverless cars. The sensor technology includes deployment of sensors and their connection; a few important sensor categories include camera, radar, LiDAR, vision sensor, optical sensor etc. Sensors are used to: realize the surroundings; detect pedestrians and other vehicles; handle the unexpected conditions; and reach the desired destination. Sensors work by converting stimuli such as light, sound, movement etc, into electronic signals to inform the situation. The more sensors the car has, the smarter it is, which means that smartness is contributed to the cars mainly by sensors. Of course, the sensors have their strengths and weaknesses related to reliability, detection range and other issues.

Now there are a few important questions that arise when we discuss the cars, such as:

- what kind of sensors are used in implementation of a driverless car?
- why is a specific type of sensor for an application to be used?
- which sensor technologies are used by the manufacturers?

Without any doubt, it is simple to understand that complex technologies are combined with AP in autonomous driving and a number of sensors are used to interpret the objects of the real world. The techniques may include appropriate signal processing methods, data acquisition and analysis system, data preprocessing etc. Driverless cars generally employ a camera, radar and LiDAR sensors as primary sensors. The cars need to see the environment to know where to go and where not to go for driving safely. Driverless cars also need to detect other vehicles and people on the road; read the street signs on the street and handle the unexpected circumstances that may occur. Therefore, many sensors are needed for various purposes so that driverless cars can run on the road without any difficulties and crashes. The sensors used in driverless cars are mainly categorized as:

- Active sensors which emit an energy wave to look for objects based on the information received back. For instance, the radar sensor which is used to find the range of an object.
- Passive sensors receive the information from the environment and the information is processed by signal processing electronic circuits, such as a camera.

Now, we will discuss common sensors which are used in driverless cars. A few important sensors are:

- GPS sensors;
- LiDAR sensors;
- video cameras;
- radar sensors;
- ultrasonic sensors.

### 10.5.1 GPS

GPS [26] is an important system employing a number of sensors, and it works as a space-based satellite navigation system. This system provides position and time data in all climate conditions. The GPS system of sensing and navigation was initiated by the US government for establishing their military navigation. Nowadays, GPS has made everyone's life easier since anyone can receive the radio signals with the help of satellite broadcasts and GPS in a mobile device, and information about location and navigation [2, 12, 21, 44, 45] can easily be obtained. We can see in figure 10.11 that satellites [2, 12] are used to capture the data and GPS sensors are made available in all mobile and other devices for capturing the data. Various types of satellites are launched in certain orbits for various purposes, one such purpose is GPS which captures the data for a wide range of applications such as navigation of a device, robot, communication and weather forecasting etc.

### 10.5.2 LiDAR

Modern digital cameras and ordinary sensors fail to sense all real-time data in all circumstances and environment conditions. LiDAR (light detection and ranging) helps driverless cars by providing location assistance with the help of light signal received from the object. The objects may be animals, human beings, plants or any living and non-living objects, which can be easily detected by LiDAR [2, 3, 8, 10, 12] since the light properties are different in all the varying types of objects being detected. LiDAR technology is developed in such a manner than it can accept objects from all directions covering all 360 degrees [11] as a rotating LiDAR on the roof of the car and for detecting all surrounding objects. LiDAR is a sensor which locates the objects by emitting a light signal and calculating how long the light takes. The light goes to the object and comes back and the time is calculated. Based on velocity of light, we can easily calculate, the distance ($d$) as:

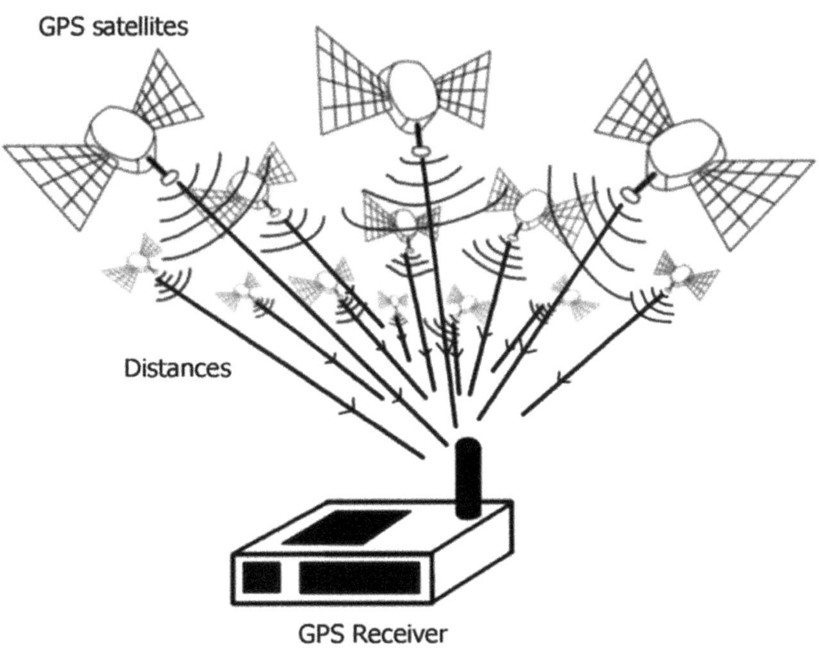


**Figure 10.11.** GPS sensor.

$$d = \text{velocity of light} \times \text{time}$$

This process is actually called time-of-flight in LiDAR, which is computed as the time for travelling to the object and coming back after getting reflected from the object. The view that can be covered by LiDAR is obviously 3D and provides shape as well as depth of the object because light signals can target at various points of the object and therefore it becomes easier to also know about the shape of the object. LiDAR sensor technology is considered as a very powerful and expensive sensor technology. It costs much to buy even a single piece of LiDAR sensor and therefore research always continues by the researchers and scientists to get upgraded and cheaper versions. The working principle of LiDAR is similar to RADAR [2, 8, 9, 11, 12, 45] except for the signal which is used for detection. The signal used in RADAR, is radio frequency, which travels and comes back and helps to locate the target or objects. LiDAR is becoming very popular in all autonomous systems these days, and uses commonly used photodetectors for ordinary applications, whereas MEMS (micro electromechanical system) types of sensors in self-driving car usage. The photodetector is used to scan and detect the pixel [27, 33, 46] of an image or an object where the pixel is actually a numerical value representing Gary scale value of the points in an object. The raw data captured using LiDAR goes to various subsequent modules for further analysis and understanding of the signal, such as preprocessing, machine learning and then post-processing. The raw LiDAR data, also called 3D point cloud, is given as input to the next module which perceives the data, with the help of pre-processing, machine learning and post-processing. The pre-processing module generally is used for:

- Transformation [47]: the raw data is converted into an absolute coordinate system since the pixels which are detected are stored in digital image format inside a computer system or similar signal processing tool.
- ROI (region of interest) selection: this is used to filter unnecessary portions of 3D point cloud raw data.
- Feature generation: finally, features are generated or extracted as a feature matrix which further goes to the next stage of machine learning.

As discussed earlier, the next task is also very important, known as machine learning model that can employ any models of machine learning for interpretation of features which are received from the previous stage. Deep learning [39, 43, 48, 49] is an obvious choice for reasons already discussed in earlier sections, in driverless cars, and the deep learning based machine learning then takes the feature matrix and produces a set of metrics for each cell of the object. Finally, a post-processing module is used which includes:

- Clustering: this produces small clusters of the objects and output metrics; and also filters unimportant clusters.
- Box building: a bounding box is generated that includes information related to height, width and length of an object.
- Tracking: the sequential frames are used to generate an object with additional information used for having perception about the object being detected.

The perception output is processed for prediction of routes and path planning [12, 38, 40, 48] in driverless car application.

Figure 10.12 shows a typical diagram of a LiDAR sensor, where the components can be seen as self-explanatory elements. Rotating mirror, laser [44] emitters, laser receivers, motor etc, are used in the sensor for various operations.

### 10.5.3 Cameras

In addition to LiDAR and RADAR sensors, a number of cameras are used in driverless cars, having good vision [21, 39, 40, 47, 50] capability to capture real-time data while the car is moving. Based on the data received through these cameras, the next stage of the driverless car pre-processes it, analyzes and then performs suitable tasks, and takes decisions. Vision-based and videos cameras help the cars to be in the right track and maintain distance from adjacent and side vehicles or pedestrians etc. The cameras are fitted in all directions of cars and also have the ability to capture data from 360 degrees by rotating and capturing facilities. The data is analyzed and interpreted using suitable machine learning and AI, and then the car takes the decision and keeps moving ahead towards the set destination. Sometimes the vision cameras face difficulties while coming across poor light illumination, bad weather conditions, heavy rain, snowfall etc. Nowadays, in ordinary cars also, vehicles use a lane departure warning system with the help of a front camera by detecting the traffic signs and painted markings on the road.

Poor illumination and bad weather conditions subject cars to constraints because the conditions might cause them not to be able to identify obstacles and nearby vehicles; although modern cameras are very powerful and versatile and can work in all possible adverse conditions, extremely bad weather conditions remain a challenging issue for camera vision sensing and capturing real-time data.

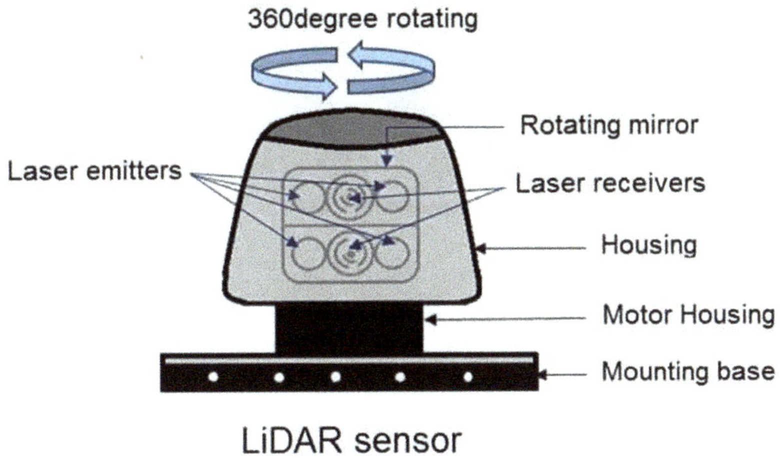


**Figure 10.12.** LiDAR sensor.

### 10.5.4 Radar sensors

Radar [2, 8, 9, 11, 12, 45] stands for radio detection and ranging and uses the principle of detection of radio waves [29] and assists cars in ranging and locating objects. The radar sensors contribute in driverless cars in significant ways since the automation is attributed by the radar controlled functions. The distance and speed of the object being tracked, can be measured using the radar sensors, which further assists the cars to maintain distance and required speed to avoid collision with other cars in the same track. The real-time data is detected and objects are located using two different radio signal pulses. The radio signals or waves travel to the object and after hitting the target object, they return and the time taken in the journey is recorded. With the help of radio frequency and time taken, the distance of the object can easily be calculated. Generally, two types of radars are used having different functions applicable in driverless cars, namely short-range radar (24 GHz) and long-range radar (77 GHz). Short-range radar assists in monitoring blind spots and also aids parking, whereas long-range radar is used in controlling automatic distance and aids braking. Unlike LiDAR, radar has no constraint related to rain, or similar adverse weather conditions. Radar operates in 2D as well as 3D; wherever 2D does not capture the range, 3D is involved so that horizontal and vertical projections can be captured. Figure 10.13 shows an example of radar being used in a driverless car.

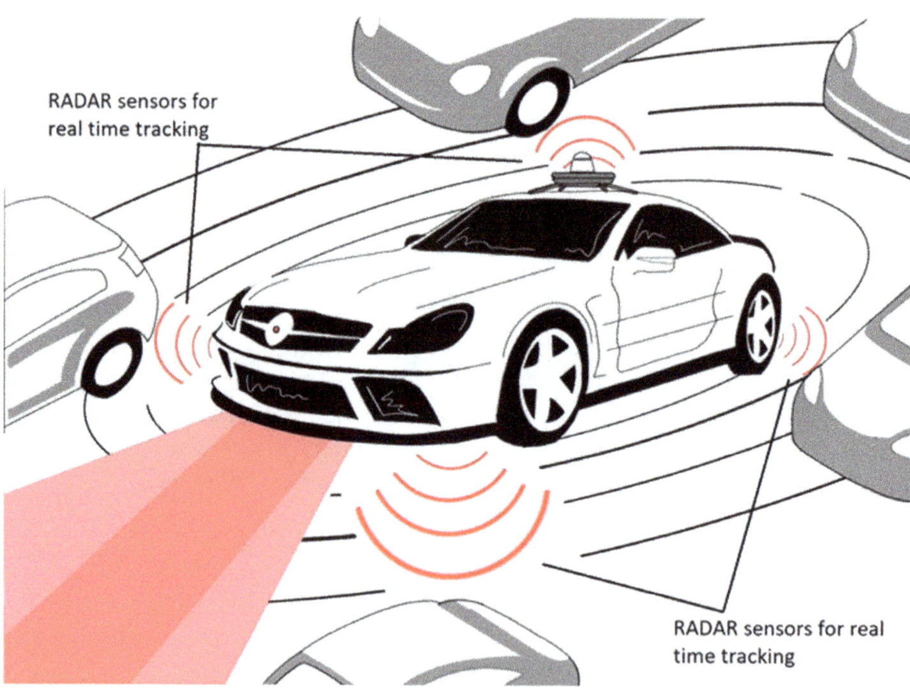


**Figure 10.13.** Radar sensor.

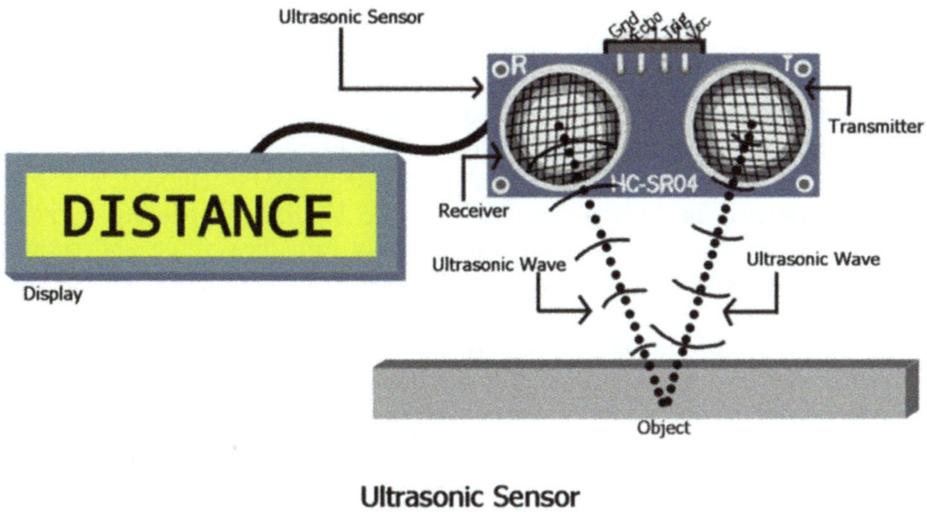


**Figure 10.14.** Ultrasonic sensor.

### 10.5.5 Ultrasonic sensors

Ultrasonic sensors, as shown in figure 10.14, are used for parking applications mainly. The time taken is recorded and ultrasonic frequency helps in estimation of distance of the target. Therefore, in ultrasonic sensors, we have two sensors, one for transmitter and another for receiver. Ultrasonic sensors [51] are not expensive and are also popularly used in ordinary cars. They can be used in driverless cars too but the sensors have the major limitation of low speed of the car. They can only detect and locate objects when the speed of the car is low.

## 10.6 Challenges and future research

Sensors are an integral part of numerous autonomous applications including self-driving cars. The detection and ranging capabilities are due to the functionalities of sensors, especially smart and intelligent sensors. The other side of the wonderful implementation of driverless cars is dealing with ethical issues and sentiments. This is also being attempted by researchers in robots as well as other humanoid autonomous vehicles so that the machines can also understand the emotions of persons and travelers and control the functions accordingly. Although, ethical and other issues are related with AI and machine learning methods, cognitive sensors are also being developed for such situations. The following are some important issues in driverless cars:

- safety and quality issues [10, 29, 52–54];
- privacy [25, 55–58];
- reliability [36, 56, 59];
- navigation [2, 12, 21, 45, 51];
- responsibility;
- security [29, 45, 58–63].

The above issues are related to AI as well as sensors, which means that reliability, for example is a major concern with sensors, and security can be enhanced with the help of better and more robust AI techniques. So, the sensors used in driverless cars should be chosen judiciously so that the issues in operation of cars are taken care of and smooth and seamless driving without any crashes can be achieved. The sensors should be designed in such a manner that the handling capability in adverse conditions can be ensured, for example in heavy rain, heavy fog and snow-like situations. Cyber-attacks are threats in systems like driverless cars, for which cognitive and intelligent sensors can be appropriately deployed and the cyber security can be enhanced. Biosensors are also modern sensors which can be used in driverless cars to avert any possibility of chemical attack or similar situations. A lot of research still is needed in the area of sensor technologies which are compatible with ethical, social, technological, cost and other issues involved in the operation of driverless cars.

## 10.7 Conclusions

This chapter discusses the role of sensors and AI techniques in design and implementation of driverless cars with more emphasis on sensor technologies. Modern, wearable and smart sensors are like the eyes and ears of driverless cars, with which the cars drive without the help of drivers smoothly on the roads. Cognitive sensors, LiDAR, radar, GPS and vision based cameras are sensors that make autonomous driving possible with AI and machine learning methods.

## References

[1] Schoettle B 2017 Sensor fusion: a comparison of sensing capabilities of human drivers and highly automated vehicles *Sustain. Worldw. Transp.* 1–42 http://umich.edu/~umtriswt/PDF/SWT-2017-12.pdf

[2] Ilci V and Toth C 2020 High definition 3D map creation using GNSS/IMU/LiDAR sensor integration to support autonomous vehicle navigation *Sensors* **20** 899

[3] Singh A K, Negi A, Azad S and Mudali S 2017 Fuzzy based controller for lidar sensor of an autonomous vehicle *Energy Procedia* **117** 1160–4

[4] Nassiri-Mofakham F 2017 Current and future developments in artificial intelligence *Intelligent Computational Systems: A Multi-Disciplinary Perspective* vol 1 (Sharjah, UAE: Bentham Science)

[5] Feigenbaum B 2018 Autonomous vehicles: a guide for policymakers *Reason Found.* 1–55 https://bit.ly/2FJDUAT

[6] Bimbraw K 2015 Autonomous cars: past, present and future: a review of the developments in the last century, the present scenario and the expected future of autonomous vehicle technology *ICINCO 2015—12th Int. Conf. Info. Contr. Autom. Robot. Proc.* 1 191–98

[7] Kornhauser A L 2013 Smart driving cars: history and evolution of automated vehicles *Florida Autom. Veh. Summit* http://teachamerica.com/favs2013/FAVS03Kornhauser/

[8] Péter G, Kiss B and Tihanyi V 2019 Vision and odometry based autonomous vehicle lane changing *ICT Exp.* **5** 219–26

[9] Surugiu M C and Stăncel I N 2019 Monitoring and warning system of a route based on roadside sensors in VANET *Procedia Manuf.* **32** 745–52

[10] Jun M and Markel T 2014 Infrastructure-based sensors augmenting efficient autonomous vehicle operations *21st World Congr. Intell. Transp. Syst. ITSWC 2014 Reinvent. Transp. Our Connect. World* https://www.nrel.gov/docs/fy18osti/62390.pdf

[11] Naveen A and Bandaru N 2016 Obstacle detection using stereo vision for self-driving cars *EEE Intelligent Vehicles Symp.* **(IV)** 926–32 https://tarjomefa.com/wp-content/uploads/2016/09/5311-English.pdf

[12] Yoneda K, Suganuma N, Yanase R and Aldibaja M 2019 Automated driving recognition technologies for adverse weather conditions *IATSS Res* **43** 253–62

[13] Ouyang Z, Niu J, Liu Y and Guizani M 2020 Deep CNN-Based real-time traffic light detector for self-driving vehicles *IEEE Trans. Mob. Comput.* **19** 300–13

[14] Fujiyoshi H, Hirakawa T and Yamashita T 2019 Deep learning-based image recognition for autonomous driving *IATSS Res.* **43** 244–52

[15] Mahdavinejad M S, Rezvan M, Barekatain M, Adibi P, Barnaghi P and Sheth A P 2018 Machine learning for internet of things data analysis: a survey *Digit. Commun. Netw.* **4** 161–75

[16] Rao Q and Frtunikj J 2018 Deep learning for self-driving cars *2018 IEEE/ACM 1st Int. Work. Softw. Eng. AI Auton. Syst.* 35–8

[17] Kurebwa J G and Mushiri T 2019 Internet of things architecture for a smart passenger-car robotic first aid system *Procedia Manuf.* **35** 27–34

[18] Friedrich M, Sonnleitner J and Richter E 2019 Integrating automated vehicles into macroscopic travel demand models *Transp. Res. Procedia* **41** 360–75

[19] Fair P and Giarda R 2018 Global driverless vehicle survey *Technical report, Baker & McKenzie International, World* p 261 https://bit.ly/2HoHqkn

[20] Zhang X, Gao H, Guo M, Li G, Liu Y and Li D 2016 A study on key technologies of unmanned driving *CAAI Trans. Intell. Technol* **1** 4–13

[21] Do T D, Duong M T, Dang Q V and Le M H 2018 Real-time self-driving car navigation using deep neural network *Proc. 2018 4th Int. Conf. Green Technol. Sustain. Dev. GTSD 2018* 7–12

[22] Center for Global Policy Solutions 2017 Stick shift: autonomous vehicles, driving jobs and the future of work

[23] Merat N, Louw T, Madigan R, Wilbrink M and Schieben A 2018 What externally presented information do VRUs require when interacting with fully automated road transport systems in shared space? *Accid. Anal. Prev.* **118** 244–52

[24] Ing P and Saponara S 2017 Measurements systems and sensors for autonomous vehicle and smart mobility IEEE http://ieee-ims.org/evts/video-tutorial/measurement-systems-and-sensors-autonomous-vehicles

[25] Allhoff F and Henschke A 2018 The internet of things: foundational ethical issues *Internet Things* **1–2** 55–66

[26] Zein Y, Darwiche M and Mokhiamar O 2018 GPS tracking system for autonomous vehicles *Alexandria Eng. J.* **57** 3127–37

[27] Liu Z, Yu S and Zheng N 2018 A co-point mapping-based approach to drivable area detection for self-driving cars *Engineering* **4** 479–90

[28] Ondruš J, Kolla E, Vertaľ P and Šarić Ž 2020 How do autonomous cars work? *Transp. Res. Procedia* **44** 226–33

[29] Zhao H 2018 Technology driving tomorrow's cars *ITU News* pp 1–45 https://bit.ly/31m63oZ

[30] Shanker R, Jonas A, Devitt S, Huberty K, Flannery S, Greene W, Swinburne B, Locraft G, Wood A and Weiss K *et al* 2013 Autonomous cars: self-driving the new auto industry paradigm *Morgan Stanley Blue Pap.* 1–109 https://bit.ly/3dEOKoj
[31] Sachdev S, MacWan J, Patel C and Doshi N 2019 Voice-controlled autonomous vehicle using iot *Procedia Comput. Sci.* **160** 712–7
[32] Colquitt J, Dowsett D, Gami A, Jaysane-Darr E, Manley C and Shad R 2017 Driverless cars: How innovation paves the road to investment opportunity *Invesco White Paper Ser.* pp 1–10 https://bit.ly/2H6jD9p
[33] Aneesh A N, Shine L, Pradeep R and Sajith V 2019 Real-time traffic light detection and recognition based on deep retina net for self driving cars *2nd Int. Conf. Intell. Comput. Instrum. Control Technol. ICICICT 2019* 1554–7
[34] Stilgoe J 2018 Machine learning, social learning and the governance of self-driving cars *Soc. Stud. Sci.* **48** 25–56
[35] Insarov W V, Sebriakov G G, Tihonova S V and Dronsky S A 2019 Autonomous unmanned aerial vehicles control system intellectualization *Procedia Comput. Sci.* **150** 376–80
[36] Nguyen V Q, Kim H, Jun S C and Boo K 2018 A study on real-time detection method of lane and vehicle for lane change assistant system using vision system on highway *Eng. Sci. Technol. Int. J.* **21** 822–33
[37] Claire P M P 2015 The Pathway to Driverless Cars: Summary Report and Action Plan (Department for Transport) pp 1–38 https://bit.ly/37qvn10
[38] Brett J A 2016 Thinking local about self-driving cars : a local framework for autonomous vehicle development in the United States *Doctoral dissertation* http://hdl.handle.net/1773/36852
[39] Ramanagopal M S, Anderson C, Vasudevan R and Johnson-Roberson M 2018 Failing to learn: autonomously identifying perception failures for self-driving cars *IEEE Robot. Autom. Lett.* **3** 3860–7
[40] Aziz M V G, Hindersah H and Prihatmanto A S 2018 Implementation of vehicle detection algorithm for self-driving car on toll road cipularang using Python language *Proc. 4th Int. Conf. Electr. Veh. Technol. ICEVT 2017* 149–53
[41] Lin W Y, Hsu W H and Chiang Y Y 2019 A combination of feedback control and vision-based deep learning mechanism for guiding self-driving cars *Proc. 2018 IEEE Int. Conf. Artif. Intell. Virtual Reality, AIVR 2018* 262–6
[42] Prabhakar G, Kailath B, Natarajan S and Kumar R 2017 Obstacle detection and classification using deep learning for tracking in high-speed autonomous driving *TENSYMP 2017—IEEE Int. Symp. Technol. Smart Cities* 3–8
[43] Ali Chishti S O, Riaz S, Bilal Zaib M and Nauman M 2018 Self-driving cars using CNN and Q-learning *Proc. 21st Int. Multi Top. Conf. INMIC*
[44] Fayjie A R, Hossain S, Oualid D and Lee D J 2018 Driverless car: autonomous driving using deep reinforcement learning in urban environment *2018 15th Int. Conf. Ubiquitous Robot. UR 2018* 896–901
[45] Elliott D, Keen W and Miao L 2019 Recent advances in connected and automated vehicles *J. Traffic Transp. Eng.* **6** 109–31
[46] Ozgunalp U and Kaymak S 2017 Lane detection by estimating and using restricted search space in Hough domain *Procedia Comput. Sci.* **120** 148–55
[47] Wu C F, Lin C J, Lin H Y and Chung H 2013 Adjacent lane detection and lateral vehicle distance measurement using vision-based neuro-fuzzy approaches *J. Appl. Res. Technol.* **11** 251–8

[48] Grigorescu S, Trasnea B, Cocias T and Macesanu G 2019 A survey of deep learning techniques for autonomous driving *J. F. Robot* **37** 362–86
[49] Shukla U, Mishra A, Jasmine S G, Vaidehi V and Ganesan S 2019 A deep neural network framework for road side analysis and lane detection *Procedia Comput. Sci.* **165** 252–8
[50] Bugeja M, Dingli A, Attard M and Seychell D 2020 Comparison of vehicle detection techniques applied to IP camera video feeds for use in intelligent transport systems *Transp. Res. Procedia* **45** 971–8
[51] Lam Loong Man C K Y, Koonjul Y and Nagowah L 2018 A low cost autonomous unmanned ground vehicle *Futur. Comput. Info.* J **3** 304–20
[52] Jing P, Xu G, Chen Y, Shi Y and Zhan F 2020 The determinants behind the acceptance of autonomous vehicles: a systematic review *Sustainability* **12** 1719
[53] Petrovi, Mijailovi R and Peši D 2020 Traffic accidents with autonomous vehicles: type of collisions, manoeuvres and errors of conventional vehicles' drivers *Transp. Res. Procedia* **45** 161–8
[54] Kurebwa J G and Mushiri T 2019 Passenger car safety and emergency healthcare: a literature review *Procedia Manuf* **35** 35–49
[55] Mihelj J, Kos A and Sedlar U 2018 Implicit aggressive driving detection in social VANET *Procedia Comput. Sci.* **129** 348–52
[56] Kaur K and Rampersad G 2018 Trust in driverless cars: investigating key factors influencing the adoption of driverless cars *J. Eng. Technol. Manag.—JET-M* **48** 87–96
[57] Martínez-Díaz M and Soriguera F 2018 Autonomous vehicles: theoretical and practical challenges *Transp. Res. Procedia* **33** 275–82
[58] Rader E and Slaker J 2017 Open access to the Proc. 13th Symp. usable privacy and security is sponsored by USENIX. The importance of visibility for folk theories of sensor data the importance of visibility for folk theories of sensor data *Thirteenth Symp. on Usable Privacy and Security {SOUPS}* pp 257–70
[59] Abdulkhaleq A, Lammering D, Wagner S, Röder J, Balbierer N, Ramsauer L, Raste T and Boehmert H 2017 A systematic approach based on STPA for developing a dependable architecture for fully automated driving vehicles *Procedia Eng.* **179** 41–51
[60] Angerholzer M, Mahaffee D, Vale M, Kitfield J, Renner H and Congress 2017 The Autonomous Vehicle Revolution: Fostering Innovation with Smart Regulation (Center for the Study of the Presidency and Congress) https://bit.ly/3jdcJMk
[61] Piao J, McDonald M, Hounsell N, Graindorge M, Graindorge T and Malhene N 2016 Public views towards implementation of automated vehicles in urban areas *Transp. Res. Procedia* **14** 2168–77
[62] Deb S, Bhattacharjee R and Deroy B 2012 Low cost versatile autonomous vehicle with intelligent control system *Procedia Eng.* **41** 965–70
[63] Bennakhi A and Safar M 2016 Ambient technology in vehicles: the benefits and risks *Procedia Comput. Sci.* **83** 1056–63

# Chapter 11

## Recent advancements in smart and wearable sensors

**Preeti Goswami, Anuj Sharma and Govind Gupta**

Sensors are transforming into an imperative part of our lives. For the environment we live in, the apparatus and the structures in our homes, vehicles, and even our bodies, splendid sensor courses of action are open, with new applications and improvements to the present ones. Sensors are fundamental for considerations regarding our condition and they give us opportunities to flourish, security, and acknowledgment, as we engage in checking of our prosperity and condition. The latest movement to sensor innovation is the development of the smart sensor, which is just a blend of distinguishing segments with an embedded digital motion processor that plays out the limits demonstrated after the data signal is received. Smart sensors can possibly be consistently applied to delicate and unpredictably molded surfaces, for example, human skin or material textures. In addition, minimum interfacing cables, high reliability, high performance and compactness are the reasons for their precedence over basic sensors. This chapter based on smart and wearable sensors (SWSs) deals with the basics of SWSs and their working principle, synthesis as well as the fabrication process. The types of SWSs and materials which are being used for the synthesis of these sensors are described in detail. Further, challenges relating to managing and protecting large data generated by sensitive sensors are discussed. Despite these challenges, in the future, SWSs will profoundly affect applications, for example, sanitation, ecological observation, and clinical diagnostics.

### 11.1 Introduction

A smart sensor is only a sensor with interfacing electronics which enables it to make decisions, perform logical functions and do two-way communications. 'Wearable' suggests an adaptable and portable sensor that can be worn constantly on the patient's body and utilized at home without the help of a physician. Nowadays, wearable sensors are gaining importance due to their skills for monitoring human

doi:10.1088/978-0-7503-2707-7ch11  

health concurrently. In a wearable sensor, a large number of physical, chemical and optical sensors are installed separately or integrated onto flexible substrates for the study of data compatibility and rotation of the display case. The data is then transmitted by wireless transmission to a computer program and monitored by a healthcare professional, as indicated in figure 11.1. The flexibility, bendability, foldability, stretchability, light weight, low cost and conformability to uneven surfaces with negligible losses to the sensor response are the attributes of a wearable sensor. The most generally utilized SWSs in medicinal services are electrocardiography sensors (electrical driving forces through heart muscles), Inertial estimation units (linear and angular movement), photoplethysmography/optical pulse sensors (blood volume changes), electroencephalography sensors (electrical action of the mind), galvanic skin reaction sensors (skin conductivity), and temperature sensors (surrounding/internal heat level) [1]. Moreover, smart sensors likewise have industrial applications, for example an optical sensor utilized for measuring exposure in optical arrays, cameras and optical angle encoders; gas sensors used in food testing (e.g. assessment of freshness in meat [2]), air detection (example $O_3$ constituents and quality of air [3]), pipeline spill recognition [4], hazard observation (e.g. explosions and toxic gas detection [5]) etc.

### 11.1.1 Basics of SWSs

A smart sensor is nothing but a sensor with appropriate interfacing electronics, as shown in figure 11.2. The vital part of a smart sensor is its microprocessors. These microprocessors complete plenty of functions such as digital processing, code change of analog to digital or digital to analog, interfacing functions, and estimation too.

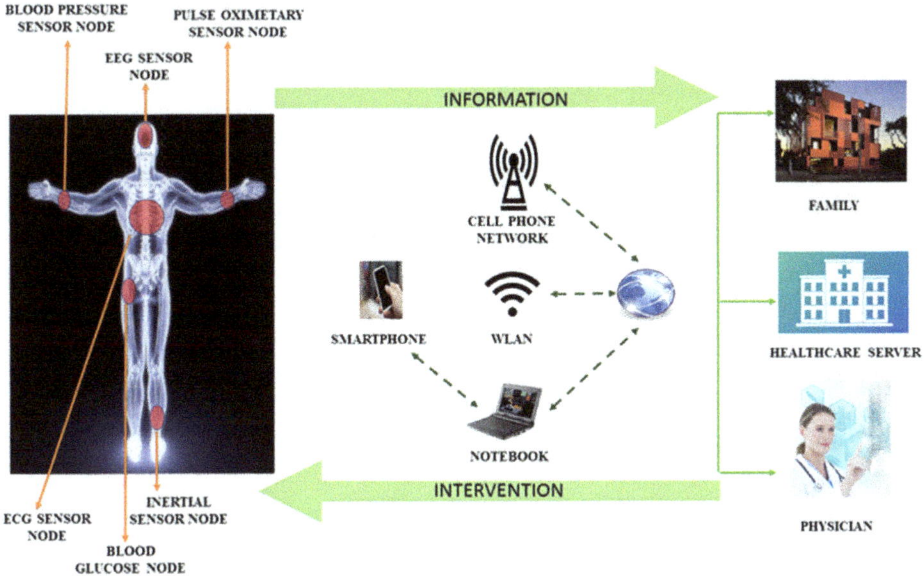


**Figure 11.1.** Framework of telemedicine based on the monitoring of smart and wearable sensors.

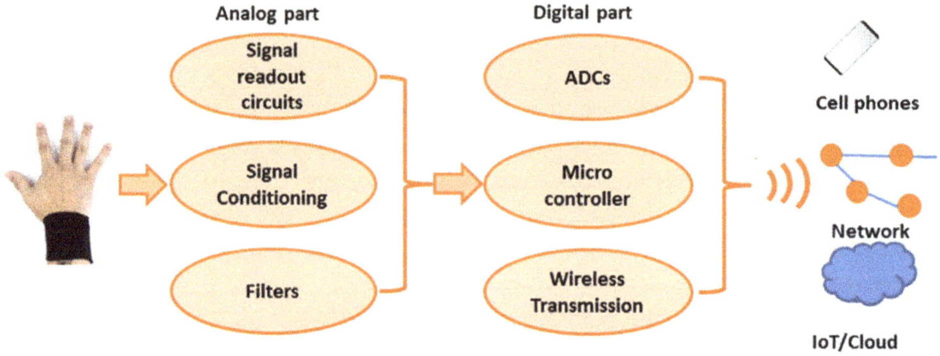


**Figure 11.2.** Schematic stream chart for estimating elements through wearable sensors and information transmission.

These even choose when they need to store or erase any part of the data as per the input along with that, and function automatically at periodic working intervals to consume less power. The presence of a controller/processor on the smart sensor has caused the amendment of various unwanted outcomes including output and timing variations, incompatibility and error sensitivity, etc. As these amendments are made to the software, no additional hardware is required, and next, the measurement changes to the electronic process. Therefore, it is possible to measure the sensor connectors during production without the need to expel the sensor from its current state or test equipment. Due to these processors within the package, it is possible to have digital communications with the standard bus and built-in self-test (BIST). These remote diagnostics can be a set of rules-based programming that works on a sensor and are therefore extremely useful for the production tests of integrated circuits.

### 11.1.2 Working principle of a smart sensor

A smart sensor is a sensor that includes a processing unit with a transmission alliance interface. Figure 11.3 shows the diagrammatical view of a smart sensor. An electrical signal corresponding to changes in parameters of sensing material upon exposure to input (like light, temperature, pressure, acceleration, position, magnetic field, radiation, gas, etc) is produced by sensor and then transferred to a signal conditioning unit that converts it into a suitable signal which further can be changed over into a computerized structure by analog-to-digital converter (ADC). At that point the significant data from the time-differing computerized signal is extricated by preparing it with a reasonable calculation and utilizing a proportionate controlling unit. The communication with the base stations is done with the help of transducers.

## 11.2 Types of wearable sensors

Wearable sensors are divided into three main sections i.e. individual sensors, multiplexed sensors and wireless sensors as illustrated in figure 11.4. Further,

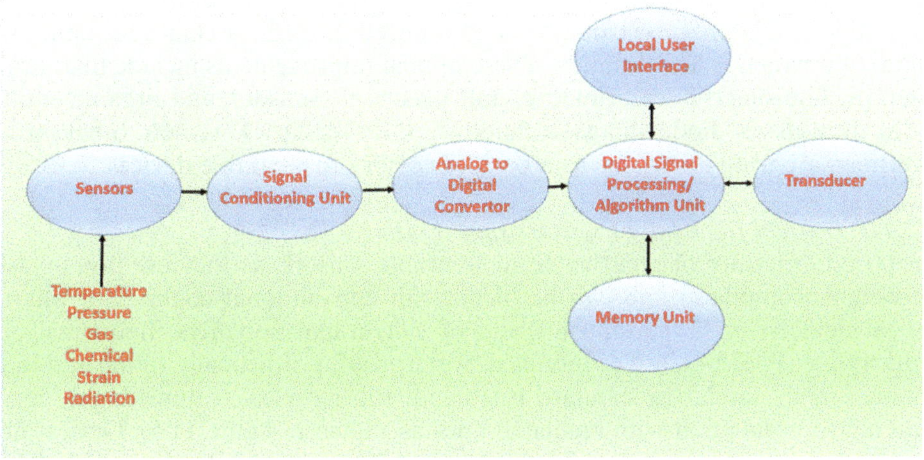


**Figure 11.3.** Basic architecture of a smart sensor.

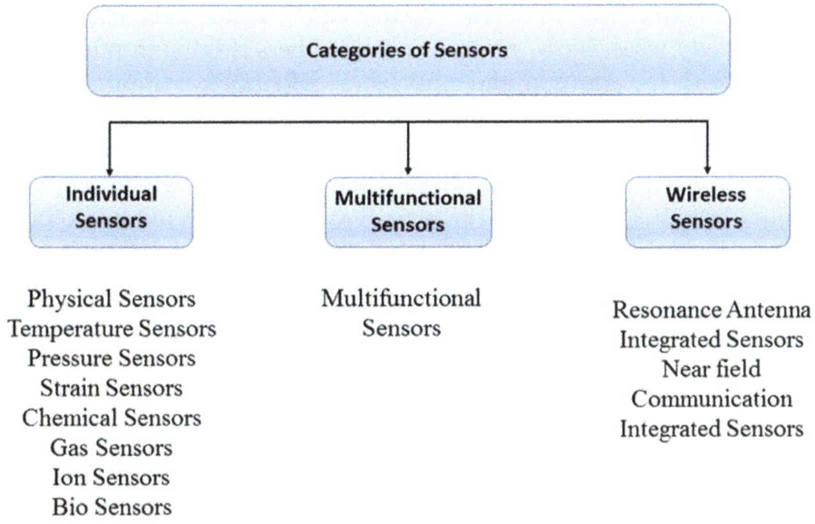


**Figure 11.4.** Classifications of wearable sensors.

individual sensors are subcategorized into optical, physical and chemical sensors, based on the species sensed by them.

### 11.2.1 Optical sensors

Optics and photonics will take the lead in creation of abstract works, as they help the most sensitive measurements of abstract information and standards related to the condition of human health and the environment. With the release of optic wearable innovations, for example, pulse rate, circulatory strain, and glucose screening, among others, individuals are becoming more endowed to produce environmental and biodiversity information, making customized medication a reality. Alternatively, this progress can also be monitored in medical clinics for concurrent lookup of a patient's health status. These optical improvements include fiber optical materials, fluorometric, calorimetric, and plasmonic sensors, and organic-emitting diodes as well as diode image technology (OPD), etc. This new breakthrough guarantees amazing assurance as a basic accessory to wearable devices.

*11.2.1.1 Optical fiber based sensors and textiles*

Catoptric fabrics are of great value to wearable sensors, as they are non-invasive, lightweight, compliant, and can be blend with conventional coatings or pharmaceutical devices [6, 7]. Catoptric fibers of silicon-like polymers, frequency polymethacrylate (PMMA), are economical, waterproof, synthetically stable and in this way are largely similar to standard treatment. Fiber optical features highly recommend active installation with regular fabrics as shown in figure 11.5). Optical fibers have much greater strength than fabric and textile fibers having the same diameter and dependency area, recommending robust incorporated textiles with thickness and surface generally not recommended for their standard counterparts [6].

The limited rays of bends and low stiffness provides a significant difference in cushioning of daily wear. Now, polymer optical fibers (POFs) in this way continue to make sense with the addition and embroidery of silica-based optical cables. The origin and seamless fiber optical coupling systems have been thoroughly investigated and evaluated based on the dynamic range of the sensors, the impact on the main bending threads, and the active coupling efficiency [6, 8–10]. In addition, POFs show some unique benefits such as visualization due to Young's modulus, high breakthrough contrast, and low cost [11]. This is sensitive to the application of pressure, small and large bends, temperatures, irregular changes in clearance inside and outside the fiber wave direction. So far, the introduction of seamless wire as sensors in ordinary objects is a great opportunity, making fiber material a viable visual platform for converging optical sensors.

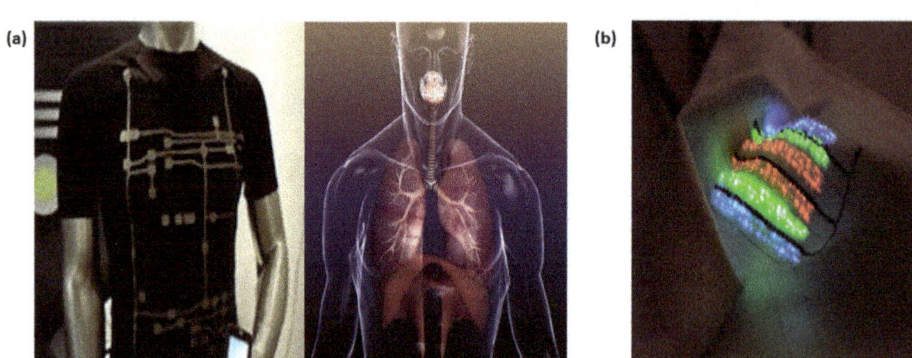


**Figure 11.5.** (a) All-respiratory screening exams using both abdominal and thoracic sensory fibers, (b) a fiber Bragg gating (FBG) sensor for respiratory tracking.

*11.2.1.2 Colorimetric sensing element*

Colorimetric sensors have a promising future as wearable implant sensors due to signal-based readings and automatic retrieval as the weight of the connection is usually transmitted to chemically modified elements or color responses. In particular, these senses enable permanent monitoring of sweat or other toxins such as drainage or spitting, and can even be used on clinical walls to prevent wound re-injury. Specifically, sweat analysis tracking applications can benefit greatly from the sensitive pH sensors of colors, by providing an underground site where installed gadgets can be developed. Understanding the pH and sweat content of some analyses, for example, sodium, potassium, and lactate is critical to a thorough assessment of physical activity and athletic performance, and in addition, it has the potential to reflect on medical issues further [12, 13].

*11.2.1.3 Plasmonic sensing substrate*

Plasmonic components can also have visual acuity sensations [14]. The main functions of these sensors include attaching incident light to plasmonic resonances, evidenced by the interaction of electrons arising from the interaction of negative and negative states. Such a combination depends largely on the shape of the nano structure and the type of print material near the field of display. In this way, the sub-fields can be built to have localized electrical potential in specific areas that can be used to selectively select analytes. The sent analyser consequently alters the direction of the ornamentation and awakens the spectacular flexibility of the plasmonic resonance which is to be evaluated in the field by way of moving or demonstrating optical readout. In addition, the enhanced localized electric field of the plasmonic structures will be able to improve the Raman distribution of the molecules. Thus, the most sensitive molecular equations will be performed similarly under those elements. Ample efforts have been made in the manufacturing of plasmonic sensors with cost effective techniques, and a diverse group has shown the design and transmitting plasmonic nerves to non-invasive substrate, for example, paper, PET, PDMS, and standard plastic tapes [15, 16]. Bacterial-cellulose (BC) paper has been effectively shown to be more cost-effective, bio-dispersed, bio-compatible, useful and flexible for plasmonic nano-particles and quantum dots [17]. Synchronized by acetobacter xylinum, a non-pathogenic microbe found in fruit, this content can be made as plain paper with a larger nano-network for mobile computers. These adaptive components show a unique plasmonic specialty, in which recovery can be performed in curved areas, for example, on the human body, with access to the skin and clothing. Thus future research on wearable plasmonic sensors is designed to develop very sensitive measurements in sensitive systems, for example, sweat ion or protein recognition, in clinical myocardial biomarker coatings, or in combination with various environmental monitoring materials around the client [18].

*11.2.1.4 Fluorometric sensors*

Fluorescence-based sensors indicate the wear potential of the coating. However, yet to be explored further, practical enhancements for synthetic materials and signal transduction recommend the potential use of fluorometric detection systems on wearable devices that detect fluorometric signals within the human body, for example, in external gadgets or blood, that emit a signal of fluorescence that transmits data modern, chemical, and organic customer experience. For example, the work of Azenbacher *et al* gave an idea of fluorescent attoreactor tangles as sensors line-up [19], in which electro-spun nanowires are mounted on amine or fluorophore predecessors shrouded in a system that is equitably dispersed to one another to keep fluorescent sources of information experienced in these channels. The detection of metal particles can be achieved by using fluorophore-polyamines that stick to the metal particles from which results the changes in their trademark fluorescence [20]. Multianalyte disconnection was uncovered by the proposed attoreactor chips, by official to three distinctive fluorophore cases (dansyl chloride, fluorescamine, furthermore, 7-chloro-4-nirobenz-2-oxa-1,3-diazole as such shaping a cross list of sensor sensors by layer of linear discrimination analysis (LDA), every one of the particles in a gathering of ten diverse metallic arrangements is separated into brilliance. Special application in is needed for hazardous coatings, or leathers. However, the little amount of these fluorescent materials gives a negative signal, which currently shows metal ion absorption with a very high concentration of about 200 microns.

*11.2.1.5 Organic LED, organic photo-diodes (OPD) and display technology*

New display has been growing fast because of the commercialization of the technology industry. OLEDs are very similar to the work of its organic part; however, they incorporate semiconductor materials which are organic, standard polymers that can provide a few benefits. For example, compared to the current flat board display, OLEDs show improved shade brightness, improved power for darker shades, lower cost due to the color change of the text, and can be fashionable on substrates [21, 22]. For the same reasons, OPDs are increasingly focused on the photovoltaic network. Their high-resolution performance enables low-cost surface production, and their excellent adoption enables ultra-thin films to reduce much of the input light. Other attractions of OLEDs too, are that OPDs have a large range of emission and absorption characteristics characterized by an appropriate polymer or semiconductor of organic material. When paired cooperatively, this is a natural combination of application enthusiasm for lightweight optical gadgets due to their ease, large scale production, and flexible environment. Current research has gone into this objective of creating an organic optical sensor for measuring photo-plethysmography and total heart rate [23]. In this case, blue-green OLEDs (532 nm and 626 nm, with 20.1 mW $cm^{-2}$ and 5.83 mW $cm^{-2}$ radios) measures pulse, rotation type, and oxygen levels. This work has shown the advantage of flexible gadgets in reduction of the potential of earthworms brought about by light interference. A 90 % decrease in device signal has been observed by using textile senors communicating with human finger, but provided all natural heartbeat oximeter should be adjusted around the perfuse object instead of unbending plane which is inclined to the spillage of surrounding light. The PPG signal obtained with the whole organic gadget was able to obtain the correct heart rate, however, it was generally of lower signal quality than the nonorganic gadget with a signal of 3 and

2.5 mVp-p as it was coupled to 26 and 16 mVp-p in green and red muscles, each [23]. Such demonstrations prepare the way for every single organic gadget that has power for one-time, expendable operation, giving more modest options in contrast to customary gadgets.

### 11.2.2 Physical sensors

Physical sensors are equipped to measure physical parameters, for example temperature, pressure, and strain. The physical parameters estimated by wearable sensors show an individual's health status by measuring body temperature and skin, heart rate, blood pressure, and skin problems. As physical sensors are able to measure these types of physical parameters, they could be utilized to look at the electronics of the human skin, the mechanical signal of the person, the individual exercises, and the applications of the individual device [24–28].

#### *11.2.2.1 Temperature sensors*

Body temperature is one of the most significant components identified with human life and body development. For example, variation in thermal changes can be characterized by physical symptoms. Similarly, constant monitoring of internal temperature is important in detecting adverse events, for example, heart rate. Alternatively, thermal views are significant for the combattant and competitor, where body development is straightforwardly identified with their achievement. Temperature sensors can be divided into several types depending on their formation process, for example, the pyroelectric heat exchanger, the heater reactors, and the thermistors. One example of heat-related research is the production of thermocouples according to the Seebeck effect [29] where thermal insulators use color-changing materials [30]. In order to use these temperature sensors to monitor individual well-being and exercise, they must be flexible and can be extended to work well on each test basis. More recently, flexibility and stretching have been made using polymers as volatile materials and by rearranging the structure.

Pyroelectric temperature examination is created utilizing the pyroelectric impact that is the result of the electric field because of the computerization of the material because of the adjustment in temperature. Barium titanate and lithium tantalite are standard pyroelectric materials, in which the crystal cross section gathering of cell bodies contains regular cerebrum. Since these pyroelectric gadgets are strong particles, it is hard to utilize them in non-straight, slanted contraptions. To conquer this issue, polymer-based pyroelectric materials, for instance, polyvinylidene fluoride (PVDF) and its copolymers, have been considered and utilized as occurrence heat sensors [31–33]. The transducer depends on the organic thin film transistor (OTFT) to identify the impact of profile [34]. Because of the material properties, charge division is made in the PVDF film when the temperature changes, which causes the present contrast in the OTFT. Another kind of temperature sensor, resistive temperature detectors (RTDs) are comprehensively used temperature sensors. Typically, ridged iron, for the most part copper or platinum, is utilized as recuperating substance. Since the serpentine structure of platinum shows a linear resistance change of 2.7 Ω/°C with the change in temperature therefore, the metal-based temperature sensor can sense the temperature variations by following the below relation

$$R = R_0[1 + a(T - T_0)] \quad (11.1)$$

In this, $R$ denotes the sensor resistance, sensor resistance at normal temperatures of 20 °C is $R_0$, $a$ is the resistance of the sensor, and $T_0$ is the temperature.

In any case, metal-based RTDs have low-temperature alterations for uses, that are utilized for screen temperatures of 30–40 °C. Not at all like metal-based RTDs, the thermistor temperature obstruction remains unaltered as shown by the comparing temperature which obeys the condition

$$R = R_0 \exp(B/T) \quad (11.2)$$

Here, $T$ is the sensor temperature and the resistance at temperature $T$ and $T = \infty$, is defined as $R$ and $R_0$ respectively. $B$ is the thermal index.

This can reveal that thermistors measure more correctly than RTDs. With this, there is an additional feature that they have high priority in bio performance as they are economical [35].

#### *11.2.2.2 Pressure sensors*

The pressure range of the wearable pressure sensors are partitioned into three classes, depending on the pressure of various parts of body to be monitored: (1) low-pressure range (<10 kPa), (2) medium-pressure range (10 to 100 kPa), and (3) high-pressure range (>100 kPa). The low-pressure range, related to intracranial as well as intraocular pressures, is a significant range since it comprises the intra-body pressure [25, 36]. On the other hand, the body pressures that come to the medium pressure range include blood pressure, beat rate, the outspread vein wave, skin modules and phonation modulation. The atmospheric pressure at high elevations or weight of individuals falls into the high-pressure range. By computing different pressures, one can screen eye illness, coronary diseases, injured vocal cords etc. The conversion of physical stimuli into corresponding electrical signal takes place by different sensing mechanisms like piezoelectric, piezoresistive, and capacitive. The electric displacement induced by pressure in a piezoelectric based pressure sensor is expressed as

$$D_i = \varepsilon_0 \varepsilon_{ij}{}^{\sigma} E_i + d_{ij} \sigma_j \text{ and } \delta_I = S_{IJ}{}^{E} \sigma_J + d_{Ii} E_i.\ D \quad (11.3)$$

where $\varepsilon_0$, $\varepsilon^{\sigma}$, $E$, $D$, $d$, $\sigma$, $S^E$, and $\delta$ represent the free-space electrical permittivity, relative electrical permittivity, the electric field, electric displacement, piezoelectric coefficient, stress by pressure, mechanical compliance matrix, and the strain by pressure. The $d_{33}$ and $d_{31}$ are labeled as the longitudinal and transverse coefficient, respectively. $i$, $j$, $I$, and $J$ are indices. So, piezoelectric effect refers to change in electric polarization produced in certain materials when they are subjected to mechanical stresses, and this change in polarization is directly related to the applied mechanical stress. Such pressure sensors are utilized widely in the recognition of dynamic pressure, for example sound vibrations, because of low response time and less power utilization. The suitable materials for the synthesis of such pressure sensors are lead titanate ($PbTiO_3$, barium titanate ($BaTiO_3$), polymers like poly-vinylidene difluoride) and copolymer P(VDF-TrFE). Preferably, thin-film based sensors using P(VDF-TrFE) are widely utilized to develop wearable piezoelectric-effect based pressure sensors [33, 37–39]. Also, a $PbZr_{0.52}Ti_{0.48}O_3$ (PZT) based stretchable and ultrathin wearable piezoelectric pressure sensor has been fabricated for observing circulatory strain; the gadget can be delicately applied on the throat, wrist, or neck [24].

Another kind of piezoresistive-pressure sensor based on the fact that the resistance of some particular material changes with an externally applied pressure is also available. The corresponding change in resistance with applied pressure can be evaluated using the below relation

$$\Delta R/R = (1 + 2v)\delta + (\Delta\rho/\rho) \tag{11.4}$$

Here $(\Delta\rho/\rho)$, $(1+2v)$ and $\delta$ represent resistivity effect, geometric effect and strain by pressure, respectively. In most cases, to create materials with piezoresistive properties conductive fillers like metal particles, carbon nanotubes or rGO are added to the elastomers, but such sensors are less sensitive .The issue of low sensitivity has been resolved by designing stretchable pressure sensors utilizing the unique geometry of interlocked micro-dome structures. Such sensors can be connected to the skin and recognize an assortment of mechanical incentives.

A capacitive pressure sensor system, as shown in figure 11.6, uses capacitors whose capacitances vary with $d$ (i.e. the thickness of dielectric material) in accordance with the following expression

$$C = \varepsilon_0\varepsilon_r(A/d) \tag{11.5}$$

Here $A$ is the area of cross section of one of the two plates, $\varepsilon_0$ and $\varepsilon_r$ are free space electrical permittivity and the relative electrical permittivity of the dielectric, respectively. As is clear from the above expression, the capacitance increases when the thickness is decreased by pressure. The suggested materials for this type of pressure sensor are those having low modulus value, for example, PDMS, ecoflex, and polyurethane (PU) etc, so as to change the thickness in accordance with applied pressure. An adaptable, capacitive pressure sensor with exceptionally high affect-ability has been reported by building an air slit between a single wall carbon nanotube (SWCNT) film and a permeable PDMS film [28]. However, these pressure sensors have constraints, for example low affectability and moderate reaction time, due to low modulus of elastomer. The affectability can be enhanced using an air dielectric layer.

#### *11.2.2.3 Strain sensors*

Slow or fast movement of the individual's body parts may cause small and enormous distortions. Therefore, wearable strain sensors have a significant role in observing the movement of various parts of the wearer, for example vocal cords vibration and joints movement. Hence, flexible-strain sensors that exhibit unique behavior from

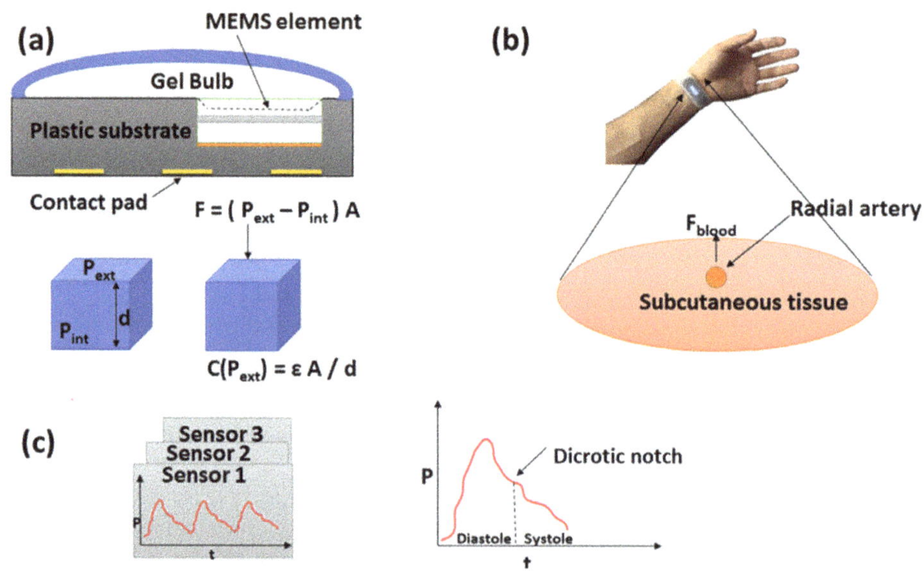


**Figure 11.6.** An overview of a sensor system. (a) The principle of operation of the sensor where the capacitance of the active element changes with outer pressure deforming the diaphragm in accordance with the displacement of gel inside the silicone gel bulb under externally applied uniform pressure. (b) The forces created by blood pressure in the radial artery are shown to the sensor array assembled on an adaptable wristband. (c) The pulse profile during one cardiac cycle in an array of sensors.

silicon based strain sensors have been produced using stretchable materials [40]. A wearable strain sensor must be more flexible, highly sensitive, mechanically reliable, and also must be able to produce a linear output signal. There are various types of strain sensors based on different mechanisms like piezoresistive, capacitive, and piezoelectric effects. In the case of piezoresistive strain sensor, the distortions are recognized mostly by the change in piezoresistivity of the active material. The resistance of a material is given as

$$R = \rho(l/A) \tag{11.6}$$

Here $\rho$, $l$, and $A$ represent resistivity, length, and area of cross section of the sensing material, respectively. As the band gap of metal and semiconductors changes during the twisting, that in turn results in the adjustment in interatomic distance [41]. Therefore, $\Delta R/R$ is known as gauge factor of strain sensors. The large change in resistivity of semiconductors leads to their high gauge factor so a significant number of semiconducting materials, for example, zinc oxide (ZnO), carbon black, and CNT are used for piezoresistive effect based strain sensors [42–45]. A strain sensor has been developed by installing SWCNTs in a conductive elastomer (PU-PEDOT: PSS) with a gauge factor of 62, 62% transmittance in visible region, and stretchability of 100% strain [42].

Capacitive strain sensors recognize strain dependence on changes in the capacitance resulting from the geometrical effect. Like a capacitor, such sensor has two electrodes separated by a dielectric material, preferably an elastomer, and the capacitance can be evaluated with the following equation

$$C_0 = \varepsilon_o \varepsilon_r l_o w_{o/} d_o \tag{11.7}$$

where $\varepsilon_o$ and $\varepsilon_r$ are already stated in equation (11.5), and $l_o$, $w_o$ and $d_o$ are length, width, and thickness of the sensing region, respectively. With the extension of material under strain $\varepsilon$, the length increases to $l = (1 + \varepsilon)l_o$ while width and thickness becomes $w = (1-v)\ w_o$ and $d = (1-v)d_o$ where $v$ is the Poisson's ratio, so by the relation (11.7) capacitive becomes $C = (1+\varepsilon)C_o$. The stretchable transparent electrodes, for example carbon nanotubes are used to ensure the stretchability in wearable sensors. A capacitive strain sensor with gauge factor 0.0348 and 700% stretchability has been demonstrated using silicon elastomers and ionic conductive fluid [46].

Also, the piezoelectric effect based strain sensors have been fabricated utilizing different oxides, for example zinc oxide, and zinc stannate or piezoelectric polymers, for example, PVDF and P(VDF-TrFE) [47–50]. Wu's group have shown zinc stannate nano/microwire-based wearable piezoelectric strain sensors with a gauge factor of 3740 at 0.35% strain [47]. These sensors typically are self-assisted, so they do not demand any extra battery or additional power source. But the mechanical properties of piezoelectric materials may limit the use of these wearable sensors as they demand good stretchability, so there are still some difficulties in the development of wearable strain sensors.

### 11.2.3 Chemical sensors

Chemical sensors are required to sense the chemical signals from clinically important biomarkers, chemical species (for example various toxic gases and volatile organic compounds) polluting the environment, or warfare agents which may be either chemical or biological [51–53]. At present, the conventional medical equipment used to extract biofluids, for example blood, cerebrospinal fluid, cyst fluid, saliva, or on the other hand tears cause pain and discomfort to patients [54]. Likewise, continuous observing of health parameters is essential for specific infections, therefore a patient's daily movements can become restricted. Hence, the development of wearable chemical sensors is a priority in healthcare services to enable minimum inconvenience for users, and to facilitate continuous observing of the health status of the wearer.

#### *11.2.3.1 Gas sensors*

There has been huge expansion in the issues related with environmental pollution caused by various toxic gases exhausted from industry directly into the surrounding air, as well as the utilization of biochemical weapons. This gives rise to great interest in having some sensors that can identify the different toxic gases, and which are also able to upgrade our capacity to ensure the safety of individuals by giving data pertaining to the presence of various harmful gases and sounding security alert signals. To develop advanced gas sensors, it is very important to consider smart and wearable technology so that wearable gas sensors which are light, cost-effective, reasonable, and successful in giving sudden alerts can be designed. Wearable gas sensors must give information about the level of the toxic gases in the surrounding environment without interrupting the activities of the wearer. Each sensor is attached to a sensing module for signal storage and data processing [55]. The improvement of wearable gas sensing frameworks is raising a high level of enthusiasm among researchers because of their huge potential in medicinal and environmental applications. Besides the strategies utilized in flexible gas detecting, screen and printing (ink-jet) empower the mass production of such frameworks. These techniques also provide an eco-friendly alternative in contrast to different systems, for example, etching, that may deliver risky synthetic waste that should be dealt with a short time later.

#### *11.2.3.1.1 Gas sensor response*

The response of a gas sensor towards a target gas is evaluated with the help of following equations:

(1) Relative sensibility

$$S[\%] = \left(R_{gas} - R_{air}\right)/R_{air} \times 100 \tag{11.8}$$

(2) Fractional sensibility

$$S[-] = R_{gas}/R_{air} \tag{11.9}$$

$R_{gas}$ represents the resistance of the sensing material in the presence of gas, whereas $R_{air}$ is sensing the material's resistance in the air.

#### *11.2.3.1.2 Sensing materials and their sensing mechanisms*

To fabricate a wearable gas sensor, the sensing material must have flexibility in order to conform to the surface of the wearer, good stretchability to prevent it from destruction under some strain and mechanical stress, large surface area for better sensitivity, low operating temperature, preferably room temperature, and good electrical properties. The suitable materials for wearable gas sensors are conducting polymers, transition metal dichalcogenides, and carbon nanomaterials, etc.

(A) *Conducting polymers*:

Various conducting polymers like polythiophene (PTh), polyaniline (PANI), and polypyrrole (PPy) as well as their derivatives have been attracting the interest of researchers who are working in the field of wearable gas sensors for many years, for many reasons like their good mechanical properties that allowing a facile fabrication of sensors, short response time, doping dependent conductivity, low operating temperature, and high sensitivity [56]. An electrochemical process is used to synthesize such conductive polymers. One can even also modify the molecular chain structure by copolymerization or structure derivations. In general, the intrinsic polymers have very low conductivity ($<10^{-5}$ s cm$^{-1}$) which can be enhanced by the doping process that is nothing but a redox reaction, especially protonation in the case of PANI. The positive charges are created on the conductive polymer's backbone as a result of doping. Thus, produced charges are known as polarons and act as the charge carriers causing an increment in the conductivity of the polymer.

*Sensing mechanism*

The conductivity of a conducting polymer-based chemo-resistive gas sensor changes autonomously with a redox reaction (i.e. the addition or withdrawal of electrons) between the polymer and the target gas. The conductivity changes upon the adsorption of gaseous molecules on the surface of the sensing layer in accordance with the nature and amount of adsorbed gaseous analytes. And subsequently, it returns to the original value after the desorption of adsorbed gaseous molecules. For example, a p-type polymer upon adsorption of ammonia should encounter an expansion in its resistance because of the addition of electrons from the ammonia gas molecules to the polymer. In any case, this is not generally the situation. Like the resistance of the PEDOT/PSS, polymer-based gas sensor decreases under ammonia gas. The main cause behind this unpredictable change is the redistribution of charges in the exposed region between the polymer and electrodes. Comparable conductivity was seen among PANI polymer and CO gas, and the clarifications were proposed for a positive charge transfer among the CO atom and amine nitrogen of the polymer, and the abatement of the charge barrier between the polymer's grains [57, 58]. In addition, a PANI detecting layer produces a decrement in conductivity in the presence of ammonia. The process responsible for the reduction in conductivity is, however, not a redox reaction, but a deprotonation process [59] shown in the below equation

$$\text{PANI H}^+ + \text{NH3} \rightleftharpoons \text{PANI} + \text{NH}_4^+ \tag{11.10}$$

(B) *Transition metal dichalcogenides*

Recently, TMDCs like $WS_2$, $WSe_2$, $MoS_2$, $MoSe_2$, $SnS_2$, $SnSe_2$ etc [60], have tremendously increased enthusiasm for gas sensing with relatively less power consumption inferable from their interesting electrical, mechanical, and chemical properties. Conventional semiconductors can likewise be utilized in wearable sensors, for example, silicon can be utilized to synthesize adaptable gadgets by scaling it down below a specific level. However, Si represents an extreme risk of deformation to the sensing layer itself because of its brittle nature. Various methods that use pure semiconductors as a sensing material were moreover researched. But still, there are some difficulties like the low ability of charge transfers and high operating voltages. The high operating voltage necessity leads to complex electronics. The mobility in the case of polycrystalline silicon is comparatively high but the extensive current leakage causes the device performance to crumble. In this way, traditional approaches have been limiting their utilization by the wearable market. But the unique properties of 2D materials, which are minimum power consumption utilization, light weight, high conductivity, high stretchability, mechanical reliability, incredible optical transparency, and great natural stability, made them magnificent candidates for adaptable gadgets. Recently, a few 2D materials, including graphene, TMDCs, and MXenes, have revealed their potential for manufacturing stretchable and adaptable gadgets that can be utilized in memory gadgets, chemical sensors, foldable displays, photodetectors, lithium-particle batteries, and supercapacitors. The huge surface-to-volume proportion promotes 2D materials as important candidates in the sensing of various chemicals.

*Sensing mechanism*

The TMDCs based gas sensor's sensing mechanism is different from that of metal-oxides based ones. In the case of these gas sensors, the sensing materials act as charge acceptors or donors depending on the nature of gaseous molecules. At the point when sensing material is exposed to various gases, the charge transfer process happens between the adsorbed molecules of gas and the detecting materials, which prompts various changes in the resistance of sensing material dependent on various exchange directions and the number of charges. Then the detecting materials are exposed to air leading desorption of gas atoms and thus resulting in the detecting material resistance back to the initial state. For example, n-type $MoS_2$ is widely used to sense various gases $NH_3$, NO, $NO_2$, CO, etc. When n-type $MoS_2$ opens to these oxidizing gases such as CO, the electrons move from $MoS_2$ to the adsorbed gaseous molecules, prompting a diminished carrier concentration in $MoS_2$. Accordingly, the resistance of n-type $MoS_2$ increases. But when $NH_3$, which is a reducing gas, is adsorbed on the $MoS_2$ layer the electrons move into the layer and thus enhance the electron's concentration of the n-type $MoS_2$sensing layer, and reduce its resistance [61].

#### *11.2.3.2 Ion sensors*

Lately, personalized health care systems have been underscored significantly and investigated broadly by various analysts, focusing on individuals who want to screen their health status persistently and demand the report of pathologies imminently. Electrolytes (like calcium, chloride, magnesium, phosphate, potassium, sodium) imbalance can be demonstrative of medical issues, for example, hypo/hyperparathyroidism, tuberculosis, anorexia nervosa, diabetic ketoacidosis, prostate cancer, syndrome of inappropriate secretion of antidiuretic hormone (SIADH), etc. Accordingly, continuous observing of these electrolytes in biofluids is significant in the diagnosis of various dangerous diseases. So, various wearable gadgets including skin patches, wristwatches, and chest bands capable of detecting these electrolytes have been developed [62, 63]. The most widely recognized analytes are pH and various clinically important ions, involving a third of all wearable electrochemical sensors. Furthermore, healthcare and medical applications represent nearly half of the whole market of wearable ion sensors. Therefore, the advancement of wearable sensors for detecting various electrolytes is fundamental in meeting the requirements of the healthcare system.

*11.2.3.3 Biosensors*

Biosensors (i.e. diagnostic gadgets) are mostly used to identify biological elements, for example, glucose, nucleic acid, lactate, enzymes, DNA, and antibodies etc, by measuring physiochemical parameters. As biological elements can demonstrate medical issues, biosensors are preferably utilized in the treatment of various medicinal problems. The wearable biosensors are capable of performing consistent and non-intrusive checking of the bio-analytes. These sensors utilize different detecting techniques like the field-effect transistor-based electrochemical technique, and optical methods [64, 65].

### 11.2.4 Multiplexed sensors

The advancements in wearable sensors have made conceivable the simultaneous sensing of different signals, for example, temperature, pressure, strain, various chemicals, and the existence of ions. So, as the result of these advancements in research, the multiplexed sensors having the capability of procuring and analyzing more data simultaneously, from the environment using different signal sensors have been fabricated. A wrist-band that consolidates a flexible printed circuit board with an adaptable multiplexed sensor has been fabricated [66]. Here, polyethylene terephthalate (PET) is used as a flexible substrate to establish a stable contact between the skin and sensor. An array of signals through human skin can be detected by using this multiplexed sensor. For example, excessive loss of sodium and potassium through the skin from sweating or burns can cause hyponatremia, muscle spasms, hypokalaemia, and diarrhea [67]; glucose in sweat has been testified to be metabolically connected with serum glucose [68]; lactate in sweat can probably function as a sensitive marker of pressure ischemia [69], and treatment of skin injuries, for example pressure ulcers, using temperature as an informative parameter [70].

### 11.2.5 Wireless sensors

Remote frameworks on wearable sensors have been concentrated to remotely work the gadgets so as to convey detecting signals to the outer detector, which prompts constant observing of various physiological parameters. For remote sensing, sensors that join resonant antennas into resistor, inductor, and capacitor (RLC) circuits have been built up [25]. Since the resonance frequency $\{f_r=1/2\pi\sqrt{(LC)}\}$ empowers the capacitive and inductive reactance to offset one another and remotely conveys the greatest measure of energy. As the resonant frequency of the antenna coil depends on physical parameters $L$ and $C$, so any changes in this frequency exhibit the changes in passive components of circuits. All these facts facilitate the miniaturization of wearable devices that can be operated wirelessly without any battery and attached to human skin.

*11.2.5.1 Bluetooth integrated sensors*

The reasons for the considerations of Bluetooth technology in the wireless wearable technology are: (1) utilization at low cost with less power; (2) transmitted data can be separated over numerous frequencies; (3) ability of Bluetooth signal to transmit through barriers like walls, so no requirement of visually checking the status of connections. Basically, the short-range Bluetooth act as a radio-frequency dependent industry standard for the interoperation of smart gadgets. A wearable multi-analytes sensor (that empowers the sensing of various important electrolytes, for example, $Na^+$ and $K^+$ ions, lactate and glucose) incorporated with the Bluetooth module has been fabricated to remotely screen the physiological data enclosed in an individual's perspiration [66].

*11.2.5.2 Near field communication (NFC) based integrated sensors*

NFC—a contact-free wireless communication with operation range near to 20 cm—has been encouraging the simultaneous exchange of both information and power for many years. This wireless technology in the field of sensors has the potential to become an advanced communication technology for wearable gadgets as a result of: its compatibility with portable electronic gadgets; reasonably high security because of short operational range; chances of scaling down of the chip; and inactive mode operation without any battery [1]. The NFC associated with wearable sensors can be operated in reader as well as writer mode. Firstly, a smartphone starts communication as a functioning device that is capable of doing both functions (reading and writing) to an NFC tag that comprises an antenna, a sensor, a chip, and other required components. With the advancement in flexible and adaptable packaging innovation, NFC-empowered sensors in a wearable form have been developed by improving the mechanical flexibility of the whole system [71–73]. At the point when attached to an individual's body, such sensors facilitate the consistent observation of the physiological parameters and parallelly computerize the sensing output in portable electronics. A wireless and battery-free optoelectronic framework has been revealed by Kim *et al* [71]. This group developed a photodevice to study the optical properties of the skin of a human, and they used multicolored LEDs which can emanate various wavelengths contingent upon the reflectance of the focused color to show the sensing results. To get the conformal contact on the uneven part of the body, the devices externally connected to the NFC chip were incorporated by producing well designed interlinks in the elastomer. This NFC innovation is capable of delivering the power to the gadgets with the help of *LC*-coupling, and the extracted output is exhibited on the screen of the smartphone.

## 11.3 Challenges in wearable chemical sensors and possible solutions

### 11.3.1 Materials-based challenges with possible solution

Wearable chemical sensors require soft and stretchable sensing materials which are able to comply with the soft and curvilinear human tissues that undergo regular multi-axle deformations, thus both sensing material and tissues must have similar mechanical properties so that a conformal contact between sensor and skin can be established [74, 75] with no interruption to the sensor response. Furthermore, in the case of implantable chemical sensors, the gadgets are required to be carefully removed after their use. But the removal process prompts unwanted extra steps.

Also, few complications may happen if the organs/tissues are harmed when the gadgets are removed or a case may happen wherein a few parts of the device are not removable. Such cases of the unintended *in vivo* breaking of the embedded gadget, causing fragments of the wrecked pieces harming nearby organs are additionally conceivable. This problem can be resolved by developing self-destroying and invisible devices that perform the essential functions and later on spontaneously deteriorate into biological components. For instance, such 'place-and-overlook' sensors will play a significant role, particularly in the case of implantable gadgets that currently demand extra careful intercession for detaching the gadget after its goal is accomplished. One can easily put on these sensors to the target tissue without the requirement of any reminder to detach the gadgets used in different healthcare services. Smart electronics with these interesting capacities have been established [76].

### 11.3.2 Operational challenges and possible solutions

With the above challenges, the chemical sensors also have some other issues obstructing the simple revolution of conventional sensors to their wearable configurations.

#### *11.3.2.1 Pre-treatment and special conditions*

The calibration of some sensors before any utilization and incubation in conditioning solutions is mandatory. These prerequisites are inconsistent with wearable sensors. Also, numerous types of sensors have drift problems resulting in significant errors in the computation of the concentration of analytes. So, sensors should be re-calibrated at regular interims. Researchers are tackling the issue of calibration by producing calibration-free sensors [77, 78]. Parallel challenges for the sensors operating on the skin are that some of them require pre-treatment of sample solution before detection. Also, few sensors do their function only under extreme conditions, for instance, metal oxide-based gas sensors sense hazardous gases at temperatures higher than room temperature. Therefore, these sensors can't be used to detect analytes directly under the same conditions felt by the wearer.

#### *11.3.2.2 Stability of the sensor*

The most significant issue encountered by most of the chemical sensors, and especially wearable sensors is stability. Biosensors depend on biological elements, so the stability of the receptors shows a significant issue of long-term storage and use of the devices. The biological elements, like catalysts, aptamers, antibodies, and DNA have very high sensitivity to their local conditions [79–81]. Other parameters like pressure, temperature, humidity, pH, ionic strength have appeared to extraordinarily influence these biomolecules. The non-ideal operating conditions can prompt the denaturation of the bioreceptors, in this manner diminishing their sensitivity. The biological recognition components must be highly stable, especially in the case of wearable sensors, since (far apart from properly regulated lab testing) wearable sensors are operated in random conditions like changing pH, temperature, concentrations of ions, pressure, humidity etc, resulting from persistent indoor and outdoor movements of the wearer. Nowadays, keeping up a stable response in random situations for long periods is incredibly burdensome. The stability of the sensor can be improved by incorporating stabilizing agents or by supporting a biocompatible environment to the bio-receptors.

#### *11.3.2.3 Sensitivity and detection limits*

High sensitivity and low detection limits are required in different applications of wearable chemical sensors, for example, the concentration of glucose in tears and in intermediary fluids extracted from the skin is two orders less than its value in blood. Low limit of detection is particularly essential in environmental and defence applications where harmful toxins, chemical, and biological warfare agents, explosives, heavy metals are required to be sensed at very low levels with high accuracy and high sensitivity. Currently, best in class incorporated benchtop analysers that meet these prerequisites are accessible [82, 83]. The most appealing part of wearable innovation is the persistent observation of various analytes. Subsequently, wearable analyte sensors should sense analytes quickly, with low time to response relating to the randomly varying concentration of the analyte. This necessity mandates that wearable sensors should have a reversible response without any reminder. In recent years, numerous protocols like incorporating nanomaterials, establishing recycling of target, and various protocols for amplification of the sensing signals have been developed to enhance the sensitivity and limit of detection of a chemical sensor.

#### *11.3.2.4 Bio-affinity sensing*

Plenty of essential chemical sensors having huge importance in the business of wearable sensors command bio-affinity examination. Thus, the detection of various physiologically important biomarkers relies upon bio-affinity based sensors [84, 85]. Also, the environmental sensors for recognizing hazardous chemicals and biological warfare agents require bio-affinity receptors. The improvement of reliable wearable bio-proclivity sensors can introduce another and significant rostrum in the IoT field for crucial disease treatment, environmental monitoring, and security. But, the realization of bio-proclivity wearable sensors is tough because of their moderate response, analytical prerequisites, and restricted re-ease of use, and henceforth such wearable gadgets have not yet been illustrated. First off, bio-proclivity receptors, for example, antibodies, DNA, RNA, PNA, and aptamers, are very liable to change and denature rather immediately when used for long periods in the conditions necessary for wearable applications. The bio-receptors depending on the lock-and-key system used to distinguish the analyte and every minute variation in the 3D structure of receptors enormously influence the capacity of perceiving the analyte, and in turn, decrease the gadget's detecting performance.

#### *11.3.2.5 Multi-analyte sensing*

In numerous situations, one must screen a few parameters parallelly so as to get the complete idea regarding the wearer or their ambiance. For instance, one must recognize an entire host of chemicals as well as physical parameters at the same time for complete profiling of an individual's health status. Also, a variety of chemicals are mandated to identify simultaneously for environmental monitoring and security purposes in industrial areas. A multiplexed wearable sensor consists of a bunch of various chemical sensors integrated on a single microchip. In such multi-analyte sensor arrays, there are more chances of insignificant cross-communication among the individual sensors. The independent status of individual sensors in such a multi-analyte sensor system can be obtained by utilizing explicit receptors, by electrically decoupling the working facts of every sensor's interface, and maintaining the spacing between the different sensors. This is the nanotechnology that has exposed the potential for creating miniaturized multi-analyte sensors with very low limits of detection [86]. Besides diminishing the dimension, analysts ought to dedicate their regard for incorporating multiple sensor modalities for recognizing multi-parameter detection; for instance, nowadays, most of the chemical sensors use only one transducer. But, if one wishes to create multi-chemical sensors, one should consolidate a few transducers on a single platform because the same transduction mechanism may not be reasonable for all different sensors to detect different pointed analytes. In comparison to a single transducer-based sensor, two totally unrelated transducers used for the detection of the same analyte estimate the concentration more precisely [22, 87].

## 11.4 Conclusion and future direction

The recent advancements and progress in the field of smart and wearable technologies are briefly discussed in this chapter. Form this discussion one may conclude that the smart and wearable sensors are used not only in a healthcare center for fast, smooth and continuous monitoring of health status but widely used in fitness tracking as well as to get the alert for various hazardous components that may be present in the working environment. To make available these sensors at an effective cost, several advancements in wearable technology have been made recently. Thus, wearable technology is showing a promising future with its applications in numerous fields like biomedical, the environment and industry etc.

## References


[1] Majumder S, Mondal T and Deen M J 2017 Wearable sensors for remote health monitoring *Sensors* **17** 130

[2] Chen J, Gu J, Zhang R, Mao Y and Tian S 2019 Freshness evaluation of three kinds of meats based on the electronic nose *Sensors* **19** 605

[3] Harrou F, Dairi A, Sun Y and Kadri F 2018 Detecting abnormal ozone measurements with a deep earning-based strategy *Inst. Electric. Electron. Eng. (IEEE) Sensors J.* **18** 7222–32

[4] Wang F, Lin W, Liu Z, Wu S and Qiu X 2017 Pipeline leak detection by using time-domain statistical features *Inst. Electric. Electron. Eng. (IEEE) Sensors J.* **17** 6431–42

[5] Bhattacharyya T K, Sen S, Mandal D and Lahiri S K 2006 Development of a wireless integrated toxic and explosive MEMS based gas sensor *Proc. IEEE Int. Conf. on VLSI Designs (Hyderabad, India)* 721–4

[6] Rantala J, Hännikäinen J and Vanhala J 2011 Fiber optic sensors for wearable applications *Personal Ubiquitous Comput.* **15** 85–96

[7] Sun Q, Wo J, Wang H and Liu D 2014 Ultra-short DBR fiber laser based sensor for arterial pulse monitoring *23rd Int. Conf. Opt. Fibre Sensors* **9157** 91572K

[8] Rothmaier M, Selm B, Spichtig S, Haensse D and Wolf M 2008 Photonic textiles for pulse oximetry *Opt. Express* **16** 448–59

[9] Augousti A T, Maletras F X and Mason J 2005 The use of a figure-of-eight coil for fibre optic respiratory plethysmography: geometrical analysis and experimental characterisation *Opt. Fiber Technol.* **11** 346–60

[10] Augousti A T, Maletras F X and Mason J 2005 Improved fibre optic respiratory monitoring using a figure-of-eight coil *Physiol. Meas.* **26** 585–90

[11] Liu H Y, Liu H B and Peng G D 2005 Tensile strain characterization of polymer optical fibre Bragg gratings *Opt. Commun.* **251** 37–43

[12] Morris D, Coyle S, Wu Y, Lau K T, Wallace G and Diamond D 2009 Bio-sensing textile based patch with integrated optical detection system for sweat monitoring *Sens. Actuators* B **139** 231–6

[13] Patterson M J, Galloway S D R and Nimmo M A 2000 Variations in regional sweat composition in normal human males *Exp. Physiol.* **85** 869–75

[14] Shir D, Ballard Z S and Ozcan A 2016 Flexible plasmonic sensors *Inst. Electric. Electron. Eng. (IEEE)Sensors J. Select. Topics Quantum Electron.* **22** 4600509

[15] Krishnamoorthy S 2015 Nanostructured sensors for biomedical applications-a current perspective *Curr. Opin. Biotechnol.* **34** 118–24

[16] Barrios C A *et al* 2015 Aluminum nanoholes for optical biosensing *Biosensors* **5** 417–31

[17] Morales-narváez E, Golmohammadi H, Naghdi T, Yousefi H, Kostiv U, Horák D, Pourreza N and Merkoçi A 2015 Nanopaper as an optical sensing platform *ACS Nano* **9** 7296–305

[18] Salinas-castillo A, Ortiz-Gómez I, Cánovas R, Salinas-Castillo A, Cuartero M and Crespo G A 2019 Wearable potentiometric ion patch for on-body electrolyte monitoring in sweat: toward a validation strategy to ensure physiological relevance *Anal. Chem.* **91** 8644–51

[19] Anzenbacher P, Li F and Palacios M A 2012 Toward wearable sensors: fluorescent attoreactor mats as optically encoded cross-reactive sensor arrays *Angew. Chem. – Int. Ed.* **51** 2345–8

[20] Prodi L *et al* 2001 Dansylated polyamines as fluorescent sensors for metal ions: photophysical properties and stability of copper(II) complexes in solution *Helv. Chim. Acta* **84** 690–706

[21] Ma R, Hewitt R, Silvernail J, Urbanik K, Hack M and Brown J J 2008 Flexible active-matrix OLED displays: challenges and progress *J. Soc. Inf. Disp.* **16** 169–75

[22] Lochner C M, Khan Y, Pierre A and Arias A C 2014 All-organic optoelectronic sensor for pulse oximetry *Nat. Commun.* **5** 5–6

[23] Khan Y, Han D, Ting J, Ahmed M, Nagisetty R and Arias A C 2019 Organic Multi-Channel Optoelectronic Sensors for Wearable Health Monitoring *IEEE ACCESS.* **7** 128114–24

[24] Dagdeviren C *et al* 2014 Conformable amplified lead zirconate titanate sensors with enhanced piezoelectric response for cutaneous pressure monitoring *Nat. Commun.* **5** 4496

[25] Chen L Y *et al* 2014 Continuous wireless pressure monitoring and mapping with ultra-small passive sensors for health monitoring and critical care *Nat. Commun.* **5** 5028

[26] Jung S *et al* 2014 Reverse-micelle-induced porous pressure-sensitive rubber for wearable human-machine interfaces *Adv. Mater.* **26** 4825–30

[27] Yeo W H *et al* 2013 Multifunctional epidermal electronics printed directly onto the skin *Adv. Mater.* **25** 2773–8

[28] Park J *et al* 2014 Tactile-direction-sensitive and stretchable electronic skins based on human-skin-inspired interlocked microstructures *Am. Chem. Soc. Nano* **8** 12020–9

[29] Park H, Kim Y, Jung E S and Kwon S 2011 Implantable hybrid chrome silicide temperature sensor for power MEMS devices *Micro Nano Lett.* **6** 895–9

[30] Gao L *et al* 2014 Epidermal photonic devices for quantitative imaging of temperature and thermal transport characteristics of the skin *Nat. Commun.* **5** 4938

[31] Ploss B and Domig A 1994 Static and dynamic pyroelectric properties of PVDF *Ferroelectrics* **159** 263–8

[32] Broadhurst M G, Davis G T, McKinney J E and Collins R E 1978 Piezoelectricity and pyroelectricity in polyvinylidene fluoride—a model *J. Appl. Phys.* **49** 4992–7

[33] Park J, Kim M, Lee Y, Lee H S and Ko H 2015 Nanomaterials: fingertip skin-inspired microstructured ferroelectric skins discriminate static/dynamic pressure and temperature stimuli *Sci. Adv.* **1** e1500661

[34] Cosseddu P, Viola F, Lai S, Raffo L and Bonfiglio A 2014 A temperature transducer based on a low-voltage organic thin-film transistor detecting pyroelectric effect *Inst. Electric. Electron. Eng. (IEEE): Electron Device Lett.* **35** 1296–8

[35] Yu C, Wang Z, Yu H and Jiang H 2009 A stretchable temperature sensor based on elastically buckled thin film devices on elastomeric substrates *Appl. Phys. Lett.* **95** 2007–10

[36] Kim J *et al* 2017 Wearable smart sensor systems integrated on soft contact lenses for wireless ocular diagnostics *Nat. Commun.* **8** 14997

[37] Zirkl M *et al* 2011 An all-printed ferroelectric active matrix sensor network based on only five functional materials forming a touchless control interface *Adv. Mater.* **23** 2069–74

[38] Lee J S, Shin K Y, Cheong O J, Kim J H and Jang J 2015 Highly sensitive and multifunctional tactile sensor using free-standing ZnO/PVDF thin film with graphene electrodes for pressure and temperature monitoring *Sci. Rep.* **5** 1–8

[39] Persano L *et al* 2013 High performance piezoelectric devices based on aligned arrays of nanofibers of poly(vinylidenefluoride-co-trifluoroethylene) *Nat. Commun.* **4** 1610–33

[40] Trung T Q and Lee N E 2016 Flexible and stretchable physical sensor integrated platforms for wearable human-activity monitoring and personal healthcare *Adv. Mater.* **28** 4338–72

[41] Smith C S 1954 Piezoresistance effect in germanium and silicon *Phys. Rev.* **94** 42–9

[42] Roh E, Hwang B U, Kim D, Kim B Y and Lee N E 2015 Stretchable, transparent, ultrasensitive, and patchable strain sensor for human-machine interfaces comprising a nanohybrid of carbon nanotubes and conductive elastomers *Am. Chem. Soc. Nano* **9** 6252–61

[43] Lu N, Lu C, Yang S and Rogers J 2012 Highly sensitive skin-mountable strain gauges based entirely on elastomers *Adv. Funct. Mater.* **22** 4044–50

[44] Yan C *et al* 2014 Highly stretchable piezoresistive graphene-nanocellulose nanopaper for strain sensors *Adv. Mater.* **26** 2022–7

[45] Lee T, Lee W, Kim S W, Kim J J and Kim B S 2016 Flexible textile strain wireless sensor functionalized with hybrid carbon nanomaterials supported ZnO nanowires with controlled aspect ratio *Adv. Funct. Mater.* **26** 6206–14

[46] Frutiger A *et al* 2015 Capacitive soft strain sensors via multicore-shell fiber printing *Adv. Mater.* **27** 2440–6

[47] Wu J M *et al* 2012 Ultrahigh sensitive piezotronic strain sensors based on a $ZnSnO_3$ nanowire/microwire *Am. Chem. Soc. Nano* **6** 4369–74

[48] Lee S *et al* 2013 Super-flexible nanogenerator for energy harvesting from gentle wind and as an active deformation sensor *Adv. Funct. Mater.* **23** 2445–9

[49] Lee S *et al* 2014 Ultrathin nanogenerators as self-powered/active skin sensors for tracking eye ball motion *Adv. Funct. Mater.* **24** 1163–8

[50] Zhong J *et al* 2015 Stretchable self-powered fiber-based strain sensor *Adv. Funct. Mater.* **25** 1798–803

[51] Lin P and Yan F 2012 Organic thin-film transistors for chemical and biological sensing *Adv. Mater.* **24** 34–51

[52] Sokolov A N, Roberts M E and Bao Z 2009 Fabrication of low-cost electronic biosensors *Mater. Today* **12** 12–20

[53] Roberts M E, Sokolov A N and Bao Z 2009 Material and device considerations for organic thin-film transistor sensors *J. Mater. Chem.* **19** 3351–63

[54] Oliver N S, Toumazou C, Cass A E G and Johnston D G 2009 Glucose sensors: a review of current and emerging technology *Diabet. Med.* **26** 197–210

[55] Feng S *et al* 2019 Review on smart gas sensing technology *Sensors* **19** 1–22

[56] Pirsa S 2017 Chemiresistive gas sensors based on conducting polymers *Mater. Sci. Eng. Conc., Methodol. Tools App.* **1–3** 543–74

[57] Watcharaphalakorn S, Ruangchuay L, Chotpattananont D, Sirivat D and Schwank J 2005 Polyaniline/polyimide blends as gas sensors and electrical conductivity response to $CO–N_2$ mixtures *Polym. Int.* **54** 1126–33

[58] Dixit V, Misra S C K and Sharma B S 2005 Carbon monoxide sensitivity of vacuum deposited polyaniline semiconducting thin films *Sens. Actuators* B **104** 90–3

[59] Crowley K *et al* 2008 Fabrication of an ammonia gas sensor using inkjet-printed polyaniline nanoparticles *Talanta* **77** 710–7

[60] Kumar R, Goel N, Hojamberdiev M and Kumar M 2020 Transition metal dichalcogenides-based flexible gas sensors *Sens. Actuators* A **303** 111875

[61] Zhao Y *et al* 2018 Low-temperature synthesis of 2D $MoS_2$ on a plastic substrate for a flexible gas sensor *Nanoscale* **10** 9338–45

[62] Bandodkar A J, Jia W, Yardimci C, Wang X, Ramirez J and Wang J 2015 Tattoo-based noninvasive glucose monitoring: a proof-of-concept study *Anal. Chem.* **87** 394–8

[63] Curto V F *et al* 2012 Real-time sweat pH monitoring based on a wearable chemical barcode micro-fluidic platform incorporating ionic liquids *Sens. Actuators* B **171–172** 1327–34

[64] Jung I Y, Kim J S, Choi B R, Lee K and Lee H 2017 Hydrogel based biosensors for *in vitro* diagnostics of biochemicals, proteins, and genes *Adv. Healthcare Mater.* **6** 1–19

[65] Kudo H, Sawada T, Kazawa E, Yoshida H, Iwasaki Y and Mitsubayashi K 2006 A flexible and wearable glucose sensor based on functional polymers with soft-MEMS techniques *Biosens. Bioelectron.* **22** 558–63

[66] Gao W *et al* 2016 Fully integrated wearable sensor arrays for multiplexed *in situ* perspiration analysis *Nature* **529** 509–14

[67] Speedy D B, Noakes T D and Schneider C 2001 Exercise-associated hyponatremia: a review *Emerg. Med. J.* **13** 17–27

[68] Talary M S, Dewarrat F, Huber D and Caduff A 2007 *In vivo* life sign application of dielectric spectroscopy and non-invasive glucose monitoring *J. Non Crystal. Solids* **353** 4515–7

[69] Derbyshire P J, Barr H, Davis F and Higson S P J 2012 Lactate in human sweat: a critical review of research to the present day *J. Physiol. Sci.* **62** 429–40

[70] Webb R C *et al* 2013 Ultrathin conformal devices for precise and continuous thermal characterization of human skin *Nat. Mater.* **12** 938–44

[71] Kim J *et al* 2016 Battery-free, stretchable optoelectronic systems for wireless optical characterization of the skin *Sci. Adv.* **2** e1600418

[72] Lee J W *et al* 2016 Soft, thin skin-mounted power management systems and their use in wireless thermography *Proc. Natl. Acad. Sci. USA* **113** 6131–6

[73] Vlahopoulos S A *et al* 2017 Dynamic aberrant NF-κB spurs tumorigenesis: a new model encompassing the microenvironment Spiros *Cytokin Growth Factor Rev.* **26** 389–403

[74] Parrilla M, Ferré J, Guinovart T and Andrade F G 2016 Wearable potentiometric sensors based on commercial carbon fibres for monitoring sodium in sweat *Electroanalysis* **28** 1267–75

[75] Jia W *et al* 2013 Electrochemical tattoo biosensors for real-time noninvasive lactate monitoring in human perspiration *Anal. Chem.* **85** 6553–60

[76] Hwang S W *et al* 2015 Biodegradable elastomers and silicon nanomembranes/nanoribbons for stretchable, transient electronics, and biosensors *Nano Lett.* **15** 2801–8

[77] Zou X U, Zhen X V, Cheong J H and Bühlmann P 2014 Calibration-free ionophore-based ion-selective electrodes with a Co(II)/Co(III) redox couple-based solid contact *Anal. Chem.* **86** 8687–92

[78] Cisternas R, Kahlert H, Scholz F and Wulff H 2015 Direct contact tungsten bronze electrodes for calibration-free potentiometric pH measurements *Electrochem. Commun.* **60** 17–20

[79] Chen C *et al* 2013 Recent advances in electrochemical glucose biosensors: a review *RSC Adv.* **3** 4473–91

[80] Gaylord S T, Dinh T L, Goldman E R, Anderson G P, Ngan K C and Walt D R 2015 Ultrasensitive detection of ricin toxin in multiple sample matrixes using single-domain antibodies *Anal. Chem.* **87** 6570–7

[81] Bankar S B, Bule M V, Singhal R S and Ananthanarayan L 2009 Glucose oxidase—an overview *Biotechnol. Adv.* **27** 489–501

[82] Baker L B, Ungaro C T, Barnes K A, Nuccio R P, Reimel A J and Stofan J R 2014 Validity and reliability of a field technique for sweat $Na^+$ and $K^+$ analysis during exercise in a hot-humid environment *Physiol. Rep.* **2** e12007

[83] Leickly E *et al* 2015 High levels of agreement between clinic-based ethyl glucuronide (EtG) immunoassays and laboratory-based mass spectrometry *Am. J. Drug Alcohol Abuse* **41** 246–50

[84] Kadimisetty K *et al* 2015 Automated multiplexed ecl immunoarrays for cancer biomarker proteins *Anal. Chem.* **87** 4472–8

[85] Skládal P 1997 Advances in electrochemical immunosensors *Electroanalysis* **9** 737–45

[86] Zheng G, Patolsky F, Cui Y, Wang W U and Lieber C M 2005 Multiplexed electrical detection of cancer markers with nanowire sensor arrays *Nat. Biotechnol.* **23** 1294–301

[87] Wang J and Pumera M 2002 Dual conductivity detection system for microchip capillary electrophoresis *Anal. Chem.* **74** 5919–23

**IOP** Publishing

Advances in Modern Sensors
Physics, design, simulation and applications
**G R Sinha**

# Chapter 12

# Design and implementation of a wearable gaze tracking device with near-infrared and visible-light image sensors

**Chih-Peng Fan**

In this chapter, a review of a proposed novel wearable gaze tracking device with near-infrared (NIR) and visible-light image sensors is given. For the NIR image sensor based wearable gaze tracking design, first of all, the computations of first-stage process, which include the process for pupil location, extracting the pupil's region of interest (ROI), the binarization process with the two-stage scheme in the pupil's ROI, and the fast scheme of ellipse fitting with the Random Sample Consensus algorithm, also called RANSAC, with binarized edge points, are described, and the proposed eye tracker device provides effective pupil center tracking performance. For the visible-light image sensor based wearable gaze tracking design, based on the fast ellipse fitting design with the RANSAC scheme and the gradients based iris center location, the proposed eye tracking device achieves effective iris center tracking capability. Besides, for the eye tracking design its deep-learning technology is also addressed. Next, the second-stage process is discussed for the function and process for calibration and the procedure of gaze tracks by the measured data from the inertial measurement unit (IMU) to compensate for the head movement effect. Finally, the experimental results and comparisons are demonstrated for performance comparisons.

## 12.1 Introduction

In order to evaluate attention disorders and people's eyesight, gaze tracking devices have been well utilized in the field [1]. In recent years, a lot of fresh consumer products have applied the gaze tracking technology embedded in their designs [2–19]. By using gaze tracking devices, the technology raises the function and application for intuitional human–computer interactions. Figure 12.1 depicts the appearance of a self-made wearable eye tracking device, and figure 12.2 also

doi:10.1088/978-0-7503-2707-7ch12

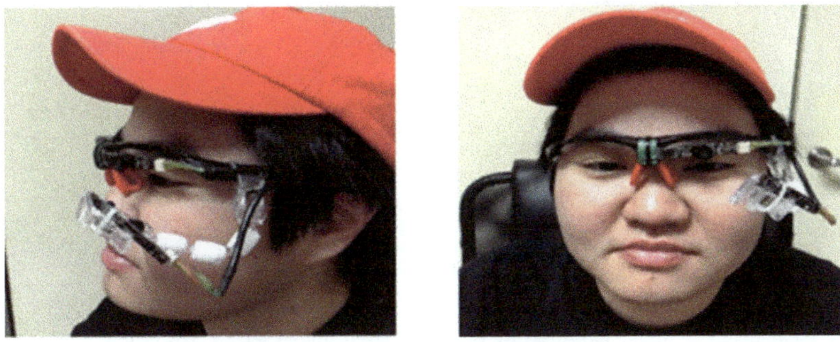

**Figure 12.1.** The appearance of wearable eye tracker.

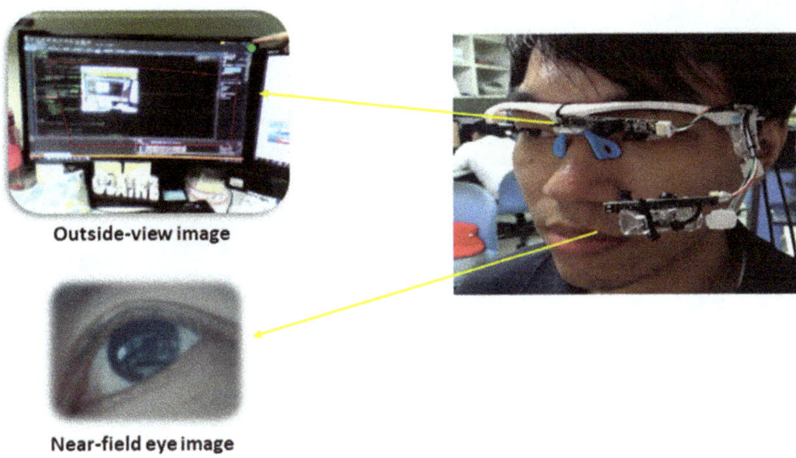


**Figure 12.2.** The outside-view scene image and near-field eye image used for the wearable device.

demonstrates the outward-view scene image and the near-field eye image, which are provided from the image sensors of the wearable eye tracker. Some previous eye tracking designs were based on high-contrast eye images, which were recorded by an active NIR image sensor. The gaze tracking designs, which used the NIR-based image sensor, were exploited in [7–14]. In [15], for the indoor application, the designer developed the visible-light image sensor based eye tracking device. In [13, 14], the wearable eye tracking devices, which used the NIR-based image sensor, displayed suitable gaze tracking capability. Although the previous NIR image sensor based designs became highly accurate, the NIR based image sensor is easily obstructed in the application field outdoors, especially under sunny conditions. Figure 12.3(a) shows the near-field eye image captured by an NIR-based image sensor, and figure 12.3(b) depicts the near-field eye image captured by a visible-light image sensor. The visible-light image sensor based methodology is adequate when the wearable eye tracking device is used properly for all of indoor and outdoor circumstances. The previous study in [15] showed it performed functionally indoors, but the device could not work well outdoors. Owning to the characteristic of the

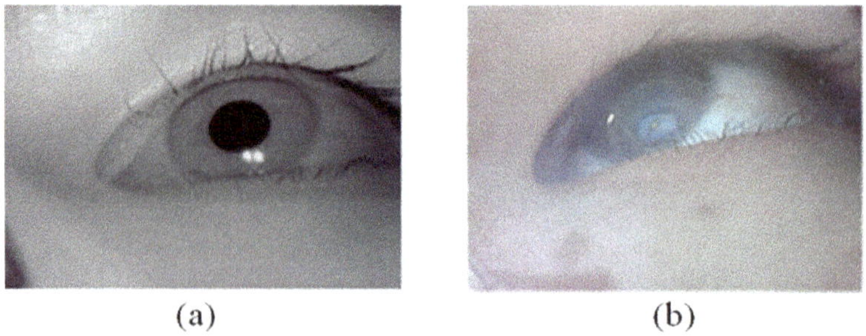

Figure 12.3. Near-field eye image captured by (a) NIR-based image sensor, (b) visible-light image sensor.

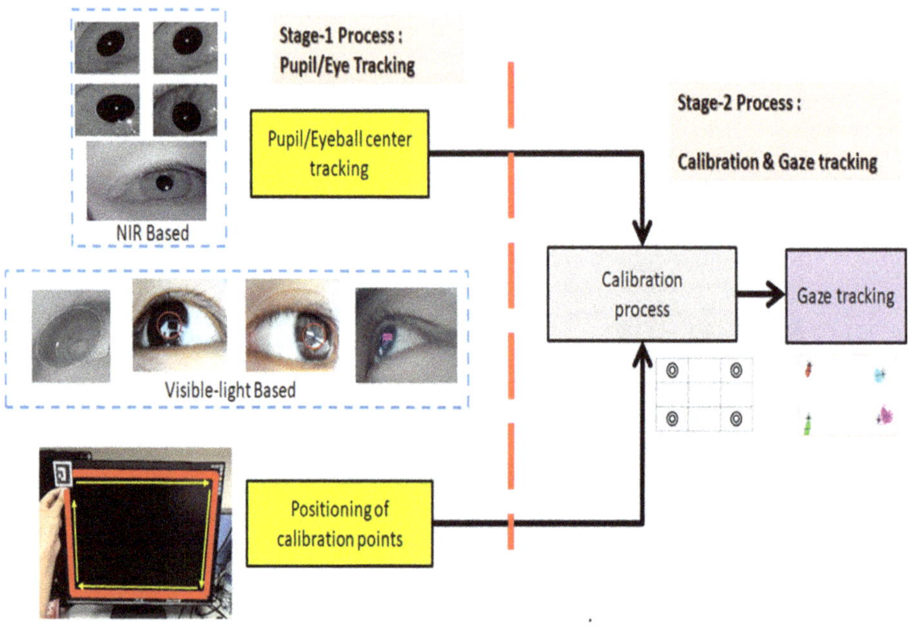


Figure 12.4. The two-stage processing flow for the design of wearable eye tracker.

chosen sensor module, the contrast of visible-light eye images may not be sufficient, and the near-field eye images may contain some random-type noises [16, 17]. In order to conquer the interferences and noises in images, the proposed visible-light eye tracker designs utilized gradients based [18] and deep-learning based [19] eye tracking technologies to solve the problem.

Figure 12.4 depicts the two-stage processing flow for the design of the proposed wearable eye tracker. The proposed wearable gaze tracking device is partitioned into two-stage computations. The first-stage operational procedure is focused on the pupil and iris center tracking technologies by using the NIR and visible-light based image sensors, respectively. Next, the second-stage process provides the function of calibration for the gaze tracking work.

The rest of this chapter is introduced as follows. In section 12.2, the first-stage computational methodologies of the developed wearable gaze tracking device are described, where the NIR image sensor based pupil center tracking design and the visible-light image sensor based iris center tracking design are depicted. Moreover, the deep-learning based iris center tracking technology is also discussed for the visible-light mode. Next, the second-stage processing flow is pointed out for the calibration and gaze tracking function by compensating for the head-movement influence. Section 12.3 shows the testing results and the comparisons. The conclusion and future work will be stated at the end of the chapter.

## 12.2 Proposed wearable gaze tracking design

The proposed wearable gaze tracking device is partitioned into two-stage computations. First, the first-stage operational procedure is focused on the pupil center and iris center tracking technologies by using the NIR and visible-light based image sensors, respectively. Secondly, the second-stage process, which provides the function of calibration for gaze tracking, is briefly discussed.

### 12.2.1 Near-infrared image sensor based wearable eye tracker design [13, 14]

Figure 12.5 depicts the proposed computational flowchart of NIR-based pupil center tracking design with fast ellipse fitting. By using the images from NIR-based image sensor, firstly the location of the eyeball center is enabled. In order to estimate the pupil region, the initial pupil's position will be predicted with the matching filter. The matching filter possesses the cross Haar feature to predict the approximate pupil center. Then the largest matching response position after the matching process is set as the possible center of pupil. For dropping down the computing operations, the center of pupil is estimated by the cross matching filter, and the pupil's region is located by an area with 160 × 160 pixels.

When the pupil's region of interest (ROI) is extracted, the pupil regional image may contain the dark pupil zone, the area of iris, and the inteferences from

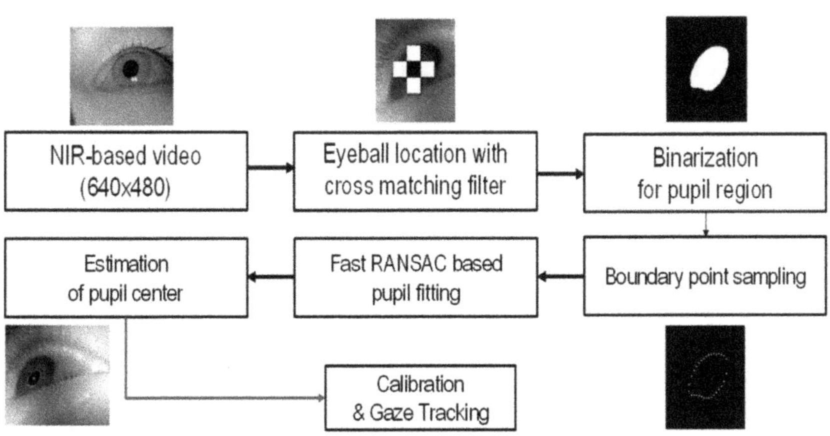


**Figure 12.5.** NIR-based pupil center tracking with fast ellipse fitting [13, 14].

eyelashes/eyelids. It is a problem that the designed system picking up the pupil region only depends on one threshold value for binarization. When a low value is assigned to the threshold, some useful information of the pupil may be cut and lost. If the threshold has a high value, several unnecessary inteferences around the eye area may destroy the pupil's binarization result. Therefore, the Otsu based [20] two-level binarization skill [14] is proposed and applied for the extraction of the pupil's zone. The design by utilizing adaptive values for threshold achieves more pupil region than that by using one fixed threshold value under the obstruction of eyelashes. By using the two-level binarization process [14], the second binarization threshold design better obtains the binarization pupil's ROI result than the the one-level binarization design in [13] with one-level adaptive binarization. Figure 12.6(a) illustrates the ellipse fitting result after one-level binarization, and figure 12.6(b) shows the ellipse fitting result after two-level binarization. After applying the proposed two-level adaptive binarization, the next-stage process for the pupil's ellipse fitting scheme will perform a better fitting result.

After the binarization process, several boundary pixels, which are located on the pupil boundary contour, are randomly selected for the fast ellipse fitting design with RANSAC [13, 14, 21]. When the ellipse model based on the pupil boundary pixels is well fitted, the fitted model will match the practical pupil's boundary. However, the matching accuracy of ellipse model by only one-stage RANSAC process may not be enough for use, then the fitted ellipse model from the first-stage estimation is utilized to remove unsuitable edge pixels at the boundary to shorten the processing time of the RANSAC process at the second stage. To enhance the accuracy of the fitting model and decrease the computational complexity, several edge pixels near the first-stage binarized ellipse will be selected together with using the radiation-like selection scheme in the ellipse fitting process at the second stage. Since the obtained ellipse at the first-stage process fits the pupil's boundary properly, the candidate edge pixels for pupil's boundary points have a high probability to match and fit a better ellipse model further. This means that the second-stage ellipse fitting model is quite similar to the practical boundary contour of the pupil. Then, the central position of the second-stage fitting ellipse pattern is predicted as the estimated central location of pupil. Figure 12.7(a) depicts the result after the boundary points sampling process, and figure 12.7(b) shows the result after the two-stage ellipse fitting scheme with

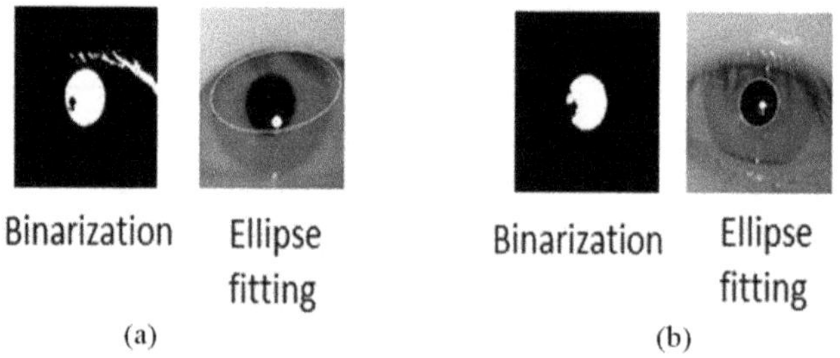


**Figure 12.6.** The results (a) after one-level binarization, (b) after two-level binarization.

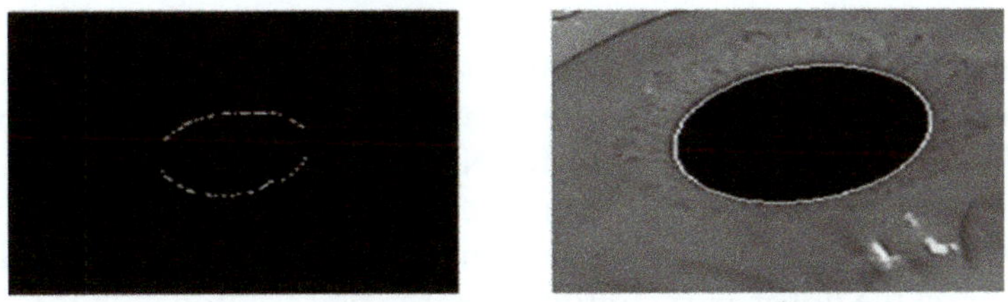

(a) (b)

**Figure 12.7.** The results (a) after the boundary points sampling, (b) after the RANSAC-based ellipse fitting.

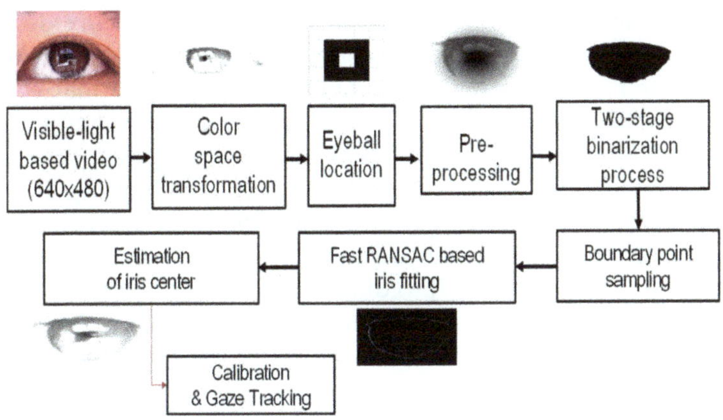


**Figure 12.8.** Visible-light based iris center tracking with fast ellipse fitting [17].

RANSAC. When the fast pupil ellipse fitting process with RANSAC is completed, the center position of the ellipse model is the estimated pupil's center.

### 12.2.2 Visible-light image sensor based wearable eye tracker design [17–19]

Figure 12.8 shows the visible-light based iris center tracking design with fast ellipse fitting technology. Firstly, the image enhancement with the color-space transformation will be active. In the expression of HSV color-space format, the 'V' means the brightness value of image, and the brightness values are independent of the change of colors. To decrease the color variation and interferences, the brightness vlaues of the visible-light based eye images are transformed and achieved from the RGB-format color images, and the image pixels with grey-level values are enhanced with a gain. After the dilation operations in eye images, the obstruction effect from both eyelashes and eyelids is reduced.

To recognize the ROI of the iris, the 2D filter with Haar characteristic is applied for detecting the iris position. In order to drop the computional complexity, the iris ROI is set to a zone with 280 × 280 pixels. Next, the proposed system will run the pre-processing operations in the ROI. To cut down the reflection effect on the iris zone and enhance the image contrast simultaneously, the design utilizes the weight mask for the pre-processing operations, and the interferences in the iris area are effectively reduced. By using the similar processes in [13, 14], when the two-level adaptive binarizations is done, several boundary pixels, which are located upon the estimated contour of iris boundary after binarization, are selected for the following fast iris ellipse fitting process with RANSAC. For the application of indoor conditions, half the candidate pixels are selected on the lower-part contour after the the binarization with the first-stage computation, and half the candidate pixels are picked out on the upper-part contour after the binarization with the second-stage process. When the fast iris ellipse fitting scheme with RANSAC is finished, the central position of the estimated ellipse contour can closely approach the estimated central location of iris zone.

Figure 12.9 depicts the visible-light based iris center tracking design with the gradients image process. Firstly, the proposed system enables the function of the pre-processing operations and iris region extraction. The RGB-format eye images are transformed into the grey-level format eye images, and the system uses the brightness information for image processing. Secondly, the proposed design utilizes the integral eye image to reduce the computations for matching by the Haar-based rectangular filter. Then the matching filter identifies the ROI of iris region for the successive processes. Next, for reflection reduction in the iris ROI, the design uses the similar weight-mask process [17] to diminish the effect of reflections on the zone of iris and also enhance the contrast of image. Thus, the interferences generated from the reflection are efficiently eliminated. Next, the gradients-based method is enabled for iris center location. To reduce the location error of iris zone, the interferences generated from the upper-part and lower-part eyelids also require to be eliminated. By using the dynamic threshold scheme, the interference of eyelids is filtered out effectively. The gradients in images, which are smaller than the threshold value, are removed. By using the gradients based estimation, the system detects the largest values from computing the inner products if the two gradient estimated vectors show as the similar trend. By the gradients property, the center position of iris zone is estimated correctly.

To intensify the value of contrast in ROI, the pixels, which are around the possible iris center and the region of iris, are enhanced with the factor for weighting.

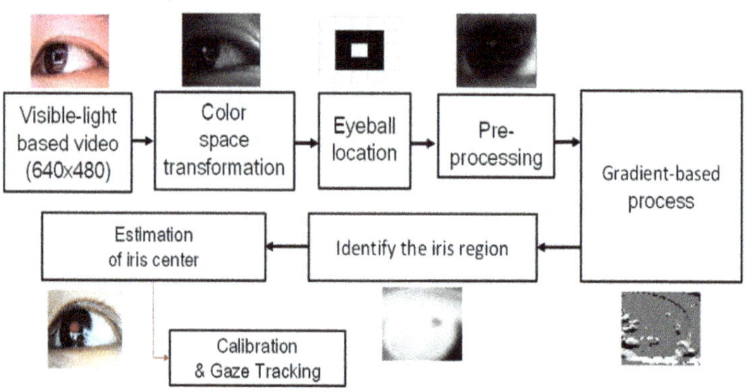


**Figure 12.9.** Visible-light mode based iris center tracking shceme with the gradient image process [18].

The applied weighting process refers the information from smoothing and reversing gray-level eye images, and then the iris ROI is smoothed by the Gaussian filter. By the weighting process, the reflective effect in the iris zone is decreased to raise the estimation accuracy of iris center. When the gradients based iris area recognition is finished, the system can estimate the correct iris center.

Figure 12.10 shows the applied methodology of the YOLOv3-tiny based iris center's tracking scheme at the visible-light mode [19]. In [19], the developed system uses the camera with the visible-light mode to obtain the near-field eye images, and then the proposed system utilizes the deep learning skill with the YOLOv3-tiny model to train the inference network and to achieve the iris center tracking function. The proposed system computes the coordinate of iris center by the infered bounding box around the pupil.

Why is the YOLOv3-tiny based deep learning network in [22] selected to develop the application? The reasons are: (1) it provides good performances of detection and identification at the same time; (2) it can compute much faster compared with other object-detection deep learning based models; and (3) the model of YOLOv3-tiny is suitable to detect small image objects because the pupil region is a small image object in the iris zone. The YOLOv3-tiny model has a small number of convolution layers, and the computational time is shortened for implementation on edge devices. The eye images are captured from eight users, and the tester utilizes the wearable eye tracking device to emulate the operations of gaze tracks, and all testers change the gaze points around the screen of the monitor, and then the near-field eye images are recorded with all probable locations of eyeball movements. The eye images without

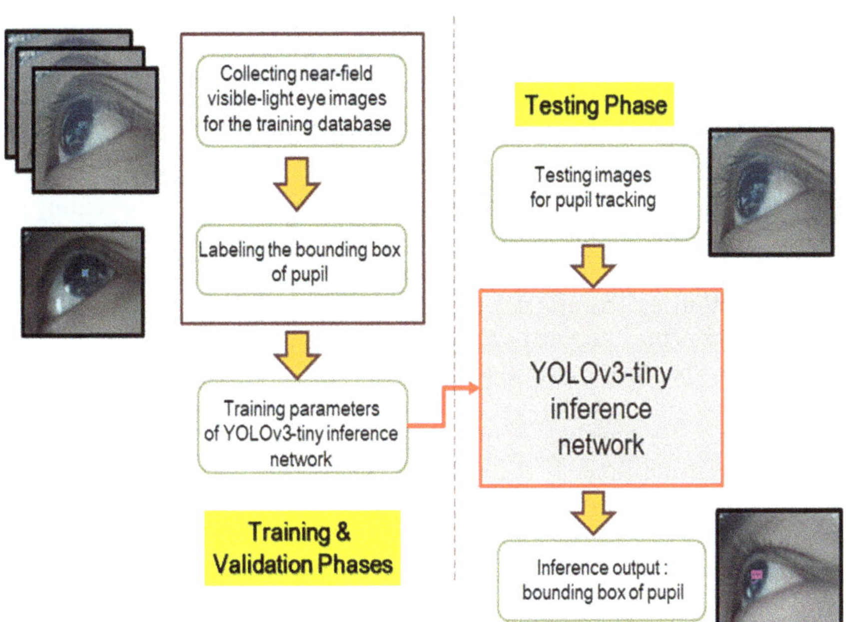


**Figure 12.10.** Visible-light pupil center tracking with YOLOv3-tiny deep-learing based classifier [19].

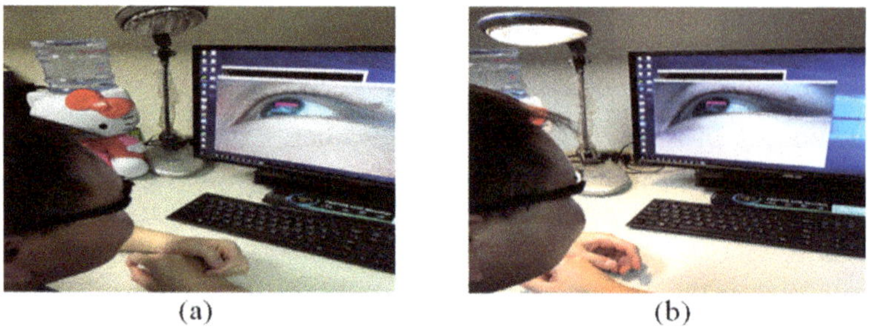

Figure 12.11. The visible-light pupil center tracking results by the YOLOv3-tiny based detector when the user (a) turns off the table lamp, (b) turns on the table lamp.

closed-eye conditions are chosen, and about 900 images per tester are utilized in the phases of training and testing processes. The overall number of the applied eye images in our dataset is 7094. When the labeling stage is active, initially the ellipse model fitting based skill in [14, 17] is applied to locate the possible center of iris previously, and the designer manually revises the coordinate of the iris center. Before the training process is active, the system designer divides all labeled near-field eye images into the training, validation, and test sections randomly, where the respective ratio is 8:1:1. Then the number of labeled images is 5676 for the training process, and the number of labeled images, which are used for the validation and the testing processes, is 709. Figure 12.11 shows the results of tracking pupil's centers with the visible-light mode by using the YOLOv3-tiny based detector. Before the deep learning model is well trained with labeled eye images, the parameters of YOLOv3-tiny network require to be set properly, and the inference model can be obtained to infer the bounding box for predicting the iris/pupil center.

### 12.2.3 Calibration and gaze tracking function for wearable eye tracking device

Before the gaze tracking function is performed, the calibration procedure needs to be processed in advance. Figure 12.12 depicts the applied calibration arrangement of the mapping process between two different coordinate systems. By calibration, the coefficients of the perspective matrix are computed to map out the pupil coordinates (i.e. centers of pupil or iris ellipse) to the scene coordinates with markers. In the calibration process, the outward camera records the four markers, and then the system calculates the center coordinates of markers. In order to drop the error of calibration, by the method in [23], the system compensates for the effect of head movement due to the parallax action between staring at the markers and head movement at the same time. If the primary process for calibration is finished, the function of gaze tracks by applying the compensation scheme for head motions is also active.

Figure 12.13 presents the utilized process of calibration by using the compensation scheme of head movement with an IMU component for the developed wearable gaze tracker system. The used perspective matrix 'T' plays the main role for mapping out the two coordinate systems from the pupil/iris center to the marker's center when

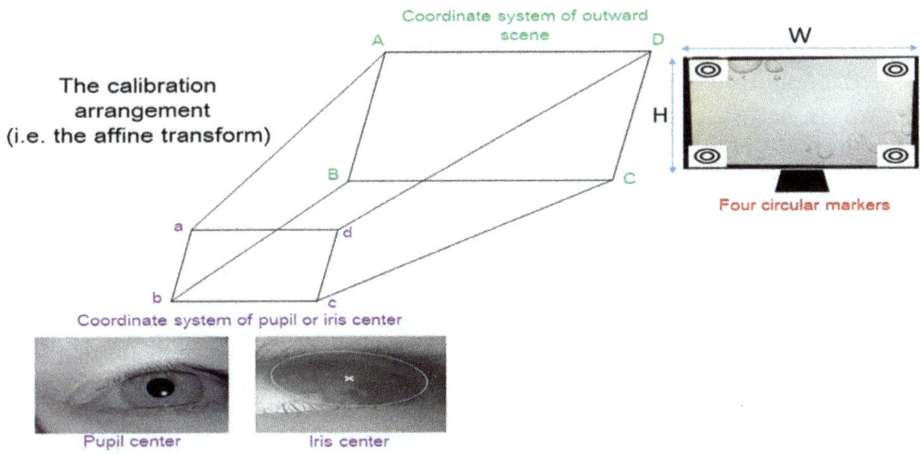


**Figure 12.12.** The calibration arrangement for mapping process between two cooordinate systems.

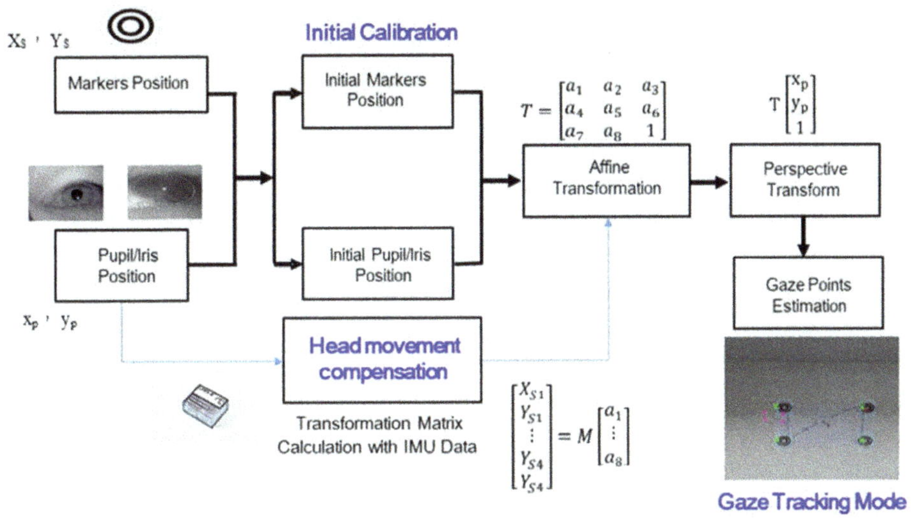


**Figure 12.13.** The applied calibration process by the head movement compensation with IMU.

the gaze tracking function is performed. In order to regulate the effect of head movement in the used perspective transform operations, the matrix for motion compensation, i.e. 'M', can be achieved by the data measured from the IMU device. Figure 12.13 reveals the adopted technology for the important calibration and the final gaze tracking function. In figure 12.13, $(X_S, Y_S)$ is the positions pf markers, and $(x_p, y_p)$ is the positions of estimated pupil/iris centers.

## 12.3 Experimental results and comparisons

In our tests, the near-field eye images are obtained by a dual-mode NIR and visible-light image sensor with resolution of 640 × 480 pixels, and wavelengths of the sensing images are ranged from the light source indoors to the light source of the NIR LED. When the predicted pupil/iris center positions are compared with the pre-estimated pupil/iris centers as the ground truth, table 12.1 lists the performance comparisons between two proposed pupil tracking designs with the NIR-based image sensor. In [14], the average tracking errors of the centers of pupil are fewer than 2 pixels value when the applied two-level binarization process and the pupil's ellipse fitting with the developed two-stage based scheme are used.

Table 12.2 lists the performance comparisons among four iris center tracking designs with the the visible-light based image sensor. Compared with the previous designs in [17] and [18], the developed YOLOv3-tiny based method [19] has fewer errors of iris central tracks in the different situations at visible-light mode. Among the four visible-light image sensor based designs in table 12.2, the YOLOv3-tiny based deep-learning classifier in [19] perfoms fewer average tracking errors for the prediction of the centers of the iris, which are fewer than the value of 4 pixels. In [19], the developed wearable eye tracking device includes two image sensing units to capture images. The camera unit of outward scene has the resolution with 1600 × 1200 pixels, and the near-field camera unit with the visible-light mode performs the resolution with 1280 × 720 pixels. Both of the two image sensing units can operate up to 30 frames per second In [19], the developed YOLOv3-tiny based inference network tracks the iris central location robustly, and the proposed design also correctly infers the eyeball's center at any position in eye images.

Figure 12.14 shows the installation of markers used for the training phase and the testing phase, respectively, and the four circle-shape markers are used for the

**Table 12.1.** Comparison of performances between two pupil's center tracking methods with the NIR mode image sensor.

| | Wu *et al* [14] | Cheng *et al* [13] |
|---|---|---|
| Image sensor mode | NIR | NIR |
| Core technology | Ellipse-fitting based | Ellipse-fitting based |
| Average tracking errors (pixels) | 1.69 | 4.15 |

**Table 12.2.** Performance comparisons among four proposed iris center tracking designs with the visible-light based image sensor.

| | Kuo *et al* [19] | Liu *et al* [18] | Liu *et al* [17] | Cheng *et al* [13] |
|---|---|---|---|---|
| Image sensor mode | Visible light mode | Visible light mode | Visible light mode | Visible light mode |
| Core technology | Deep learning based | Gradient-based | Ellipse-fitting based | Ellipse-fitting based |
| Average tracking errors (pixels) | 4 | 5.99 | 6.8 | >10 |

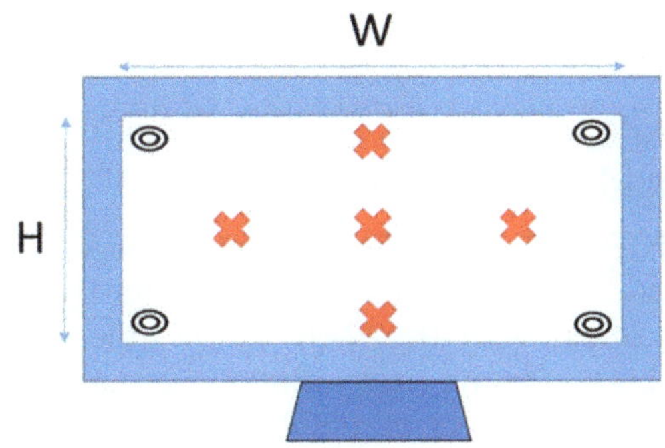


**Figure 12.14.** The locations of the applied markers for training and testing processes.

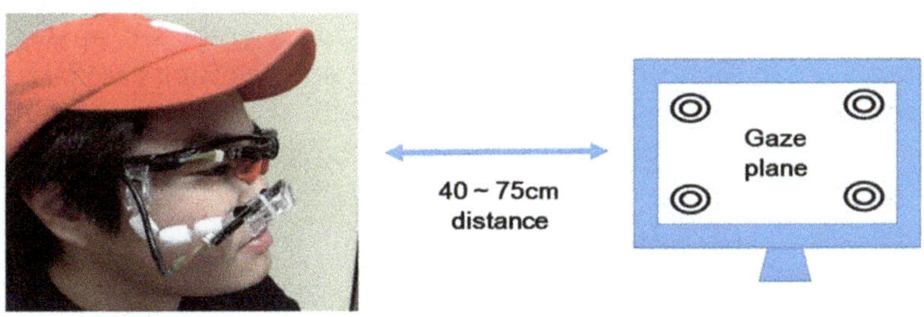


**Figure 12.15.** The setup and application of the wearable eye tracker.

**Table 12.3.** The gaze tracking error comparisons between two proposed eye tracker designs.

| | Liu *et al* [17] | Wu *et al* [16] |
|---|---|---|
| Image sensor mode | Visible-light | Visible-light |
| Head movement compensation | Yes | Yes |
| Average tracking errors at training mode (degrees) | 1.59 | 2.74 |
| Average tracking errors at testing mode (degrees) | 1.96 | N/A |

calibration process as the training phase, and the other five cross-shape markers are applied for the testing phase. Before the gaze tracking mode is enabled, the used calibration procedure only requires to be processed once, and the real-time function of wearable gaze trackers is not impacted by the calibration process. To calibrate the outward scene images for the developed gaze tracking device, in our experiments, the value of the width '*W*' is equivalent to the value of 450 pixels. Besides, the value of the height '*H*' is equivalent to the value of 280 pixels. Figure 12.15 depicts the setup and application circumstances of the wearable eye tracking device, where the application length of distance within the user and gaze plane with markers is ranged from 40 cm to 75 cm. Table 12.3 lists the gaze tracking error (i.e. degrees) comparisons between the two proposed gaze tracking designs. In table 12.3, at the testing phase, by the visible-light based image sensor with the compensation of head movement effect, the average gaze tracking errors of the design in [17] are less than 2.0 degrees.

## 12.4 Conclusion and future works

In this chapter, the NIR and visible-light image sensors based eye gaze tracking methodologies are introduced for wearable applications. By using the compensation scheme of head movement with the IMU device, the developed wearable eye tracking design achieves effective executing performance. In our studies, the averaged values of tracking errors of pupil centers are less than 2 pixels value when the NIR image sensing mode [14] is active, and the averaged values of tracking errors of iris central positions can be fewer than 4 pixels value when the visible-light image sensing mode [19] is enabled. By using the visible-light based image sensor for gaze tracking, the errors of gaze tracking values on average are smaller than 2.0 degrees value [17]. In future studies, the developed deep learning based iris and pupil center tracking technologies can be combined with the process of calibration to examine the performance of gaze tracks for the wearable device at visible-light conditions.

## Acknowledgments

This work was supported by Ministry of Science and Technology, Taiwan, R.O.C. under Grant MOST 104–2220-E-005–002, MOST 105–2221-E-005–078, and MOST 107–2218-E-005–014. The author also would like to thank Chun-Wen Cheng, Wei-Liang Ou, Jia-Hao Wu, Ting-Lun Liu, Chih-Hsuan Fang, and Tzu-Ling Kuo for providing the technical support.

## References

[1] Chennamma H R and Yuan X 2013 A survey on eye-gaze tracking techniques *Indian J. Comput. Sci. Eng. (IJCSE)* **4**(5) 388–93

[2] Li D, Babcock J and Parkhurst D J 2006 Open eyes: a low-cost head-mounted eye-tracking solution *ACM Eye Track. Res. App. Symp.* 95–100

[3] Drewes H 2010 Eye gaze tracking for human computer interaction *PhD thesis* (Media Informatics Group, LMU University of Munich)

[4] Lienhart R and Mayd J 2002 An extended set of haar-like features for rapid object detection *Int. Conf. Image Proc.* 900–3

[5] Świrski L, Bulling A and Dodgson N 2012 Robust real-time pupil tracking in highly off-axis images *ACM Proc. Symp. Eye Track. Res. App.* 173–6

[6] Chen Y and Su J 2012 Fast eye localization based on a new haar-like feature *10th World Congr. Intell. Contr. Auto. (WCICA)* 4825–30

[7] Ohno T, Mukawa N and Yoshikawa A 2002 FreeGaze: a gaze tracking system for everyday gaze interaction *Eye Track. Res. App. Symp.* 125–32

[8] Morimoto C H and Mimica M R M 2005 Eye gaze tracking techniques for interactive applications *Comput. Vis. Image Understand.* **98** 4–24

[9] Li D, Winfield D and Parkhurst D J 2005 Starburst: a hybrid algorithm for video-based eye tracking combining feature-based and model-based approaches *IEEE Comput. Vis. Pattern Recogn.—Workshops* 1–8

[10] Sugita T, Suzuki S, Kolodko J and Igarashi H 2007 Development of head-mounted eye tracking system achieving environmental recognition ability *SICE Annu. Conf.* 1887–91

[11] Chen S and Julien J E 2014 Efficient and robust pupil size and blink estimation from near-field video sequences for human-machine interaction *IEEE Trans. Cybernet.* **44** 2356–67

[12] Zhang X, Sugano Y, Fritz M and Bulling A 2015 Appearance-based gaze estimation in the wild *IEEE Conf. Comput. Vis. Pattern Recogn.* 1–10

[13] Cheng C W, Ou W L and Fan C P 2016 Fast ellipse fitting based pupil tracking design for human-computer interaction applications *IEEE Int. Conf. Consumer Electron. (ICCE 2016)* 445–6

[14] Wu J H, Ou W L and Fan C P 2017 NIR-based gaze tracking with fast pupil ellipse fitting for real-time wearable eye trackers *IEEE Conf. Depend. Sec. Comput.* 93–7

[15] Kao W C, Chang W T, Wu S J, Liu C H and Yin S Y 2013 High speed gaze tracking with visible light *Int. Conf. Syst. Sci. Eng. (ICSSE)* 285–8

[16] Wu J H, Ou W L and Fan C P 2016 Fast Iris ellipse fitting based gaze tracking with visible light for real-time wearable eye trackers *5th IEEE Global Conf. Consumer Electron. (GCCE 2016)* 1–2

[17] Liu T L and Fan C P 2017 Visible-light based gaze tracking with image enhancement pre-processing for wearable eye trackers *6th IEEE Global Conf. Consumer Electron. (GCCE 2017)* 1–2

[18] Liu T L and Fan C P 2018 Visible-light wearable eye gaze tracking by gradients-based eye center location and head movement compensation with IMU *Int. Conf. Consumer Electron. (ICCE 2018)* 1–2

[19] Kuo T L and Fan C P 2020 Design and implementation of deep learning based pupil tracking technology for application of visible-light wearable eye tracker *IEEE Int. Conf. Consumer Electron. (ICCE 2020)* 1–2

[20] Bangare S L, Dubal A, Bangare P S and Patil S T 2015 Reviewing Otsu's method for image thresholding *Int. J. Appl. Eng. Res* **10**(9) 21777–83

[21] Fischler M A and Bolles R C 1981 Random sample consensus: a paradigm for model fitting with applications to image analysis and automated cartography *Commun. ACM* **24**(6) 381–95

[22] Joseph R and Farhadi A 2018 YOLOv3: an incremental improvement arXiv: 1804.02767

[23] Kocejko T, Bujnowski A, Ruminski J, Bylinska E and Wtorek J 2014 Head movement compensation algorithm in multidisplay communication by gaze *7th Int. Conf. Human Syst. Interact. (HSI)* 88–94

**IOP** Publishing

Advances in Modern Sensors
Physics, design, simulation and applications
**G R Sinha**

# Chapter 13

# Vibration powered wireless sensor networks—harvesting energy from good vibrations

**K P Venkat Vivek, Abhishek Choubey and Shruti Bhargava Choubey**

We use a vast number of batteries. Most of the batteries we use in daily life are not rechargeable and cause environmental pollution. Wireless sensors should be made in such a way that they should survive for longer periods of time without any service and should have the capability to survive on alternative methods of obtaining the energy, for example any renewable source. Wireless sensor network (WSNs) refers to a group of spatially dispersed and dedicated sensors for monitoring and recording the physical conditions of the environment and organizing the collected data at a central location. WSNs measure environmental conditions like temperature, sound, pollution levels, humidity, wind, and so on. The modern approach is to use vibrations present in the environment for energy harvesting. Trees, being the astounding living beings, are known to carry a unique frequency of energy, or vibration (about 100 Hz). As can be seen in our modern world, there is dynamic change. But if we look at the general topic from a wide angle, we can see that all electronic devices have one conundrum in common: energy. Therefore, harvesting energy in this way can provide a lot of potential for powering small devices.

## 13.1 Introduction

Wireless sensor networks have the capability to analyse the environmental parameters like intensity of sound, levels of pollutants, temperature of surroundings, wind speed, humidity, and many more by means of spatially dispersed and allocated sensors which are essentially used for observing and retrieving the said parameters of any place and organizing the traversed information at a centralized node. On a broader scale, any sensor is a module, or sub-system that is intended to detect any variation in its environment and communicate that information with another electronic device, generally a computer processor. In order to get electricity from the environmental ambient energy, a piezoelectric cantilever is used which is

doi:10.1088/978-0-7503-2707-7ch13

specifically shaped to detect any vibrations that are in the frequency range from 50 Hz to 200 Hz. A power-module is used to collect the electrical energy, store it and deliver it to the electronic devices. A microcontroller is used so that the variations of the environment in the form of signals are retrieved from a sensor and then to perform analysis, converts that signal to an understandable form which can be communicated with the help of a radio link. Finally, this link in the RF range is used so that the processor's information is sent to the host receiver, which can also be termed as a data collection point [1–5]. Normally, WSN comprises 'nodes' ranging from a small number to several, where every node is joined to any one of the sensors and they are in a topology—mesh, star, cluster etc. WSNs are usually deployed in several cities to monitor the environmental conditions.

## 13.2 literature survey

### 13.2.1 Piezoelectric sensors

An element called transduction element which is composed of material made from piezoelectric substances is generally signalized in piezoelectric sensors. To obtain an output signal, there is no need of external energy in the case of active sensors, here piezoelectric sensors. Typically, many passive sensors show a difference in the transduction element and the sensing element, but not in piezoelectric sensors. The measured quantity is directly transmitted on to the said element in a piezoelectric sensor designed to produce charge under mechanical stress. This element produces a comparable charge of electricity as yield which is induced by mechanical stress through the piezoelectric effect.

### 13.2.2 Modeling and analysis of a bimorph piezoelectric cantilever beam for voltage generation

The capability of acquiring electricity as a reaction to the employment of mechanical stresses is presented inherently by the piezoelectric materials. Actuators and sensors using this property of acquiring electricity from mechanical stress can be considered as one of many large implementations. This study which has been published in the said paper aims at modelling of a material made for piezoelectric purposes which has a combination of generation of power and voltage from the ambient vibrations by converting them to electricity. Thus, we can say that the said technology can be operated for longer periods of time in maintenance free conditions and has the prospect of superseding the use of the battery which can yield power in microwatt range, sufficient for the operation of sensors along with their data transmission.

### 13.2.3 Feasibility of structural monitoring with vibration powered sensors

The civil engineered structures can be kept on a tab by employing wireless sensors and networks comprising these sensors. Predominantly, the endurance and the usefulness of said sensors are terribly used in a minimal condition due to decrease in the amount of power which can be stored. An abundant energy of ambience can be converted to useful electrical power which has the capability of eliminating the conundrum of wastage or not enough electrical supply to allow indefinite tracking. The emphasis of this study is on vibrations from civil engineered monuments as the primary origin of ambient energy; the crucial query being, do the vibrations provide enough electrical energy? Some of the real-time origins of said vibrations can be from earthquakes, wind moving trees or any other structures or load from traffic. Theoretically, maximum levels of energy which are to be captured from these potent sources can be enumerated [6]. A mathematical model can be used to measure the energy which is being generated by the said potent loads which are subjected to any piezoelectric generator. The requirement of energy of many sub electronic systems has been compared with the energy that has been collected. A research found that the sensor node of volume 5 $cm^3$, which is normally the volume of a small stone, has the capability of providing enough energy to contemporary electronic chips under the influence of earthquakes which can be considered an acute event. It was concluded that the ideal power which is generated can be increased by linearly increasing the mass of the sensor. By using modern techniques, by employing a sensor with mass between 99 and 999 g under load of traffic, we can self-power the sensor-node. An effort is being made to lower the consumption of energy by the electronic components [7]. As the advancements are being made in the field of electronics, it can be possible in the nearby future to use a generator of volume less than 6 cm to power the node by harvesting the vibrations.

### 13.2.4 Vibration powered wireless sensor networks

A network having the characteristics of an architecture which is distributed and has electronic devices having the capability to collect data regarding the physical environment and provide automation, tracking and other applications is termed as a wireless sensor network. These wireless networks have found their way into specific types of industries with unlimited uses and applications which require permanence, more reliability, increased life of battery, varied topologies, data throughput and sensor use.

## 13.3 Existing methodology

Typically, there are five basic components/elements in an extreme low powered wireless sensor: (1) for the detection and quantification of numerous parameters of the environment like proximity, motion, pressure, temperature, light, pH, vibration etc, a sensor is used; (2) for the conversion of ambient source of energy to electricity, an energy producing transducer is used; (3) to deliver the stored electrical form to the electronic devices, a power module is used; (4) the signal received from the sensor must be converted into a more useful form for the sake of analysis and communication through a radio link for which a microcontroller is used; (5) the host receiver and data-collection point must receive the information from a processor on an event-driven basis or periodic or continuous for which a radio link is required [8–10]. The block diagram of a wireless sensor network is shown in figure 13.1. It consists of a transducer, power module, sensors, processor and radio link.

The major challenge is the powering of nodes, especially when they are arranged at dangerous or inaccessible locations or in harsh environments. The engineers of the

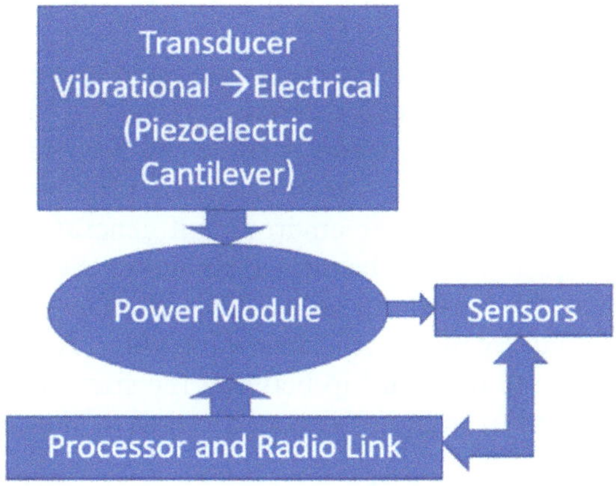


**Figure 13.1.** Block diagram of a wireless sensor network.

contemporary world have been attracted by some of the revolutionary energy harvesting technologies which are responsible for the conversion of microwatt or mille-watt level of power from the environment [11]. This sort of implementation achieves a monitoring of various conditions which is free from maintenance. This main purpose of this review is to investigate, stimulate and evaluate the energy resources and their applications, which aids in the improvement of their systems.

Traditionally, for transferring the information from transducers to devices that acquire this information requires a plethora of cables or wires in any system monitoring the said conditions. But, the main loophole of such systems is that they demand high costs and there is difficulty in installation along with low operational reliability [12–15]. In order to overcome such drawbacks, an effective and efficient solution must be developed and deployed in the latest sensors working wirelessly.

The power unit in a network of sensors poses a problem as the conventional batteries are said to have a finite lifespan, capacity and limited energy density [16, 17]. Additionally, the batteries' performance hasn't shown the necessary improvement to support the required power consumption of the electronic devices. When the nodes are placed remotely or in inaccessible locations, it is not advisable to replace and recharge the empty batteries as it can be an expensive and difficult task. Fortunately, due to the advent of new energy harvesting (EH) methods, the ambient source of energy in any form can be transformed to electrical energy from many wasted energy sources to power the sensor nodes [18–26]. Hence, this approach can be deemed as one of the best as it reduces the maintenance cost and avoids environmental contamination due to batteries.

Sources of energy which are being wasted (like vibration, motion, magnetic energy and heat energy) are being used for harvesting by employing many EH techniques. These techniques are targeted primarily at ultra-low and small power devices. For any mechanical system, it has been observed that the power transformation and transmission results in energy losses during its operation. Therefore, mechanical efficiency is always observed to be below 100%.

### 13.3.1 Proposed methodology

A phenomenon in which mechanical energy is converted into electrical energy due to inherent polarization characteristics of certain crystals is known as piezoelectric effect. An example of piezoelectric effect is shown in figure 13.2. The beam in the proposed design called a cantilever beam made of piezoelectric crystal is set in resonance, when any excitation in the environment generates a frequency which is nothing but the self frequency of the beam, which makes the yielding voltage as well as the amplitude reach the maximum values [27, 28]. The frequency of excitation including the circumstance by the force is determined first. A theoretical formula can be determined based on the relationship between the size and the self-frequency of the beam, which is derived from finite element analysis of the cantilever beam made of piezoelectric material.

Pressure sensors have a sensing element, usually a diaphragm, which is responsible for transmission of the pressure of the fluid to the element of transduction. The force transmitted to the transduction element is directly proportional to the acting pressure in a well-designed sensor as the effective area of the diaphragm is made constant, which again is converted into an electric charge proportionally. The sensing material-mass, here termed as seismic mass, is attached to the acceleration sensors which are also a type of force sensors. Due to the inertia, some force is exerted on the sensor when accelerated [29–34]. According to Newton's second law, we can deduce that this seismic mass remains constant and the force which generates the charge of electricity is proportional to the acceleration.

Practically, the deformation of an elastic sensing element of any force sensor depends on the acting force. The said deformation must be measured in passive sensors. Hence, to produce a significant deformation a high compliance sensing

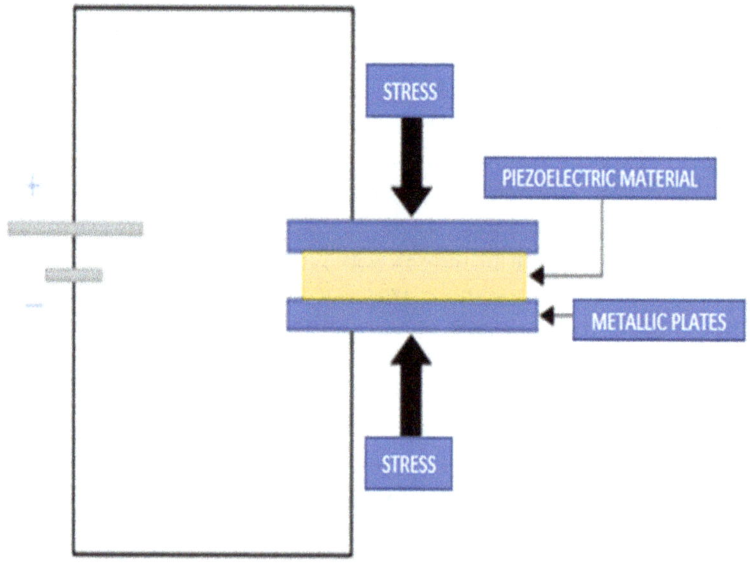


**Figure 13.2.** Piezoelectric effect.

element must be used to obtain an adequate sensitivity. The key reason why the sensing element is to be kept tinier by many orders of magnitude is because it is similar to the transduction element which gives output directly where the measuring of such deformation is not necessarily required in piezoelectric force sensors [35–39].

To design a piezoelectric energy harvester, the physics of a 3D model of a cantilever is defined as a piezoelectric device which includes both solid mechanics and electrostatics. After adding physics, the cantilever beam is designed according to the dimensions required as shown in figure 13.3.

According to the requirements of environmental vibration, a micropiezoelectric harvester, which has sensitivity towards the vibrations, can generate energy for which the working frequency of the harvester and the vibrations of the environment are to be synced, and it should also have the simple structure necessary for fabrication [40–42]. The charge that is produced by the piezoelectric material can be collected by the upper and lower electrode layer. The loss of charge by elastic layer is prevented by the insulation layer. It is essential for resonance of the cantilever because the frequency of the cantilever should be made closer to the environmental vibration with which we can achieve the small amount of mass which decreases the natural frequency of cantilever [43–46]. On the other hand, the input energy, which is in mechanical form, and output power increase accordingly as the seismic mass also can increase the inertia force and deformation. Modal analysis of the cantilever can determine the modes of vibration and natural frequencies which are termed as vibration characteristics. Therefore, the size of the structure and seismic mass at the end of the cantilever has great impact and can be easily determined by the simulation.

A base station, also called a sink in WSNs, is said to act like an interface between users and the network. By injecting queries, one can retrieve required information from the network. Usually, a wireless sensor network has hundreds of thousands of sensing nodes which communicate among themselves using radio EM waves and are also equipped with power components, radio transceivers and sensing and computing devices. The individual nodes are resource constrained inherently, i.e. they are limited by their capacity of storage, communication's bandwidth and processing power. Sensor nodes self-organize in a flexible network organization and infrastructure after their deployment, often using multi-hop communication between them, after which, the on-board sensors start collecting information of interest [47–53]. Any specific

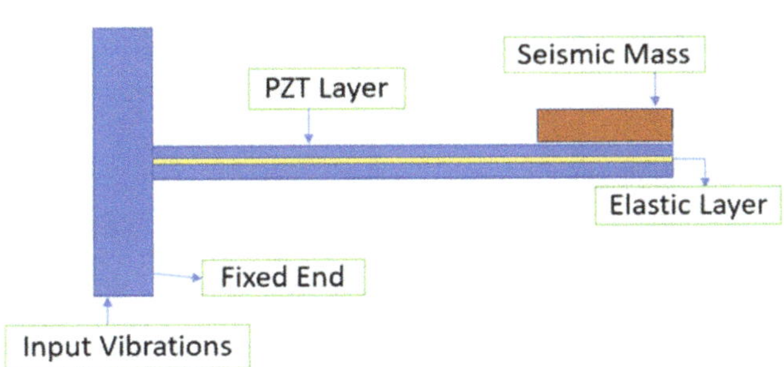


**Figure 13.3.** Piezoelectric cantilever beam.

instructions can be performed or sensing samples can be provided by the wireless sensor devices powered by vibrations which respond to questions or errors sent from a 'control site'. The working mode may either be on the same path or event driven and the location or information of the position can be obtained by global positioning system (GPS) and algorithms that position locally. Actuators make these wireless sensors 'act' upon certain conditions also if equipped [54–59].

The node structure of a network may be viewed physically or can be interpreted logically as a usage of graph theory where we are trying to communicate with devices which are used as nodes and the connections between the devices are made through links or lines between the selected nodes. There are various types of topologies, i.e. star, bus,ring, point to point, mesh and tree, which are shown in figure 13.4. One of the burning issues when operating these sensor nodes is the security. It has been observed that many kinds of threats are detected against a sensor node at varying levels. To make it more understandable, a duplicate node can be planted by an unknown source and can be manipulated to provide distorted information. Further, this duplicate node can not only distort but also provide wrong information about the position to the dumb node. If this node uses multiliteracy with this wrong position information as the input, the information which is calculated for the dumb node shows error, resulting in misrepresenting the location by the sensor wing [60–63]. This conundrum can be solved if we authenticate each and every node which actually is an effective mechanism for their communication. All these issues and security related algorithms must be included while deploying the network.

The scarcest resource of wireless sensor nodes is energy which determines the lifetime of WSNs which are deployed in huge numbers in various environments where ad hoc communications are the vital components. Considering the above reason, necessary protocols and algorithms need to address the flexibility and error tolerance, increased life span and the ability to reconfigurate.

Many man-made fabrications of wireless sensor nodes have proven that they are not always suitable because the network can be deployed in regions where humans cannot reach, considering that including GPS installation isn't feasible in every wireless sensor network because it is not economical and it does not support indoor setup. The mentioned disadvantages made engineers develop local algorithms or codes, in which particular nodes referred to as main nodes, will make us aware of

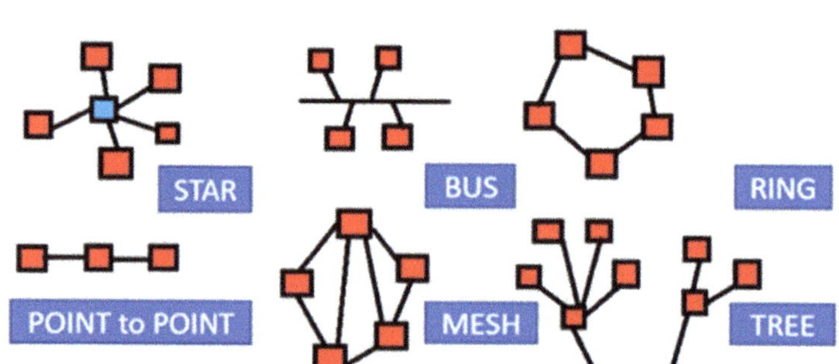


**Figure 13.4.** Network topologies.

their positions at a prior time, which enables the other nodes to estimate their own positions by taking the help of main nodes. The mentioned techniques of localization can be termed as in-network localization. Further, to understand the need of ranging based information for local nodes helps us to design a network with less cost as it requires less hardware material.

### 13.3.2 Comparison of proposed methodology with existing methodology

Economically cheaper and small sensor nodes must be produced which can be considered as a major challenge in a WSN. The situation of the commerce of 1970s-related home computing can be compared to the noticed increment of the companies which produce WSN hardware. Acquisition of data and communicated in the RF range can be achieved by deploying power methods of lesser amounts which are adopted inherently by the sensor networks in which the nodes are being continuously researched and are in the stage of development [64].

The problem of drained electrical supply can be eliminated by the ability of this methodology by the conversion of a cornucopia of ambient energy to electricity which will aid in increased amount of monitoring. Thus, we can say that the emphasis of this chapter is on structures which are vibrating as a primary source of ambient energy. But speculation arises whether sufficient energy is produced from such vibrations [65–67]. Wind, earthquake and significant traffic are some of the real-time sources of such vibrations.

In many applications, the data which has been stored must be processed by many devices with many resources, which is done by establishing communication between a WSN and a LAN or WAN via a device called a gateway, which is usually a bridge between the WSN and the other networks. A wireless wide area network used primarily for low-power devices is known as a low-power wide-area network (LPWAN) which is the most typical network deployed in real-time.

Also, we know that India is a place of numerous temples and devotional places where huge gatherings are present. The idea is to use piezoelectric material which we are going to establish on railway and temple premises to generate power. Imagine the huge number of people on railway platforms and temples; if we are generating power by just walking, then why not switch to this idea and rethink how to really implement it? Piezoelectric materials can be kept between the two layers to make use of the energy generated due to the weight of the people moving or walking on it. As we know that individuals will move over the first layer, they will exert a force into the piezomaterial and thus these crystals deform. This deformation will absorb the force and thus power is generated [68–70].

In third world countries like India, the number of speed breakers observed on roads belonging to different zones and accident-prone areas are considerably fewer than elsewhere. So, we can deduce that to prevent the occurrence of accidents and make roads more secure, we can build and implement these synthetic based speed breakers at the said zones. They can be designed in such a way that there are essentially three parts in a speed breaker—an upper and a lower part which are separated from each other by the part which has the piezoelectric crystal, while the lower layer is fixed to the road. [71–74]. Whenever a vehicle goes by such fabricated speed breakers, a deformation is formed on them vertically by the pressure of weight of the automobile, which is converted to electricity directly using the piezoelectric effect and this method can be considered as more effective as there is less wastage of energy.

This technique produces power with the usage of electric components that create use of the power from human footsteps. The purpose of super capacitors is to store and discharge energy while not intense abundant energy. Thus, the emerging need of constant increase of power will be met by putting in these systems in closely packed places. This may doubtless not solely overcome energy crises, but rather conjointly build up a healthy encompassing.

It is a known fact that an Internet of Things (IoT) system provides us with an ecosystem where we have the facility to communicate with a wide range of devices in a single go. Less power consumed and complex networks consist largely of integrated nodes (with constrained processing power and memory usage). These routers are interconnected by a number of links which support typically only lesser data rates which are known to be more unreliable with relatively low packet delivery rates. Also, the noise or traffic is not point to point but from point to multipoint.

The primary characteristics of any WSN include: compact size, easily usable for others, nature of heterogeneity of different nodes, cross-layer design, ability to find solutions to overcome node failures, ability to combat adverse environmental situations, homogeneity of nodes, scalability to large scale of deployment and power consumed by nodes using various types of batteries or energy harvesters.

### 13.3.3 Advantages

In harsh and hostile environments like in a dense forest, wireless sensor nodes powered by vibrations can be dropped from the air as it is difficult to deploy wired networks. The WSNs are known for their scalability because in certain scenarios like monitoring of structures based on their health a dense amount of deployment is required. Otherwise, a setup made of wires if densely deployed may result in a chaotic scenario at the time of deployment. Additionally, such deployment of wired networks is a costlier procedure when compared to wireless deployment which is usually hassle-free and can stay maintenance-free for over longer periods of time. Despite certain disadvantages mentioned below, these networks are widely used in IoT and can be treated as their backbone.

### 13.3.4 Disadvantages

The only disadvantages that encompass wireless sensor networks are the less advancement in computation and the resources used for communication. This less advancement will result in not enough battery power, not enough storage, and less capability of computation, greater susceptibility to security threats and limited communication due to less bandwidth.

## 13.4 Conclusion

The need for energy is increasing with the aim of becoming more powerful in the world for countries. As can be seen in our modern world, each sector is changing dynamically. But if we look at the general topic from a wide angle, we would see that they all have one conundrum in common: energy. Therefore, harvesting energy in this way can provide a lot of potential for powering small devices.

The times when energy harvesting using piezoelectric materials was treated as unreliable, as the output was typically less, in the range of μW, can be considered long gone, as there are many advancements which have demonstrated improvement in the development of ultra-low power electronics. This has increased the output power to the range of mW tremendously in devices since 2018.

Energy harvesting is a critical technology for the expansion of self-governing, self-powered electronic devices. The requirement of energy by various electronic devices which require lesser power by using energy harvesting systems has acquired great predominance and is considered as a promising technology in the long run. This transition of battery powered small electronic devices to self-governed as well as self-powered systems has improved by such energy harvesting technology.

There are more than 250 piezoelectric materials that can be used for the generation of electric impulses. Therefore, appropriate ones can be used for this purpose. Barium titanate is the first ceramic that showcased the principle of piezoelectricity. Also, the ceramic lead zirconate titanate is the most commonly used piezoelectric material. In fact, there are materials which have the flexibility to supply electricity as a result of conversion of mechanical energy into electricity. These devices are units usually named as energy harvesters and can be utilized in applications where outside power is unattainable and batteries are not a viable alternative. Whereas recent experiments have shown that these materials have the potential to be portayed as power generators, the joules of energy made by this continue to be very low, hence the need to optimize them.

Certain challenges like providing supply continuously to the wireless sensor nodes which are deployed remotely, using any relevant device which has the capability to store required energy and operate all day long are targeted in this chapter. In addition to this, these vibration powered devices can be used to face the said challenges but difficulty comes when they cannot be deployed everywhere. Thus, sometimes wireless sensor nodes can be rendered useless in certain instants which effect the overall operation.

As shown and demonstrated here, the study focuses on exploration of ideas which help us to achieve multi-functionality of piezoelectric vibration energy harvesters. It can be conceived that these harvesters show great potential in powering of sensor nodes which are also governed by them. To elaborate, we can say that the electromechanical coupling exhibited by the said materials in the piezoelectric harvesters pose a great application idea by the employment of a procedure in which there is conversion of on-going disturbances in the environment of a system which requires electrical energy for its functioning and survival. With the ongoing advancement in the field of piezoelectricity, the craze for small devices with no batteries at all, and with more battery efficiency has created a challenge.

## References


[1] Ajitsaria J, Choe S, Shen D and Kim D 2007 Modeling and analysis of a bimorph piezoelectric cantilever beam for voltage generation *Smart Mater. Struct.* **16** 447

[2] Gautschi G 2002 Piezoelectric sensorics: force, strain, pressure, acceleration and acoustic emission sensors, materials and amplifiers *Sens. Rev* **22** 363–4

[3] Elvin N, Lajnef N and Elvin A 2006 Feasibility of structural monitoring with vibration powered sensors *Smart Mater. Struct.* **15** 977

[4] Fuh Y-K, Huang Z, Wang B and Li S 2017 Self-powered active sensor with concentric topography of piezoelectric fibers *Nanoscale Res. Lett.* **12** 44

[5] Liu M, Cui T and Wang L 2007 Fabrication and characterization of the piezoelectric microcantilever integrated with PZT thin-film microforce sensor and actuator *TRANSDUCERS and EUROSENSORS '07—4th Int. Conf. Solid-State Sensors, Actuat. Microsyst* 10.1109/SENSOR.2007.4300618

[6] Uddin M N, Islam M S, Sampe J, Wahab S A and Ali S H M 2016 Vibration based T-shaped piezoelectric cantilever beam design using finite element method for energy harvesting devices *2016 IEEE Int. Conf. Semiconduct. Electron. (ICSE) (Kuala Lumpur)* 137–40 doi:10.1109/SMELEC.2016.7573610

[7] Berringer K July 2008 High-performance mixed-signal MCUs in low power applications *EPN* **37** 12

[8] Vieira M A M Jr, Coelho C N, da Silva D C Jr and da Mata J M 2003 Survey on wireless sensor network devices *Proc. 2003 IEEE Conf. on Emerging Technologies and Factory Automation (EFTA 2003)* 1 537–44

[9] Alippi C, Anastasi G, Di Francesco M and Roveri M April 2009 Energy management in wireless sensor networks with energy-hugry sensors *IEEE Instr. Measure. Mag.* **12** 16–23

[10] Want R and Marayanaswami K F January–March 2005 Energy harvesting and conservation, guest editor's introduction *IEEE Pervasive Comput.; Mob. Ubiquitous Syst.* **4** 14–7

[11] Rosing T, Todd M, Farrar C and Hodgkiss W 2008 Energy harvesting for structural health monitoring sensor networks *J. Infrastruct. Syst.* 14

[12] Chalasani S and Conrad J 2008 A survey of energy harvesting sources for embedded systems *Conf. Proc.—IEEE SOUTHEASTCON* 442–7 10.1109/SECON.2008.4494336

[13] Wang K, Wang Y, Sun Y, Guo S and Wu J Dec. 2016 Green industrial internet of things architecture: an energy-efficient perspective *IEEE Commun. Mag.* **54** 48–54

[14] Jayakumar H, Lee K, Lee W S, Raha A, Kim Y and Raghunathan V 2014 Powering the internet of things *Proc. 2014 Int. Symp. Low Power Electron. Design. (New York)* (ACM) pp 375–80

[15] Raghunathan V and Chou P H 2006 Design and power management of energy harvesting embedded systems *Proc. 2006 Int. Symp. Low Power Electron. Design (New York)* (ACM) pp 369–74

[16] Pimentel D and Mus´ılek P 2010 Power management with energy harvesting devices *Electrical Comput. Eng. (CCECE), 2010 23rd Canadian Conf. IEEE* 1–4

[17] Seah W K, Eu Z A and Tan H-P 2009 Wireless sensor networks powered by ambient energy harvesting (wsn-heap)-survey and challenges *Wireless Commun. Vehi. Technol. Info. Theor Aerospace Electron. Syst. Technol. Wireless VITAE 1st Int. Conf. IEEE* pp. 1–5

[18] Sudevalayam S and Kulkarni P 2011 Energy harvesting sensor nodes: survey and implications *IEEE Commun. Surv. Tutor.* **13** 443–61

[19] MachinaResearch 2016 Press release: global internet of things market to grow to 27 billion devices, generating USD3 trillion revenue in 2025 [Online]

[20] Shirvanimoghaddam M, Dohler M and Johnson S J 2017 Massive nonorthogonal multiple access for cellular IoT: potentials and limitations *IEEE Commun. Mag.* **55** 55–61

[21] Prasad R V, Devasenapathy S, Rao V S and Vazifehdan J 2014 Reincarnation in the ambiance: devices and networks with energy harvesting *IEEE Commun. Surv. Tutor.* **16** 195–213

[22] Safak M 2014 Wireless sensor and communication nodes with energy harvesting *J. Commun., Navig., Sens. Serv. (CONASENSE)* **1** 47–66

[23] Kamalinejad P, Mahapatra C, Sheng Z, Mirabbasi S and Guan Y 2015 Wireless energy harvesting for the internet of things *Commun. Mag. IEEE.* **53** 102–8

[24] Tervo J, Manninen A, Ilola R and Hänninen H 2009 *State-of-the-art of Thermoelectric Materials Processing* (Finland: Ju lkaisija-Utgivare Publisher) pp 6–7

[25] Zhao Y, Dyck J S, Hernandez B M and Burda C 2010 Enhancing thermoelectric performance of ternary nanocrystals through adjusting carrier concentration *J. Am. Chem. Soc.* **132** 4982–3

[26] Barnes M, Conway C, Mathews J and Arvind D K 2010 ENS: an energy harvesting wireless sensor network platform *Proc. ICSNC* 83–7*(August 2010)*

[27] Eu Z, Tan H and Seah W 2011 Design and performance analysis of MAC schemes for wireless sensor networks powered by ambient energy harvesting *Ad Hoc Netw.* **9** 300–23

[28] Chen Y, Wang Q, Gupchup J and Terzis A 2010 Tempo: an energy harvesting mote resilient to power outages *Proc. IEEE LCN 2010* 933–4*(October 2010)*

[29] Rezaei Z and Mobininejad S February 2012 Energy saving in wireless sensor networks *Int. J. Comput. Sci. Eng. Surv. (IJCSES)* **3**

[30] Bao X and Ding G 2016 An routing algorithm for maximizing network performace in energy harvesting wireless sensor network IEEE. doi:10.1109/ICISCE.2016.271

[31] Fujii C and Seah W K G 2011 Multi-tier probabilistic polling in wireless sensor networks powered by energy harvesting *2011 Seventh Int. Conf. on Intelligent Sensors, Sensor Networks and Information Processing* 383–8

[32] Iannello F, Simeone O and Spagnolini U 2012 Medium access control protocols for wireless sensor networks with energy harvesting *IEEE Trans. Commun.—TCOM.* **60** 1381–9

[33] Mitcheson P D, Yeatman E M, Rao G K, Holmes A S and Green T C 2008 Energy harvesting from human and machine motion for wireless electronic devices *Proc. IEEE* **96** 1457–86

[34] Benasciutti D and Moro L 2013 Energy harvesting with vibrating shoemounted piezoelectric cantilevers *Advances in Energy Harvesting Methods* (Berlin: Springer) 141–62

[35] Discenzo F M, Chung D and Loparo K A 2006 Pump condition monitoring using self-powered wireless sensors *Sound Vib* **40** 12–5

[36] Sterken T, Fiorini P, Baert K, Puers R and Borghs G 2003 An electret based electrostatic\spl mu\-generator *TRANSDUCERS, Solid-State Sensors, Actuators and Microsystems, 12th Int. Conf. 2003* 2 (Piscataway, NJ: IEEE) 1291–4

[37] Sazonov E, Li H, Curry D and Pillay P 2009 Self-powered sensors for monitoring of highway bridges *IEEE Sensors J.* **9** 1422–9

[38] Kymissis J, Kendall C, Paradiso J and Gershenfeld N 1998 Parasitic power harvesting in shoes *Digest of Papers. Second Int. Symp. Wearable Comput.* (Piscataway, NJ: IEEE) pp 132–9

[39] Park K-I, Xu S, Liu Y, Hwang G-T, Kang S-J L, Wang Z L and Lee K J 2010 Piezoelectric batio3 thin film nanogenerator on plastic substrates *Nano Lett.* **10** 4939–43

[40] Kwon J, Seung W, Sharma B K, Kim S-W and Ahn J-H 2012 A high performance PZT ribbon-based nanogenerator using graphene transparent electrodes *Energ. Environ. Sci.* **5** 8970–5

[41] Hwang G-T, Park H, Lee J-H, Oh S, Park K-I, Byun M, Park H, Ahn G, Jeong C K and No K *et al* 2014 Self-powered cardiac pacemaker enabled by flexible single crystalline PMN-PT piezoelectric energy harvester *Adv. Mater.* **26** 4880–7

[42] Roundy S J 2003 Energy scavenging for wireless sensor nodes with a focus on vibration to electricity conversion *Ph.D. dissertation* (Berkeley, CA: University of California)

[43] Franciscatto B 2014 Design and implementation of a new low-power consumption DSRC transponder *Ph.D. dissertation* (Universite Grenoble Alpes)

[44] Tanaka T, Suzuki T and Kurihara K 2014 Energy harvesting technology for maintenance-free sensors *Fujitsu Sci. Technol. J.* **50** 93–100

[45] Becker T, Kluge M, Schalk J, Tiplady K, Paget C, Hilleringmann U and Otterpohl T 2009 Autonomous sensor nodes for aircraft structural health monitoring *IEEE Sensors J.* **9** 1589–95

[46] Torfs T, Sterken T, Brebels S, Santana J, van den Hoven R, Spiering V, Bertsch N, Trapani D and Zonta D 2013 Low power wireless sensor network for building monitoring *IEEE Sensors J.* **13** 909–15

[47] Whelan M J, Gangone M V and Janoyan K D 2009 Highway bridge assessment using an adaptive real-time wireless sensor network *IEEE Sensors J.* **9** 1405–13

[48] Park G, Rosing T, Todd M D, Farrar C R and Hodgkiss W 2008 Energy harvesting for structural health monitoring sensor networks *J. Infrastruct. Syst.* **14** 64–79

[49] Bai H, Atiquzzaman M and Lilja D 2004 Wireless sensor network for aircraft health monitoring *Broadband Netw. 2004 BroadNets 2004 Proc. First Int. Conf. IEEE* 748–50

[50] Nelson C A, Platt S R, Albrecht D, Kamarajugadda V and Fateh M 2008 Power harvesting for railroad track health monitoring using piezoelectric and inductive devices *15th Int. Symp. Smart Struct. Mater. Nondestructive Eval. Health Monitor. Int. Soc. Opt. Photon.* 69280R–69280R

[51] Alemdar H and Ersoy C 2010 Wireless sensor networks for healthcare: a survey *Comput. Netw.* **54** 2688–710

[52] Mascareñas D, Flynn E, Farrar C, Park G and Todd M 2009 A mobile host approach for wireless powering and interrogation of structural health monitoring sensor networks *IEEE Sensors J.* **9** 1719–26

[53] He C, Arora A, Kiziroglou M E, Yates D C, O'Hare D and Yeatman E M 2009 Mems energy harvesting powered wireless biometric sensor *2009 Sixth Int. Workshop Wearable Implantable Body Sensor Netw.* (Piscataway, NJ: IEEE) 207–12

[54] Hawkes A M, Katko A R and Cummer S A 2013 A microwave metamaterial with integrated power harvesting functionality *Appl. Phys. Lett.* **103** 163901

[55] Qi S, Oudich M, Li Y and Assouar B 2016 Acoustic energy harvesting based on a planar acoustic metamaterial *Appl. Phys. Lett.* **108** 263501

[56] Drozdenko B, Zimmermann M, Dao T, Chowdhury K and Leeser M 2018 Hardware-software codesign of wireless transceivers on ZYNQ heterogeneous systems *IEEE Trans. Emerg. Topics Comput.* **6** 566–78

[57] Cid-Fuentes R G, Naderi M Y, Chowdhury K R, Cabellos-Aparicio A and Alarcón E 2016 On the scalability of energy in wireless RF powered internet of things *IEEE Commun. Lett.* **20** 2554–7

[58] Mishra D, De S, Jana S, Basagni S, Chowdhury K and Heinzelman W 2015 Smart RF energy harvesting communications: challenges and opportunities *IEEE Commun. Mag.* **53** 70–8

[59] Mishra D, De S and Chowdhury K R 2015 Charging time characterization for wireless RF energy transfer *IEEE Trans. Circ. Syst. II: Exp. Briefs* **62** 362–6

[60] Chen Z-G, Han G, Yang L, Cheng L and Zou J 2012 Nanostructured thermoelectric materials: current research and future challenge *Prog. Nat. Sci. Mater. Int.* **22** 535–49

[61] Satyala N, Norouzzadeh P and Vashaee D 2014 Nano bulk thermoelectrics: concepts, techniques, and modeling *Nanoscale Thermoelectrics.* (Berlin: Springer) 141–83

[62] Lu X and Yang S-H 2010 Thermal energy harvesting for WSNS *IEEE Int. Conf. Syst. Man Cybernet. (SMC)* (Piscataway, NJ: IEEE) 3045–52

[63] Leonov V, Torfs T, Fiorini P and Van Hoof C 2007 Thermoelectric converters of human warmth for self-powered wireless sensor nodes *IEEE Sensors J.* **7** 650–7

[64] Wang Z, Leonov V, Fiorini P and Van Hoof C 2009 Realization of a wearable miniaturized thermoelectric generator for human body applications *Sens. Actuators* A **156** 95–102

[65] Cuadras A, Gasulla M and Ferrari V 2010 Thermal energy harvesting through pyro-electricity *Sens. Actuators* A **158** 132–9

[66] Ferrari M, Ferrari V, Marioli D and Taroni A 2006 Modeling, fabrication and performance measurements of a piezoelectric energy converter for power harvesting in autonomous microsystems *IEEE Trans. Instr. Measure.* **55** 2096–101

[67] Harb A 2011 Energy harvesting: state-of-the-art *Renew. Energ.* **36** 2641–54

[68] Singh H and Lalchand C M 2012 Self powered wearable health monitoring system *Adv. Mater. Res.* **403** 3839–46

[69] Ramsay M J and Clark W W 2001 Piezoelectric energy harvesting for bio-mems applications *Smart Structures and Materials 2001: Industrial Commercial Applications of Smart Structures Technology* vol 4332 (Bellingham, WA: SPIE) 429–39

[70] Zurbuchen A, Pfenniger A, Stahel A, Stoeck C T, Vandenberghe S, Koch V M and Vogel R 2013 Energy harvesting from the beating heart by a mass imbalance oscillation generator *Ann. Biomed. Eng.* **41** 131–41

[71] Dagdeviren C, Yang B D, Su Y, Tran P L, Joe P, Anderson E, Xia J, Doraiswamy V, Dehdashti B and Feng X *et al* 2014 Conformal piezoelectric energy harvesting and storage from motions of the heart, lung, and diaphragm *Proc. Natl Acad. Sci.* **111** 1927–32

[72] Toh W Y, Tan Y K, Koh W S and Siek L 2014 Autonomous wearable sensor nodes with flexible energy harvesting *IEEE Sensors J.* **14** 2299–306

[73] De Mil P, Jooris B, Tytgat L, Catteeuw R, Moerman I, Demeester P and Kamerman A 2010 Design and implementation of a generic energyharvesting framework applied to the evaluation of a large-scale electronic shelf-labeling wireless sensor network *EURASIP J. Wireless Commun. Netw.* 1 2010

[74] Prieto M D, Millan D Z, Wang W, Ortiz A M, Redondo J A O and Martinez L R 2016 Self-powered wireless sensor applied to gear diagnosis based on acoustic emission *IEEE Trans. Instr. Measure* **65** 15–24

# Chapter 14

# Comprehensive review on brain–computer interface sensor-based smart home appliances control system

**Praveen Kumar Shukla, Rahul Kumar Chaurasiya, Mridu Sahu and Shrish Verma**

In India, 2.21% of the total population is disabled. According to Census 2011, in India, 20% have movement disability, 19% have disability in seeing, and 8% have multiple disabilities. Physically disabled people need constant supervision. It is the responsibility all over the world for people to care for these physically disabled people, but this is not possible in our busy schedules to care for everybody. It is very important to have a system to constantly monitor physically disabled people to improve their quality of life. Brain–computer interface (BCI) allows people to communicate with a personal computer without undermining the body part's actions. BCI is characterized by a combination of hardware and software that allows the brain to monitor portable devices or even computers. The goal is to help people with severe disabilities live their lives as normal as possible. Therefore, this type of system is useful for people who are completely disabled and suffer from conditions like ALS (amyotrophic lateral sclerosis), spinal cord injury, brain stroke, etc. Such systems do not depend on any ordinary medium of interchanges like physical touch, talking, etc. In fact, BCI is an effective communication tool for people suffering from multiple motor disorders but with strong cognitive abilities. The collected signal is recorded by sensors in the form of amplitude and frequency. This system has the cognitive ability to control any of the devices like wheelchair, robot, keyboard, cursor control, music system, etc.

This chapter gives a comprehensive review of the sensor-based smart home automation system. A smart home is where the machines, for example, TV, telephones, music system, fans, and so on can be controlled just by brains. The thought-based BCI systems increase the convenience, comfort level of human beings, providing a better life quality, for those suffering from severe motor disabilities that are unable to do physical activities for their day-to-day living requirements. The

doi:10.1088/978-0-7503-2707-7ch14  

chapter gives an all-inclusive review of the different types of control signals, artifacts, and various sensor-based methods for single and multiple home appliances control.

## 14.1 Introduction

BCI is a technology that gives correspondence between the brain and the outside world. The BCI-construction method does not rely on any ordinary communication medium such as physical, talking, etc. It is the most recent innovation that gathers brainwaves and responds accordingly. A general BCI-based system is shown in figure 14.1. BCI depends on the translating of electroencephalographic (EEG) reactions produced by incitement of a subject. EEG is a primary method for recording and analyzing brain activities [1]. It is a noninvasive technology that measures the electrical activity of the brain [2]. The general range of the EEG signal is 0–100 microvolt, with frequency ranges from 1 Hz to 100 Hz. The EEG can be used for disease diagnosis in the brain including various investigations like alertness monitoring [3], device control, etc [4]. EEG signals can be used as control signals once classified signals are converted into command signals.

A BCI system passes several phases, including pre-processing, extraction of features, signal recognition and ultimately control. Several preprocessing and classification methods have been used for BCI-based systems. For example preprocessing methods like filtering methods such as energy combination (MEC) [5], canonical correlation analysis (CCA) [6], common average reference (CAR) [7], principal component analysis (PCA) [8], independent component analysis (ICA) [9], autocorrelation (AU) and single value decomposition (SVD) are also applied. Further extracted signals are transferred for feature extraction, namely wavelet-based element extractions [10], power spectral density [11], band power [12], autoregression [13], etc. The most common method for feature extraction for

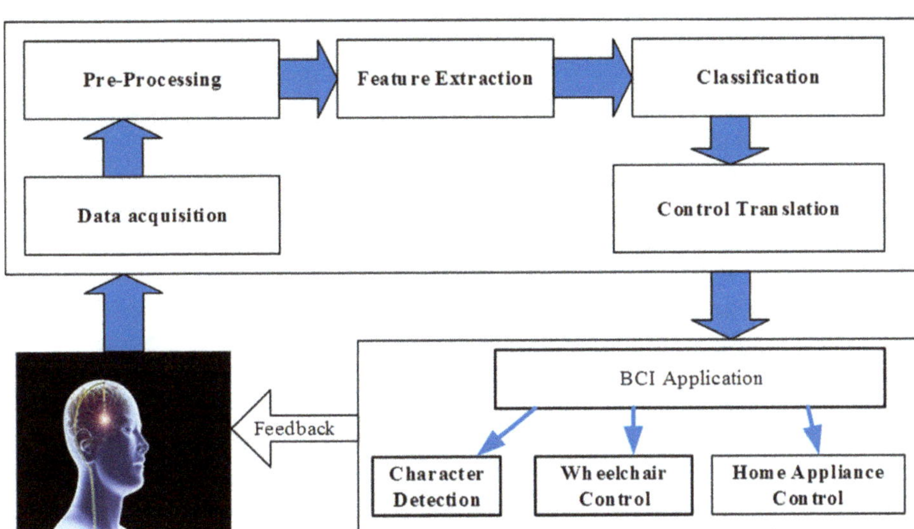


**Figure 14.1.** General brain–computer interface system.

EEG signal includes fast Fourier-transform (FFT) [14], independent component analysis (ICA) [15], discrete wavelet transform [16] (DWT) and Heiberg–Hung transform (HHT). Further, different linear and nonlinear classifiers like linear discriminant analysis (LDA) [17], hidden Markov model (HMM) [18], support vector machine (SVM) [19], K-nearest neighbors (KNN) [20] and Bayesian classifier [21] etc, have been used to classify signals for a BCI-based system.

Numerous BCI innovations have been proposed for disabled individuals, viz. application to enter telephone number [22], control their surroundings [23], and voice-based system [24], screen control through a cursor [25], robot (motor imagery-based) control [26], 3D game [27], smart wheelchair, assistive home mechanization system, home computerization system and web-based system [28–30], another system like FPGA-based engineering to control a clinical nursing in bed [31], implementation of a reduced cost FPGA-based architecture [32] have been developed. Recently, research and development of the EEG-based home automation system have received great attention because it brings mobility back to people and improves their quality of life by making disabled and elderly persons self-sustained. Different appliance controls based on a Bluetooth remote transmission [33], infrared [34], UPnP (Universal plug and play) [35], eye movement [36], LED (light emitting diode), control using Arduino [37], embedded system [38] etc, have been developed.

This chapter presents a comprehensive review covering all aspects, key techniques for controlling appliances, underlying challenges, and issues in the BCI-based smart home system. The main goal of this review is to provide a comparative and comprehensive study of controlling single and multiple appliances just by thoughts. Regarding this, we searched 45 papers related to BCI-based home environment systems from 2008 to 2018. We searched the keywords BCI-based home automation system or P300 in the Scopus database as a search engine. In figure 14.2 a bar plot of year-wise published papers for BCI-based home automation systems is depicted.

### 14.1.1 Motivation and requirement

The brain–computer interface communicates with people whose motor cells are damaged and who are unable to perform any muscle activity. Therefore, by using BCI, they can interact with the external environment and monitor such items as a wheelchair, home appliance, 2D–3D cursor control, etc. Some of the requirements for disableds and elderly people are as follows:

1. To provide quality of life for physically disabled people and elderly people.
2. To provide an easy and economical BCI-based controlling home appliances system for disabled and elderly people.
3. To provide an accurate system for physically disabled and elderly people.

The rest of the chapter is organized as follows: section 14.2, describes the background including EEG in detail. Section 14.3 provides a detailed step-by-step description for a BCI-based home automation system. Section 14.4 provides a detailed description of the controlling methods; section 14.5 is dedicated to outlining limitations, challenges and discussion. Finally, section 14.6 concludes the paper.

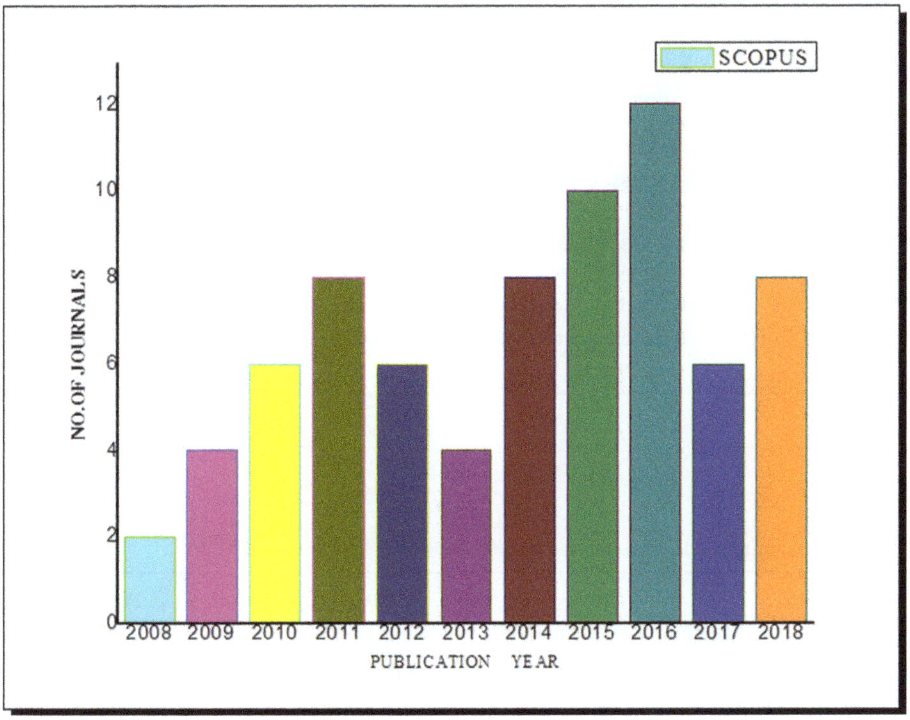


**Figure 14.2.** Bar plot of year-wise published papers for BCI-based home automation systems.

## 14.2 Background

### 14.2.1 Electroencephalography (EEG)

The EEG signal represents human brain activity and therefore EEGs are known as complex brain signals. These signals depend upon the mental state of the subjects. The EEG data recording system comprises four items: electrodes, amplifiers, converters from analog to digital (A/D) and recording devices. The EEG signals are obtained by attaching electrodes to the scalp and amplifying the recorded signals afterward. The A/D converter transforms those amplified analog signals into digital form. Such amplified analog signals are translated into digital form by the A/D converter. Finally, the digital signals are stored in a recording device like a computer in discrete form. The EEG-related brain waves are divided into different bands, depending on the frequency range. These bands are described in the next subsection.

### 14.2.2 Brain waves

EEG consists of a set of waves, categorized according to their frequencies. Such frequency bands are categorized from low to high as delta (δ), theta (θ), alpha (α), bBeta (β), and gamma (γ) [39, 40]. In table 14.1 various types of waves are described in detail.

**Table 14.1.** Types of waves.

| Brain waves | Frequency-band | Applications |
|---|---|---|
| Delta | 0.5–4 Hz | It is correlated with deeper sleep. |
| Theta | 4–7 Hz | It is associated with emotional stress, light sleep, creative motivation, and relaxation. |
| Alpha | 8–13 Hz | It is associated with a closed eye, physical relaxation, and cognitive inactivity. |
| Beta | 13–30 Hz | It is related to the skills of alertness, concentration, and problem-solving. |
| Gamma | 30–70 Hz | It is related to hyper-alertness and integration as well as subjective alertness. |

**Table 14.2.** Types of artifacts.

| S. no | Study | Year | Method | Artifacts removal |
|---|---|---|---|---|
| 1 | Flexer *et al* [41] | 2005 | ICA | Ocular |
| 2 | Gao *et al* [42] | 2010 | PDAIC | Ocular |
| 3 | Guerrero-Mosquera and Vasquez [43] | 2012 | ICA-RLS | Ocular |
| 4 | Li *et al* [44] | 2013 | DWICA | Ocular |
| 5 | Yang and He [45] | 2014 | ICA-RLS | Ocular |
| 6 | Maki and Ilmoniemi [46] | 2011 | TMS | Muscles |
| 7 | Chen *et al* [47] | 2013 | EEMD-ICA | Muscles |
| 8 | Fitzgibbon *et al* [48] | 2016 | ICA | Muscles |
| 9 | Gerla *et al* [49] | 2017 | PCA | Ocular + Muscles |
| 10 | Chen *et al* [50] | 2017 | IVA | Ocular + Muscles |
| 11 | Hamaneh *et al* [51] | 2014 | ICA–WT | Cardio |
| 11 | Navaro *et al* [52] | 2015 | EMD-AF | Cardio |
| 12 | Patel *et al* [53] | 2017 | EMD+REG | Cardio |

### 14.2.3 EEG artifacts

The human mind is not a static system; indeed, it is actually very dynamic. EEG signals are noisy as they are also influenced by the surrounding environment and are non-stationary by nature. This makes BCI-based communication involving decoding EEG commands a challenging task.

The artifacts origins include eye blinks, muscle movement, and cardio artifacts. Some examples, different types of artifact and their removal techniques are depicted in table 14.2.

As per the analysis single method, ICA is the most commonly used method for the removal of artifacts. Overall most of the study used a hybrid method for removing and identification of artifacts.

### 14.2.4 Control signal of BCI

Our brain generates different types of control signals depending upon the situation. There are different studies that identify brain signal as a control signal [54–56].

#### *14.2.4.1 Evoked potential*

The subject generates these signals unconsciously when he/she receives external stimuli. They concentrate on the external stimulation that may be the snug, awkward and stressful thing about the subject. The few possible signals that have been evoked are described below.

i. **State evoked potential**: When the subject sees pictures, listens to sound or experiences some sensation, it creates steady-state evoked potential signals.
ii. **P300**: After almost 300 ms, these signals are produced, when the subject is accessible to the unusual or surprising task. These signals are created by unusual paradigms where the consumer is asked to go on the stimuli's random sequence.

#### *14.2.4.2 Spontaneous signal*

The subjects generate these signals voluntarily and do not require any external stimulation. These signals are generated when frequency belongs to $\leqslant$1 Hz.

i. **Motor and sensorimotor rhymes**: Such signals are mostly found in the front and middle of the cortex.
ii. **Slow cortical potential**: These are low recurrence signal that shows the low variety, positive and negative of the cortical movement and may last from a millisecond to a few seconds.

#### *14.2.4.3 Hybrid signals*

Such signals are the synthesis of the brain signals of two or more types. The main objective of the BCI to use these hybrid signals is to make the system effective and to avoid demerits of every form of signal. Examples of these hybrid signals include SSVEP and EMG, P300 and SSVEP, SSVEP and Alpha wave, etc. Some examples of control signals and their applications are described in table 14.3.

**Table 14.3.** Control signals and their BCI applications.

| S. No. | Study | Year | Control signals | Application |
|---|---|---|---|---|
| 1 | Thulasidas *et al* [57] | 2006 | P300 | Character detection |
| 2 | Trejo *et al* [58] | 2006 | SSVEP | 1D, 2D cursor control |
| 3 | Rebsamen *et al* [59] | 2007 | P300 | Wheelchair control |
| 4 | Müller-Putz and Pfurtscheller [60] | 2008 | SSVEP | Electrical Prosthesis |
| 5 | Sirvent *et al* [61] | 2010 | P300 | Internet browsing |
| 6 | Edlinger *et al et al* [62] | 2011 | P300+SSVEP | Home appliance control |
| 7 | Bordoloi *et al* [63] | 2012 | MI | Gaming control |
| 8 | Reshmi and Amal [64] | 2013 | MI | Wheelchair control |
| 9 | Long *et al* [65] | 2013 | SSVEP + P300 | Wheelchair control |
| 10 | Ma *et al* [66] | 2014 | EEG +EOG | Robot control |
| 11 | Aydin *et al* [67] | 2016 | P300 | Home appliance control |
| 12 | Belwafi *et al* [68] | 2016 | MI | Home appliance control |
| 13 | Martišius and Damaševičius [69] | 2016 | SSVEP | Gaming system |
| 14 | Anindya *et al* [70] | 2016 | SSVEP | Home appliance control |

## 14.3 Step involved in BCI-based controlling home appliances system

The step of the BCI-based home automation system consists of several phases like data collection, preprocessing, feature extraction, classification and hardware translation. These steps are as follows.

### 14.3.1 Data acquisition framework

Several researchers have focused on the BCI home appliances control system using various data acquisition techniques. The data acquisition framework has specific specifications like sampling frequency and channel length. Table 14.4 represents the data acquisition instrument used in the BCI-based home appliances control system. Several researchers worked on different channel number headsets like 60, 16, 16, 14, 8, 3, 1 and sampling frequency 512, 500, 256, etc. These data acquisition frameworks can be utilized on the specific application. Nowadays for intentional control like home appliances control and keyboard control emotive, NeuroSky headsets are used widely because of low cost, low power, and easy availability.

### 14.3.2 Preprocessing and feature extraction

Real-time acquired EEG signals are very noisy. There are various types of artifacts affected by these signals. For this situation preprocessing of signals plays a major role. In the experiment of a BCI–based home appliances control system, various preprocessing and feature extraction methods include Butterworth low pass filter

**Table 14.4.** Summary of data acquisition instruments used in a BCI-based home appliances control system.

| Study | Data acquisition instruments | Electrode | Application | Sensor |
|---|---|---|---|---|
| Lin *et al* (2012) [71] | Headband | 1 | Home appliance | Ski sensor |
| Alshbtat *et al* (2014) [72] | Emotive EPOC headset | 14 | Home appliance | Semi-dry polymer |
| Miralles *et al* (2015) [33] | g.Nautilus headset | 60 | Home appliance | 3-axis acceleration sensor |
| Aydin *et al* (2016) [67] | actiCAP Xpress brain products | 16 | Home appliance | Dry electrode sensors |
| Achanccaray *et al* (2017) [73] | g.tech | 16 | Home appliance | Monitoring sensor |
| Virdi *et al* (2017) [74] | g.GAMMA cap | 8 | Home appliance | Ladybird and scarabeo |
| Uma and Sheela (2017) [75] | NeuroSky mind wave headset | 1 | Home appliance | Think gear AM (TGAM) EEG sensor |
| Tabbal *et al* (2018) [76] | Open BCI kit | 3 | Home appliance | MyoWare sensor |

**Table 14.5.** Summary of applied preprocessing and feature extraction methods in a BCI-based home appliances control system.

| Author | Year | Preprocessing methods | Feature extraction method |
|---|---|---|---|
| Hoffman *et al* [77] | 2008 | 6th order Butterworth filtering | Feature vector |
| Corralejo *et al* [78] | 2014 | Moving average filter | Stepwise method |
| Aydin *et al* [17] | 2015 | Baseline correction, Butterworth low pass filter, down sampling | Create feature vector |
| Anindya *et al* [79] | 2016 | Band pass filter | Fast fourier transform |
| Achanccaray *et al* [73] | 2017 | IIR zero-phase band pass filter | Feature vector |
| Uma and Sheela [75] | 2017 | Butterworth bandpass filter | Extract feature as an eye blink signal |
| Masud *et al* [80] | 2017 | – | – |
| Belwafi *et al* [68] | 2017 | Kaisewin, Butterworth, Chebyshev 1, 2, Elliptic | CSP (common special pattern) |

[51], Kaiser win filter [58], band pass filter [73], impulse response FIR filter [76], whereas feature extraction methods like common spatial pattern (CSP) [58], FFT [55, 76] have been widely used. A summary of preprocessing and feature extraction methods is shown in table 14.5.

### 14.3.3 Classification results

Several researchers have focused on the BCI-based home appliances control system using various data acquisition techniques that have used the artificial neural network (ANN), random forest, LDA, and adaptive neuro-fuzzy inference system (ANFIS) classification methods. As reported in EEG-based home automation system research works, Corralejo *et al* [78] reported classification accuracy of 74.4% using a stepwise linear discriminant analysis (SWLDA) classifier. Achanccaray *et al* [73] reported above 80% classification accuracy for an EEG-based home automation system using an ANFIS classifier. Anindya *et al* [79] reported 83% classification accuracy using a redial base function-support vector machine-redial base function (RBF-SVM). In a similar context, for EEG-based home automation system. Masud *et al* [80] reported an increased, 87% classification accuracy using a random forest classifier, however, this may not be considered to be an excellent system. Uma and Sheela [75] reported 90% accuracy using an ANN classifier for multiple appliance control. Aydin *et al* [17] reported 95% classification accuracy for the EEG-based home appliances control system. Hoffmann *et al* [77] reported 100% classification accuracy for an EEG-based home control system using (SWLDA) and Fisher-linear discriminant (FLDA) classifier. The classification accuracy of that recorded in various research

**Table 14.6.** Typical classifiers applied in the BCI-based home environment system.

| Author | Subjects | Type of paradigms | Classifier | Total selection (session) | Average accuracy (%) | ITR (bit/mint) |
|---|---|---|---|---|---|---|
| Hoffman *et al* [77] | 4 | 3 × 2 images | Stewise-LDA, Fisher-LDA | 24(2) | 100 | 15.9 |
| Aloise*et al* [81] | 7 | 4 × 4 matrix images | | 75(2) | 72.7 | |
| Corralejo *et al* [78] | 15 | 10 matrices 3 × 4, 4 × 4 and 5 × 5 Images | Stepwise linear discriminant analysis (SWLDA) | 134(3) | 74.4 | 16.6 |
| Aydin *et al* [17] | 5 | Main menu 6 × 6, Submenu 6 × 6 | Linear discriminate analysis (LDA) | 20(2) | 95 | 22.1 |
| Anindya *et al* [79] | 4 | | SVM-RBF | | 83.26 | |
| Achanccaray *et al* [73] | 8 | Matrix 2 × 3 | Adaptive neuro-fuzzy inference system (ANFIS) | | Above 80 | |
| Uma and Sheela [75] | 5 | Main menu 2 × 3 GUI based menus | Artificial neural network | | 90 | |
| Masud *et al* [80] | 3 | Matrix 4 × 6 | Random forest | | 87 | |

works 74.4%, 80%, 83%, 87%, 90%, 95%, and 100% was achieved by SWLDA, ANFIS, SVM-RBF, Random forest, ANN, LDA and (SWLDA, FLDA) classifiers, respectively. It can, therefore, be concluded that according to the research published on BCI home appliances control, LDA classifier and its versions such as SWLDA, FLDA has used greater accuracy than others. Table 14.6 provides a description of the results of various classifiers implemented in the BCI-based home environment system.

## 14.4 Controlling methods based on single and multiple appliances

Once the classified signals are converted into a command the same will be given to home appliances. Various researchers have worked to develop control EEG-based smart home control systems. This section describes system design methods to control single and multiple appliances using BCI. The section covers BCI systems design to control lights, fan, door, buzzer and music individually and multiply at a time.

### 14.4.1 Single appliance control

*14.4.1.1 Light control*

The artifact, for example, eye flicker can be utilized as a part of the automation of the home through BCI. There are three forms of reflexive, voluntary, and spontaneous eye blinking [52]. Reflexive eye blinking does not rely on cortical structure, whereas voluntary eye blink depends upon the movement of eyelids due to the previous condition. No outer encouragement is required for spontaneous eye blinking. The most commonly used method for an automation system is voluntary eye blinking.

Researchers [82] gave an idea about EEG-based voluntary eye blink signals that was used three times over a four-second period [82, 83]; the EEG signals were accumulated by setting three electrodes on the subject's frontal, ear ligament, and occipital regions. The RS-232 link was utilized for associating three terminals. Module KL-72001 fundamental unit and KL_75004 EEG module were associated with the PC by means of an RS-232 link. The generated EEG signals have been stored in an excel file. A flowchart of the light control system is shown in figure 14.3.

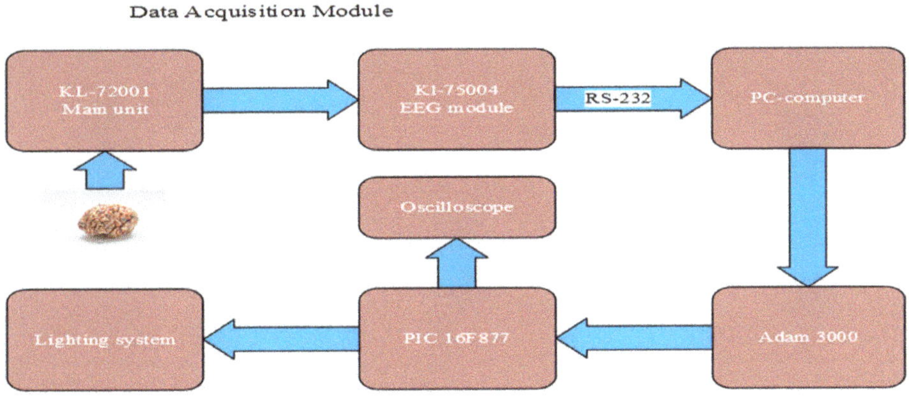


**Figure 14.3.** Light control system.

Instead, the stored EEG signals were transmitted through a data acquisition system (DAQ) to the microcontroller, which was connected to a computer. The test system, which was made up of the PIC16F877A microcontroller and lighting circuit, consisted of a diode, transistor, transfer circuit, and light bulb. The microcontroller detected voluntary eye blinks, and a light bulb was switch on through a relay circuit. To exchange the EEG signals from the PC to the PIC16F877A microcontroller, visual basic software (mikroC) was employed. As the program gets three eye squints inside the duration of four seconds, it will pass one to the microcontroller, which drives the relay to turn the light on.

*14.4.1.2 Fan control using think gear Am (TGAM) EEG sensor*

Wang *et al* [84] describe a BCI-based fan control system by utilizing steady-state visually evoked potential (SSVEP control signal) of different frequency values to generate the EEGs. For data collection, the NeuroSky chip was used. The EEG-input pin was connected to the occipital lobe for acquiring the VEP signal. The REF-pin was connected to the frontal lobe, which generates the spontaneous signal. An EEG–GND pin is associated with the human ear for zero potential references. A db4 wavelet was utilized to break down the EEG signals in order to remove noise from them. The new EEG signals were without noise. For evoked potential signal, an FFT was utilized to decode the signals. The application was developed on an android pad. This application had four flash squares producing frequencies of 6 Hz, 7 Hz, 8 Hz, and 11 Hz, which represent four activities 'turn on the fan', 'pivot with quick speed', 'pivot with low speed', 'turn off the fan'. The android pad was made as a mobile control terminal. A Bluetooth module was utilized to transmit information between the portable control terminal and the EEG acquisition module. The control terminal worked as a simulator receiver and analyzer of SSVEP. It also generated control codes. The control codes were sent to the fan control circuit. The control circuit, which has an STC89C52R single-chip engine, derived the L298N circuit and PWM was utilized to change the speed of the fan. A flow chart of the fan control system is depicted in figure 14.4.

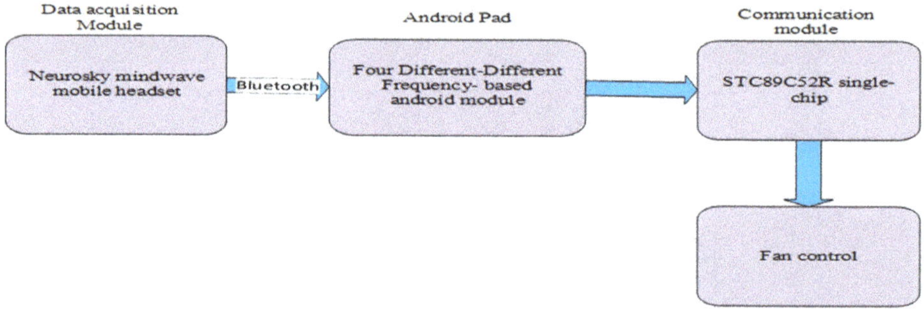


**Figure 14.4.** Fan control system.

*14.4.1.3 Door control using semi-polymer sensor*

Emotive EPOC has the facility to record three-types of suite: (1) cognition suite (2); performance matrix suite; and (3) facial expression suite. The cognitive suite offers various cognitive actions, such as left, right and push–pull action, as well as cognitive action, which involves the imagination. The facial expression suite detects the facial expression and events resulting from facial expression. A performance matrix suite that includes the user's pressure, meditation, motivation, attention, and excitement etc, can detect the emotional and subconscious dimension of the user brain signal. A BCI-based door control system [85] consisted of three main components, viz. emotive headset, base smart home (BSH) system, and thirdly Arduino. The emotive (EPOC) headset is used to distinguish the EEG signal, whereas sensors detect brain electrical activity when the user is relaxed and the eyes are shut [86, 87]. After that, a signal is passed to the BSH system through wireless communication. The base smart home system (BSH) comprises two subsystems called a cognitive and facial subsystem. A cognitive subsystem is responsible for emotive cognitive signal and the facial subsystem is responsible for emotive facial expression signals. Both are subsystem-received signals from the EPOC+ headset. In the psychological (cognitive) subsystem, select push–pull subjective activity was chosen to open and shut the door. In facial expression subsystem select clasping is chosen to open and shut the door [88]. Both of the sessions required instructional training, which is stored in the client profile. After that, the BSH signal comes to the Arduino functioning as a microcontroller, a servomotor associated with the Arduino is used to open and shut the door.

*14.4.1.4 Buzzer control using dry electrode sensor*

A brain concentration-based assistive buzzer system was designed by Chowdhury *et al* [89]. It was designed by using a noninvasive EEG device for acquiring the brain waves. For data collection, a noninvasive EEG device with the specification of 64 channels and 160 Hz examining frequency was used with BCI2000 software [90]. The data collected was loaded into MATLAB for signal processing. The signals were passed through a finite impulse response (FIR) filter for removing noise. Spectral analysis was done using the FFT algorithm. Brainwaves were calculated after this based on the peak value of the brainwave ($\delta$, $\theta$, $\alpha$, $\beta$, $\gamma$). The transition of signals was grouped into five categories. Physically disabled individuals require consideration when waking up, in this condition alpha waves are generated. When the user has a thought, the wave will change from bringing down alpha to higher alpha, in this condition the signal is not initiated. Theta and alpha waves also represent drowsy and sleepy state so in these conditions buzzers will not activate. The buzzer will be on when amplitude crosses 8 Hz frequency. We need users to think in hyper condition, waves turn gamma or higher beta from alpha [91].

*14.4.1.5 Music control*

A brain–computer interface for a music control system designed by Tseng *et al* [92] contained a wireless multichannel data acquisition module and mobile headset. The multichannel data collection module was used to acquire EEG signals and transmit them to the mobile headset, remotely. The mobile headset features a multimedia control program (MIDlet), which is a cognitive state-based detection algorithm. The MIDlet program was used to acquire EEG signals and stored the evaluated psychological (cognitive) state ceaselessly. Four kinds of music (variety 1, variety 2, variety 3, and variety 4) were saved on the computer in four separate folders. At the stage where the playing music is towards the end, the mental (cognitive) states of the user should select the appropriate sort for the next music. In this experiment, the wireless multichannel device is used for data collection. After data collection, signals were down-sampled at 64 Hz applied and FFT was applied to get different EEG spectra. In starting the first few minutes when EEG has recorded users, it should be in an alert state so alpha and theta rhythms can match the alert model. Late in a drowsy state, the user may diverge from the alert mode. As per the alert model, the resultant Mahalanobis distance (MD) index is high, the user's cognitive state trends to drowsy and if the MD index is small the user's cognitive state is alert. Music can be selected according to the user's current situation of four levels of the cognitive state. For example, generally, people in a relaxing situation or sleep suggested variety 3, while folk and modern are set at variety 1 [93, 94]. Levitin and McGill [95] recommended in a walking situation it should be set to variety 2 because this state requires little alertness, the music is close to few beats per minute (BPM) or higher.

### 14.4.2 Multiple appliance control

#### *14.4.2.1 Eye blinks-based method using think gear AM (TGAM) EEG sensor*

Another design, by Uma and Sheela [75], has multiple accessing of apps like wheelchair control, movie player, personalized music player, light and fan control, web control, sending SMS through a mobile device, etc. The main panel of the GUI (graphical user interface) was designed in a button form in a matrix (row and column). The flashing time interval was fixed to 3 s. The subject was able to desire an option by blinking his/her eyes twice. A hardware device called a NeuroSky headset that has a portable sensor capable of extracting EEG data works was used to acquire the eye blink commands. For this purpose, the MindRec tool was employed to which a filtered eye could blink data from EEG data. In the whole process, the EEG data pattern was examined and eye blink information was extracted. After the data collection, the extraction process diminishes the complexity of recorded raw EEG data and extracted feature. The artificial neural network method was used for classification target and non-targets in the stimuli with a 2N hidden layer. Once users select the desired option by eye blinks, the sub-GUI menu was redirected to the main menu and the user was able to select the desired option in the sub-GUI menu again by a double eye blink. A flow chart of the eye blink-based home automation system is shown in figure 14.5.

Some of the other concepts for the BCI-based control framework for smart home appliances are Masood *et al* [96], and Rani and Wahidah [97] using the eye blink method.

#### *14.4.2.2 Microcontroller-based method using semi-dry polymer sensor*

Researchers Alshbatat *et al* [72], came up with an idea regarding the BCI-based home automation system using a microcontroller. The system had two blocks: one

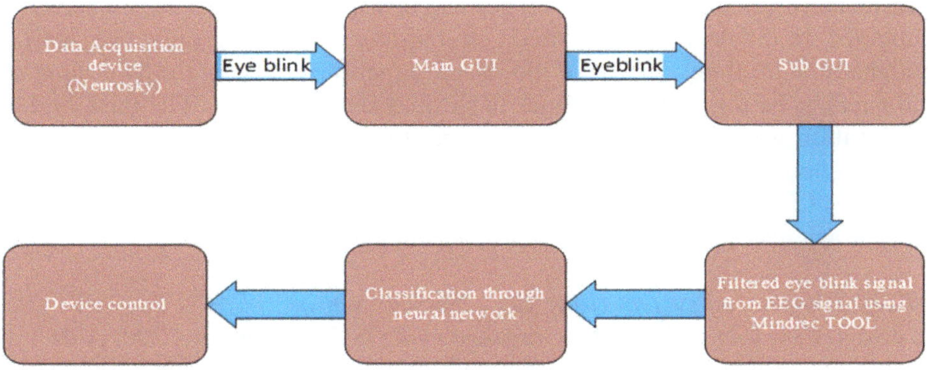


**Figure 14.5.** Flowchart of the eye blink-based home automation system.

block for the transmitter and one block for the receiver. A block of transmitters has six sub-blocks, viz. emotive epoch headset sub-blocks, three wireless modules, a computer, and one converter. The receiver block had eight six sub-blocks, three for the wireless module, two for the PIC microcontroller, one for the GSM module and two for the appliance module. The design working principle of this system was as follows, an EMOTIV EPOC Neuroheadset with 14 electrodes was used to capture thoughts through EmoEngine application software. The application software of EmoEngine was used to process the data for matching the trained pattern from the brain. In this experiment, the test cognitive suite (push, pull, left-right movement of the virtual cube) was classified after training by similar data. Emokey is a tool of EMOTIVE EPOC headset, which can read user thought and send it as a command through the Wi-Fi module directed to the microcontroller. Two microcontrollers were used at the receiver end, the first microcontroller used as a decoder and runs appliances like the light system, the second microcontroller was used to run another appliance and send text messages after cognitive thought detected by a subject.

*14.4.2.3 Virtual reality-based system*

Researchers Carabalona *et al* [98] designed a system for physically impaired persons to control home appliances in a real environment. For the user a 6 × 6 matrix is provided in the form of icons. Each icon represents a command related to a day-to-day life device. The icons flash one by one on the computer screen. Once the desired icon is achieved, there is a peak signal observed. In this study four participants participated, tested accuracy varies from 33% to 100% among different users.

Other designs are given by Holzner *et al* and Guger *et al* [99, 100]. BCI links to a virtual reality control system for home appliances. Guger *et al* designed a P300-based bird's eye view virtual 3D picture of smart home appliances. All seven-control masks were created in this design, namely: light, music, telephone, temperature, TV, moving mask and a go-to mask. The user can do a variety of tasks like switch on or off lights, control TV sets, play music, create video cameras to walk around the house, etc. For the collection of data, 12 subjects participated; each person needed to select seven icons and seven control commands for training purposes. Subjects were asked to focus on the specific icons, for the chosen devices. All of the icons were highlighted in single character mode in random order, i.e. only one icon was highlighted at a time. In the P300-based experiment there are two important parameters to be taken into account, flash time and dark time. The flash and dark time duration was long enough for a subject to detect flashes without overlapping and short enough to maintain high command speed [99]. For training purposes, each icon flashes 15 times, in this experiment, 8, 4 and 2 flashes of each icon were checked, and accuracy and speed achieved were measured.

#### *14.4.2.4 P300-based system using dry electrode sensor*

The next system was designed by Corralejo *et al* [78] based on the P300. A BCI 2000 tool was used for controlling smart home electronics devices through an infrared (IR) emitter device. In this BCI, disabled users were able to control eight devices. In this system, there is one main screen with a $3 \times 4$ matrix menu. In the first two rows, there are eight options representing eight different devices while the first element in the third row represents a phone menu, whereas the last three elements in the third row represent three different commands, namely pause, resume and stop. The user can access the submenu matrix when directed from the main menu. Its size will depend upon the device functionality and the choice to switch to the main menu. There are different sizes of the submenu like $3 \times 4$, $4 \times 4$ and $5 \times 5$, etc. In this study, the stimulus dim time was defined as 62.5 ms and the screen remained active unvarying for 125 ms. Thus the total time for this whole process was 187.5 ms. The user was asked to concentrate on a particular item on the menu and count the number of flashes. All the devices were worked on a computer-connected infrared (IR) emitter (RedRAT Ltd, Cambridge, UK). In this RedRat analysis, the software development key (SDK) was used for storing and transmitting each device's control commands.

#### *14.4.2.5 Web-server based using dry electrode sensor*

The web server-based home appliance control system is an application over region-based paradigms, which is an oddball paradigm based simulation requiring the selection of the target in two levels [67]. In the RB-based paradigm, characters are recognized at two stages. The first stage is used to select the target region, and the second stage is used to select an exact character in that region. In the first stage, the subject focuses on the specific region where the target is located. The target region is a TV, AC; the telephone is the command group at the first level. The entire seven regions have seven commands, which are related to each other.

In this system, EEG signals were recorded employing a cap (actiCAP cerebrum items) which is embedded with sixteen active electrodes and the signal was then amplified by a V-amp amplifier then digitized at the rate of 250 Hz. After, the baseline correction signal was filtered with a low pass filter followed by 30 HZ. The down-sampled data was reconstructed into a feature vector. In completion of the target prediction, a control statement was generated using the target command. HTML language was used to design the web page. In order to provide security, there is a provision of a username, and password authentication also popped up on the web page. The hardware control module consists of an ethernet module, microcontroller, and relay group. The input interface was associated with the control device (module) through LAN, whose yield interface with the outer ethernet module (ENC28J60) utilized TCP/IP connection. The ethernet module worked to receive a command from the client and send it to the microcontroller PIC18F4620, where mikroC language was used for the microcontroller firmware. The web server always monitors the system and checks whether the demand has arrived or not. Once a demand is obtained from the client, the control module checks port number (80), and after the login secret key, the control statement from the BCI customer is activated. All the appliances and control devices are connected through the wired connection of a 5 V relay input with a driving operating voltage rated up to 250 V [67].

*14.4.2.6 Real sound source-based system*

In this experiment, BCI utilized the real source of sound as stimuli. Two separate sounds were simultaneously played on both the left and right sides of the headphone, to produce evoked responses [101]. Every preliminary has two phases: if the subject listens to the auditory stories played through the left earphone in the first stage and right in the second stage, the outcome after this check will be 'LR'. There are four outcome code combination commands that would be obtained in each system simulation screen: RR, RL, LL, LR. Every screen has separate sound commands, for the main screen, if the subject has to monitor the 'TV screen', focus on the left ear and left ear (LL), the 'control AC' focus on the left ear and right ear (LR), the 'make emergency call', focus on the right ear and left ear (RL), the 'For Quit', focus on the right ear and right ear (RR).

When one of the framework control screens is displayed, the system begins a sound guide to let the subject know instantly how to choose one of the four commands in the current screen. Thus, when the subject cannot see the computer, the subject must focus on the instructions for auditory sound RR, RL, LL, LR, accordingly to trial test for controlling purposes. Afterwards, the subject's brainwaves command is sent to the smartphone via the HC-05 Bluetooth module, which is inbuilt in the Arduino. When the smartphone gets the command signal via Bluetooth, the app is executed in the smartphone. The user's phone becomes an infrared transmitter that would transmit the signal to the home devices. The different domestic devices work on different wavelengths, so the system needs to change wavelength frequency according to the domestic device [101].

*14.4.2.7 Universal plug and play networking based system using ski sensor*

A smart living auto adjustment system (BSLEACA)-based brain–computer interface is the universal plug and play (UPnP) home networking technique for smart home applications [71]. BSLEACS mainly consists of four components, viz. wireless data acquisition module, an embedded module, host system, and simple and basic control convention/control line correspondence convention. In this system, the headband is used for data acquisition. The wireless acquisition module acquires the EEG signal from the brain and transmits it via Bluetooth to the embedded acquisition module. An embedded signal processing module contains a blackfin embedded processor which is used in signal processing with an operating frequency of 600 MHZ. The third system is a host system, which is intended for UPnP control point to get an answer from the UPnP control gadget and the fourth system is a power line communication (PLC) environmental controller /simple control protocol (SCP) that is utilized to control all the home gadgets like day-night knob and aeration and cooling systems and so on.

A similar concept was designed by Pradeep and Padmajothi [102] using non–invasive data acquisition machines to control home appliances. They also used drowsiness or alertness state of the user's signals and converted into a command signal to activate different home appliances.

#### *14.4.2.8 FPGA–based system*

This system was designed by Belwafi *et al* [68], using the P300-BCI tool for controlling smart home electronics devices through an FPGA-based embedded system. The FPGA-based home automation deals with two issues, viz. increasing the motor imagery classification accuracy through designing an adaptive filter and designing hardware and software embedded architecture. The motor imagery processes of right- and left-hand movement detection require different preprocessing techniques based on time and a frequency domain transform such as FFT or on DWT. To remove the electrooculography (EOG) artifact from the EEG signal, filtering techniques such as finite impulse response (FIR), an infinite impulse response (IIR) filter, an adaptive filtering technique, and independent component analysis (ICA) were employed. Once the filtering was completed, a common spatial pattern (CSP) method was used for feature extraction. The classifier based on the Mahalanobis distance was then used for classifying the hand and foot movements. To interact with external devices like TV, fan, control switch, and AC, the client needed to envision left- or right-hand development as per the stated machine mechanism. After completion of the filtering, feature extraction, and classification, EEG-based MATLAB code was dumped into the embedded system architecture, which had hardware and software components. Verilog designed the hardware components, and the software portion was written in ANSI-C language. In the hardware core section, NIOS-II fast version common instruction set architecture was used. ANSI-C language was used to prepare a BCI soft-core application and initially developed on the Nios-II-ISS instruction set simulator. The developed code was compiled; GSL library and ANSI-C code were integrated into the eclipse environment of the Altera, which was uploaded on Stratix-IV EP4SGX230KF40C2 kit. After the connection of this kit, the relevant appliance was activated.

As per the discussion and descriptions in previous sections, a number of smart home appliances like light, TV, fan, music system, door, cooling system, buzzer etc, were controlled by different modules like Bluetooth, Arduino board microcontroller, FPGA kit, and infrared etc. The architecture of a BCI–based home environment system is depicted in figure 14.6.

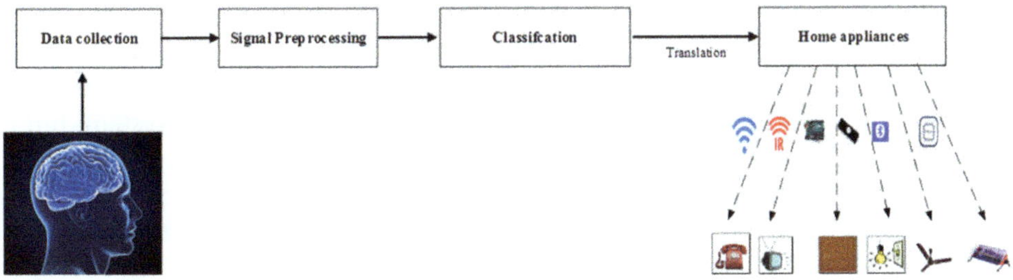


**Figure 14.6.** The architecture of a BCI-based home appliances control system.

## 14.5 Observation and discussion

### 14.5.1 Hardware analysis

- This study was investigated based on the research status of several avenues in this area for the EEG-based smart home appliances control system. Tseng *et al* [92] gave a design for multimedia controller through Bluetooth, but the system was able to control only one device. Miralles *et al* [33] designed an environmental control application through Bluetooth, but its limitation is the number of the control devices it can control. However, the system resolved the cable complexity problem (appliances connection through cable only).
- For Cincotti *et al* [34] the TCP/IP communication approach was an alternative approach for local wireless communication. This technology increased the number of devices but the limitation of this system is the lack of internet communication. Lin *et al* [71] designed a UPnP-based wireless signal acquisition module to handle an environmental control system. This is an alternative approach to the internet (TCP/IP) access. This technology was able to control a number of devices, but this technology failed because the UPnP network setting is quite complex and electrical devices are not compatible with this UPnp technology.
- The next significant technology was the infrared-based control home control system. Corralejo *et al* [78], designed an idea for eight appliances to be controlled through an infrared (IR)-emitter device. Using this, Hsieh *et al* [101] designed a control system for TV and AC through the infrared module. This technology required direct communication between the transmitter and the receiver. This technology had a limitation that a device can only be controlled in a very restricted area. This problem was solved by Aydın *et al* [67], as they designed and implemented an embedded web server (internet-based) wireless control of EEG-based home environment. This system not only removed the distance limitation complexity. It was also cost-effective, low power consumption, and had a small dimension device compatible with the home appliance.
- An Arduino board kit was also utilized for EEG-based appliances control. Anindaya *et al* [79] worked to overcome this limitation by designing a three LED control system through Arduino UNO [85] for a door control system

using Arduino kit, however, the system was able to control only a single appliance. Still, there is huge scope to design a convenient, fast, cost-effective and generalizable system to control home appliances using BCI.

- Shyu *et al* [32] designed an FPGA-based low-cost multimedia system, but this design was able to control only a single appliance. Belwafi *et al* [68] also designed a system through an FPGA-based home control system. This system controlled appliances by Stratix-IV EP4SGX230KF40C2 alterakit, which is very costly. This system used hardware components designed in Verilog and software components designed in C-language, which made the overall system very critical and costly.
- By using the eye blink, Wahy and Mansor [82] designed a system using three eye blink signals (in the duration of four seconds) using a PIC16F877A microcontroller, however, the system was designed only for one appliance. Researchers. Uma and Sheela [75] designed a system using double eye blink to control a home appliance, but they did not focus on the mentioned hardware components. The work has been verified for a healthy person, but it may not be useful for disabled persons.
- An EEG-based home control system using a microcontroller was designed by Alshbatat *et al* [72] but this system was complex and costly.

There remains huge scope to design a convenient, fast, cost-effective and generalizable system to control home appliances using BCI, with different modules utilized to send a command to appliances. Very few research works were actually performed with disable persons. Corralejo *et al* [78] verified their EEG-based home control system for disabled persons. However, a large number of research articles have been reported with an experiment on healthy subjects. Various researchers [17], such as Uma and Sheela [75], and Belwafi *et al* [68] have verified their EEG-based home control system only for healthy persons. Some of the researchers have experimented both with healthy and disable participants. Hence, a research gap exists in terms of checking the physicality and convenience of an EEG-based home control system for actually disable persons. A summary of cited articles for physically disable persons or healthy persons is provided in table 14.7.

**Table 14.7.** Summary of cited articles for physically disabled persons or healthy persons.

| S. No. | Participant | Article |
|---|---|---|
| 1. | Physically disable people (yes) | Corralejo *et al* [78] |
| 2. | Healthy people (yes) | Aydin *et al* [17, 67], Uma and Sheela [75], Belwafi *et al* [68] |
| 3. | Both healthy and physically disabled people | Hsieh *et al* [101] |

### 14.5.2 Discussion

In this chapter, we analyzed most of the widely used algorithms for implementing hardware using the BCI-based home automation system. Figure 14.7 shows the percentage of articles published with different hardware algorithms. Arduino-based (29%) and eye blink classification-based (17%) algorithms are used largely for a BCI-based home control system. Very few researchers worked on FPGA (3%)-based hardware for a home appliances control system.

The research and development of an EEG-based smart home appliances control system can help bring autonomy back to disabled people and research and development of an EEG-based smart home appliances control system will help bring power back to people with disabilities and thereby boost their standard of living and make them self-sustaining. In this chapter, we represented a comprehensive review of different EEG-based home control systems, a key technique about controlling and limitation of the smart home control system. The major difference between the brain-controlled smart home control system and other brain-controlled systems is that these smart home appliances require greater protection because they are used to transport people with disabilities. Numerous analysts have designed various BCI-based home appliances control systems using different techniques. Utilization of other techniques such as intelligent techniques and shared control techniques has made these home control systems safer. Moreover, there is still much work that needs to be done before the BCI-based home system is put into practice, including discovering approaches to enhance the performance of the BCI framework, to establish a standard assessment strategy to encourage the innovative work of BCI–based home appliances control systems.

Firstly, improving the performance of the BCI system (especially robustness) is critical to making the brain-controlled smart home system functional in real-world conditions. One viable direction is to adopt the hybrid BCI system for a better smart home control system.

The BCI mechanisms used in all brain-controlled systems rely on only one form of adequate mind signal (e.g. P300, ERD, or SSVEP) to monitor the intent of the

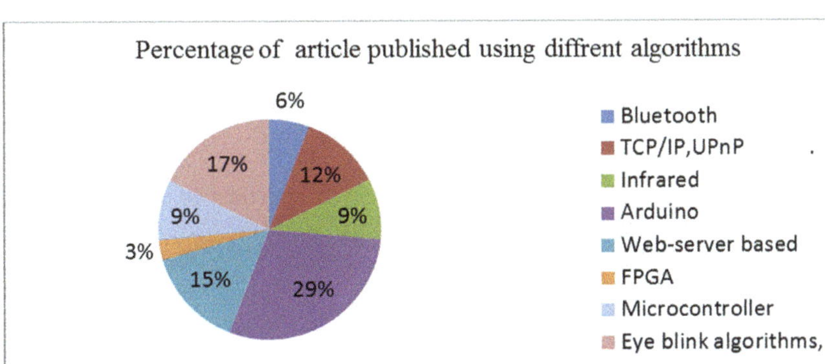


**Figure 14.7.** A pie chart showing the percentages of papers published using different algorithms.

user. However, those BCI systems which supported a single type of signal did not work for all users. Hence, the hybrid BCI approach combining the two approaches should be used to boost accuracy. Second, most of the papers used healthy subjects to check and suggest their method. Moreover, this is an open research question as to whether a healthy person's good performance of the BCI program automatically means good performance for the disabled, whether such connections rely upon the particular disabilities and also particular BCI frameworks and applications. These need to be developed and validated by the targeted disabled community to ensure that the emerging EEG-based smart home automation system can be used by disabled people.

## 14.6 Conclusion

This chapter aims to review different EEG-based home control systems. It was focused on reviewing home control systems using BCI. The performance of EEG-based appliance control systems is judged for the accuracy and ITR of the designed system. In the majority of reviewed articles, researchers have recruited healthy or older people as participants. Only a few articles have reported with disable participants. Different alterations to the client interface plan, the control instruments, and assessment techniques are required before the framework can be practically utilized for physically disabled people. Nowadays there are many low-cost EEG devices on the marketplace. Further clinical research will help to validate the use of these new technologies as a means of communication for people with physical disabilities. Research on a BCI-based home environment system has achieved several vital accomplishments. The success and advancement of the BCI-based home environment system research his going to lead to the development of technologies that disabled people can use, thus improving the standard of living and making them self-dependent.

## References


[1] Huang G *et al* 2012 Effect of duty cycle in different frequency domains on SSVEP based BCI: a preliminary study *Eng. Med. Biol. Soc. (EMBC), 2012 Annu. Int. Conf. IEEE* pp 5923–6

[2] Muller K-R and Blankertz B 2006 Toward noninvasive brain–computer interfaces *IEEE Sign. Process Mag.* **23** 126–8

[3] Hu J-f 2010 Biometric system based on EEG signals by feature combination *Measur. Technol. Mechatron. Auto. (ICMTMA), 2010 Int. Conf.* pp 752–5

[4] Gupta S and Singh H 1996 Preprocessing EEG signals for direct human-system interface *Intell. Syst., 1996, IEEE Int. Joint Symp.* pp 32–7

[5] Nan W *et al* 2011 A comparison of minimum energy combination and canonical correlation analysis for SSVEP detection *Neural Eng. (NER), 2011 5th Int. IEEE/EMBS Conf.* pp 469–72

[6] Wang Y *et al* 2010 Visual stimulus design for high-rate SSVEP BCI *Electron. Lett.* **46** 1

[7] Casagrande A *et al* 2015 PolyMorph: increasing P300 spelling efficiency by selection matrix polymorphism and sentence-based predictions arXiv:1502.04485

[8] Pouryazdian S and Erfanian A 2009 Detection of steady-state visual evoked potentials for brain–computer interfaces using PCA and high-order statistics *World Congr. Med. Phys. Biomed. Eng. (September 7–12, 2009) (Munich, Germany)* pp 480–3

[9] Zhou B *et al* 2014 Robust spatial filters on three-class motor imagery EEG data using independent component analysis *J. Biosci. Med.* **2** 43
[10] Bostanov V 2004 BCI competition 2003-data sets Ib and IIb: feature extraction from event-related brain potentials with the continuous wavelet transform and the t-value scalogram *IEEE Trans. Biomed. Eng.* **51** 1057–61
[11] Millan J R and Mouriño J 2003 Asynchronous BCI and local neural classifiers: an overview of the adaptive brain interface project *IEEE Trans. Neural Syst. Rehabil. Eng.* **11** 159–61
[12] Pfurtscheller G *et al* 1997 EEG-based discrimination between imagination of right and left hand movement *Electroencephalogr. Clin. Neurophysiol.* **103** 642–51
[13] Pfurtscheller G *et al* 1998 Separability of EEG signals recorded during right and left motor imagery using adaptive autoregressive parameters *IEEE Trans. Rehabil. Eng.* **6** 316–25
[14] Oikonomou V P *et al* 2016 Comparative evaluation of state-of-the-art algorithms for SSVEP-based BCIs arXiv:1602.00904
[15] Alomari M H *et al* 2014 Wavelet-based feature extraction for the analysis of EEG signals associated with imagined fists and feet movements *Comput. Info. Sci.* 17
[16] Shukla P K and Chaurasiya R K 2018 An experimental analysis of motor imagery EEG signals using feature extraction and classification methodologies *2018 Int. Conf. Comput., Power Commun. Technol. (GUCON)* pp 796–9
[17] Aydin E A *et al* 2015 Region based brain computer Interface for a home control application *Eng. Med. Biol. Soc. (EMBC), 2015 37th Annu. Int. Conf. of the IEEE* pp 1075–8
[18] Rabiner L R 1989 A tutorial on hidden Markov models and selected applications in speech recognition *Proc. IEEE* **77** 257–86
[19] Li Y *et al* 2008 A self-training semi-supervised SVM algorithm and its application in an EEG-based brain computer interface speller system *Pattern Recognit. Lett.* **29** 1285–94
[20] Cecotti H 2015 Toward shift invariant detection of event-related potentials in non-invasive brain–computer interface *Pattern Recognit. Lett.* **66** 127–34
[21] Kindermans P-J 2014 A Bayesian machine learning framework for true zero-training brain–computer interfaces *PhD Thesis* Ghent University
[22] Cheng M *et al* 2002 Design and implementation of a brain–computer interface with high transfer rates *IEEE Trans. Biomed. Eng.* **49** 1181–6
[23] Helal S *et al* 2005 The gator tech smart house: a programmable pervasive space *Computer* **38** 50–60
[24] Yerrapragada C and Fisher P S 1993 Voice controlled smart house *Consumer Electron., 1993. Digest Techn. Papers. ICCE., IEEE 1993 Int. Conf.* pp 154–5
[25] Long J *et al* 2012 Target selection with hybrid feature for BCI-based 2-D cursor control *IEEE Trans. Biomed. Eng.* **59** 132–40
[26] Pfurtscheller G and Neuper C 2001 Motor imagery and direct brain–computer communication *Proc. IEEE* **89** 1123–34
[27] Lalor E C *et al* 2005 Steady-state VEP-based brain–computer interface control in an immersive 3D gaming environment *EURASIP J. Adv. Sign. Process.* **2005** 706906
[28] Nijholt A *et al* 2008 Brain–computer interfacing for intelligent systems *IEEE Intell. Syst.* **23** 72–9
[29] Bi L *et al* 2013 EEG-based brain-controlled mobile robots: a survey *IEEE Trans. Human-Mach. Syst* . **43** 161–76
[30] Simpson R C 2005 Smart wheelchairs: a literature review *J. Rehabil. Res. Dev.* **42** 423

[31] Gao X *et al* 2003 A BCI-based environmental controller for the motion-disabled *IEEE Trans. Neural Syst. Rehabil. Eng.* **11** 137–40
[32] Shyu K-K *et al* 2010 Development of a low-cost FPGA-based SSVEP BCI multimedia control system *IEEE Trans. Biomed. Circ. Syst.* **vol. 4** 125–32
[33] Miralles F *et al* 2015 Brain computer interface on track to home *Sci. World J.* **2015** 623896
[34] Cincotti F *et al* 2008 Non-invasive brain–computer interface system: towards its application as assistive technology *Brain Res. Bull.* **75** 796–803
[35] Bayliss J D 2003 Use of the evoked potential P3 component for control in a virtual apartment *IEEE Trans. Neural Syst. Rehabil. Eng.* **11** 113–6
[36] Gupta S S *et al* 2012 Detecting eye movements in EEG for controlling devices *Comput. Intell. Cybernet. (CyberneticsCom), 2012 IEEE Int. Conf.* pp 69–73
[37] Mahajan R and Bansal D 2017 Real time EEG based cognitive brain computer interface for control applications via Arduino interfacing *Procedia Comput. Sci.* **115** 812–20
[38] Kais B *et al* 2014 An embedded implementation of home devices control system based on brain computer interface *Microelectron. (ICM), 2014 26th Int. Conf.* pp 140–3
[39] Ramadan R A *et al* 2015 Basics of brain computer interface *Brain–Computer Interfaces* (Berlin: Springer) 31–50
[40] Ramadan R A and Vasilakos A V 2017 Brain computer interface: control signals review *Neurocomputing* **223** 26–44
[41] Flexer A *et al* 2005 Using ICA for removal of ocular artifacts in EEG recorded from blind subjects *Neural Netw.* **18** 998–1005
[42] Gao J F *et al* 2010 Automatic removal of eye-movement and blink artifacts from EEG signals *Brain Topogr.* **23** 105–14
[43] Guerrero-Mosquera C and Vazquez A N 2009 Automatic removal of ocular artifacts from EEG data using adaptive filtering and independent component analysis *2009 17th European Sign. Process. Conf.* pp 2317–21
[44] Li M *et al* 2013 Automatic removal of ocular artifact from EEG with DWT and ICA method *Appl. Math. Info. Sci.* **7** 809
[45] Yang B and He L 2014 Removal of ocular artifacts from EEG signals using ICA-RLS in BCI *2014 IEEE Workshop Electron. Comput. App.* 544–7
[46] Mäki H and Ilmoniemi R J 2011 Projecting out muscle artifacts from TMS-evoked EEG *Neuroimage* **54** 2706–10
[47] Chen X *et al* 2014 A preliminary study of muscular artifact cancellation in single-channel EEG *Sensors* **14** 18370–89
[48] Fitzgibbon S *et al* 2016 Automatic determination of EMG-contaminated components and validation of independent component analysis using EEG during pharmacologic paralysis *Clin. Neurophysiol.* **127** 1781–93
[49] Gerla V *et al* 2017 Automatic identification of artifacts and unwanted physiologic signals in EEG and EOG during wakefulness *Biomed. Sign. Process. Contr.* **31** 381–90
[50] Chen X *et al* 2017 Simultaneous ocular and muscle artifact removal from EEG data by exploiting diverse statistics *Comput. Biol. Med.* **88** 1–10
[51] Hamaneh M B *et al* 2013 Automated removal of EKG artifact from EEG data using independent component analysis and continuous wavelet transformation *IEEE Trans. Biomed. Eng.* **61** 1634–41
[52] Navarro X *et al* 2015 Denoising preterm EEG by signal decomposition and adaptive filtering: a comparative study *Med. Eng. Phys.* **37** 315–20

[53] Patel R *et al* 2017 Common methodology for cardiac and ocular artifact suppression from EEG recordings by combining ensemble empirical mode decomposition with regression approach *J. Med. Biol. Eng.* **37** 201–8

[54] Bera T K 2015 Noninvasive electromagnetic methods for brain monitoring: a technical review *Brain–Computer Interfaces* (Berlin: Springer) pp 51–95

[55] Amiri S *et al* 2013 A review of hybrid brain–computer interface systems *Adv. Human-Comput. Inter.* **2013** 187024

[56] Polich J 2007 Updating P300: an integrative theory of P3a and P3b *Clin. Neurophysiol.* **118** 2128–48

[57] Thulasidas M *et al* 2006 Robust classification of EEG signal for brain–computer interface *IEEE Trans. Neural Syst. Rehabil. Eng.* **14** 24–9

[58] Trejo L J *et al* 2006 Brain-computer interfaces for 1-D and 2-D cursor control: designs using volitional control of the EEG spectrum or steady-state visual evoked potentials *IEEE Trans. Neural Syst. Rehabil. Eng.* **14** 225–9

[59] Rebsamen B *et al* 2007 Controlling a wheelchair using a BCI with low information transfer rate *2007 IEEE 10th Int. Conf. Rehabil. Robot.* pp 1003–8

[60] Muller-Putz G R and Pfurtscheller G 2007 Control of an electrical prosthesis with an SSVEP-based BCI *IEEE Trans. Biomed. Eng.* **55** 361–4

[61] Sirvent J L *et al* 2010 P300-based brain–computer interface for internet browsing *Trends in Practical Applications of Agents and Multiagent Systems* (Berlin: Springer) pp 615–22

[62] Edlinger G *et al* 2011 A hybrid brain–computer interface for smart home control *Int. Conf. Human-Comput. Interact.* pp 417–26

[63] Bordoloi S *et al* 2012 Motor imagery based BCI for a maze game 2012 *4th Int. Conf. Intell. Human Comput. Interact. (IHCI)* pp 1–6

[64] Reshmi G and Amal A 2013 Design of a BCI system for piloting a wheelchair using five class MI Based EEG *2013 Third Int. Conf. Adv. Comput. Commun.* pp 25–8

[65] Long J *et al* 2012 A hybrid brain computer interface to control the direction and speed of a simulated or real wheelchair *IEEE Trans. Neural Syst. Rehabil. Eng.* **20** 720–9

[66] Ma J *et al* 2014 A novel EOG/EEG hybrid human–machine interface adopting eye movements and ERPs: application to robot control *IEEE Trans. Biomed. Eng.* **62** 876–89

[67] Aydın E A *et al* 2016 Implementation of an embedded web server application for wireless control of brain computer interface based home environments *J. Med. Syst.* **40** 27

[68] Belwafi K *et al* 2017 A hardware/software prototype of EEG-based BCI system for home device control *J. Sign. Process. Syst.* **89** 263–79

[69] Martišius I and Damaševičius R 2016 A prototype SSVEP based real time BCI gaming system *Comput. Intell. Neurosci.* 2016 3861425

[70] Anindya S F *et al* 2016 A prototype of SSVEP-based BCI for home appliances control *2016 1st Int. Conf. Biomed. Eng. (IBIOMED)* pp 1–6

[71] Lin C-T *et al* 2012 Brain computer interface-based smart living environmental auto-adjustment control system in UPnP home networking *IEEE Syst. J.* **8** 363–70

[72] Alshbatat A, Vial P, Premaratne P and Tran L C 2014 EEG-based brain–computer interface for automating home appliances. *J. Comput.* 9 2159–66

[73] Achanccaray D *et al* 2017 A P300-based brain computer interface for smart home interaction through an ANFIS ensemble *Fuzzy Syst. (FUZZ-IEEE), 2017 IEEE Int. Conf.* pp 1–5

[74] Virdi P, Syal P and Kumari P 2017 Home automation control system implementation using SSVEP based brain computer interface *2017 Int. Conf. Inv. Comput. Info. (ICICI) (Nov 23 2017 IEEE)* pp 1068–73
[75] Uma M and Sheela T 2017 Analysis of collaborative brain computer interface (BCI) based personalized GUI for differently abled *Intell. Auto. Soft Comput.* 1–11
[76] Tabbal J, Mechref K and El-Falou W 2018 Brain computer interface for smart living environment *2018 9th Cairo Int. Biomed. Eng. Conf. (CIBEC) (Dec 20 2018 IEEE)* pp 61–4
[77] Hoffmann U *et al* 2008 An efficient P300-based brain–computer interface for disabled subjects *J. Neurosci. Method* **167** 115–25
[78] Corralejo R *et al* 2014 A P300-based brain–computer interface aimed at operating electronic devices at home for severely disabled people *Med. Biol. Eng. Comput.* **52** 861–72
[79] Anindya S F *et al* 2016 A prototype of SSVEP-based BCI for home appliances control *Biomed. Eng. (IBIOMED), Int. Conf.* pp 1–6
[80] Masud U *et al* A P300 brain computer interface based intelligent home control system using a random forest classifier *Comput. Intell. (SSCI), 2017 IEEE Symp. Ser.* pp 1–5
[81] Aloise F *et al* 2011 Asynchronous P300-based brain–computer interface to control a virtual environment: initial tests on end users *Clin. EEG Neurosci.* **42** 219–24
[82] Wahy N and Mansor W 2010 EEG based home lighting system *Comput. App. Ind. Electron. (ICCAIE), 2010 Int. Conf.* pp 379–81
[83] Esmaeli M, Jabalameli M H and Moghadam Z 2007 A new scheme of EEG signals processing in brain–computer interface systems *2007 IEEE Int. Conf. on Granular Computing (GRC 2007)* p 522
[84] Wang M *et al* 2016 A fan control system base on steady-state visual evoked potential *Comput., Consumer Contr. (IS3C), 2016 Int. Symp.* pp 81–4
[85] Alrajhi W *et al* 2017 Smart home: toward daily use of BCI-based systems *Info., Health Technol. (ICIHT), Int. Conf.* pp 1–5
[86] Simoens P *et al* 2014 Vision: smart home control with head-mounted sensors for vision and brain activity *Proc. Fifth Int. Workshop Mob. Cloud Computing Services* pp 29–33
[87] Lee W T *et al* 2013 A brain computer interface for smart home control *Consumer Electron. (ISCE), 2013 IEEE 17th Int. Symp.* pp 35–6
[88] Graimann B *et al* 2009 Brain–computer interfaces : a gentle introduction *Brain–Computer Interfaces* (Berlin: Springer) 1–27
[89] Chowdhury A M *et al* 2017 Brain controlled assistive buzzer system for physically impaired people *Electrical, Comput. Commun. Eng. (ECCE), Int. Conf. on* pp 666–9
[90] Schalk G *et al* 2004 BCI2000: a general-purpose brain–computer interface (BCI) system *IEEE Trans. Biomed. Eng.* **51** 1034–43
[91] Popper K R and Eccles J C 2012 *The Self and Its Brain* (Berlin: Springer)
[92] Tseng K C *et al* 2012 Brain computer interface-based multimedia controller *Intell. Info. Hiding and Multimedia Sign. Process. (IIH-MSP), 2012 Eighth Int. Conf.* pp 277–80
[93] Cassidy G and MacDonald R 2009 The effects of music choice on task performance: a study of the impact of self-selected and experimenter-selected music on driving game performance and experience *Musicae Sci.* **13** 357–86
[94] Holm J *et al* 2010 Associating emoticons with musical genres *NIME* 383–6
[95] Juslin P N and Sloboda J A 2001 *Music and Emotion: Theory and Research* (Oxford: Oxford University Press)

[96] Masood M *et al* 2016 Brain computer interface based smart home control using EEG signal *Sci. Int.* **28** 2219

[97] Rani M S and Wahidah B T 2009 Detection of eye blinks from EEG signals for home lighting system activation *2009 6th Int. Symp. Mechatron. App. ISMA 2009*

[98] Carabalona R *et al* 2010 Home smart home: brain–computer interface control for real smart home environments *Proc. 4th Int. Conv. Rehabil. Eng. Assistive Technol* **51**

[99] Holzner C *et al* 2009 Virtual smart home controlled by thoughts *Enabling Technol.: Infrastruct. Collaborative Enterprises, 2009 WETICE'09 18th IEEE Int. Workshops* pp 236–9

[100] Guger C, Holzner C, Grönegress C, Edlinger G and Slater M 2008 Control of a smart home with a brain–computer interface (unpublished)

[101] Hsieh K *et al* 2017 Home care by auditory brain computer interface for the blind with severe physical disabilities *Appl. Syst. Innov. (ICASI), 2017 Int. Conf.* pp 527–30

[102] Pradeep S and Padmajothi V 2012 Brain controlled and environmental auto-adjustment smart home network *Int. J. Adv. Eng. Sci. Res* **8** 363–70

# Chapter 15

## Future directions: smart sensors for green Internet of Things (green IoT)

**K C Raveendranathan**

The theme of implementing an energy-efficient green Internet of Things (green IoT) emerged with the pervasive growth of the IoT and its offshoot technologies like smart sensors and wireless adhoc networks. The green IoT focuses on reducing the carbon footprint of IoT devices and applications, globally. It is an accepted practice that the 17 United Nation's Sustainability Development Goals (SDGs) are to be complied with worldwide by implementing sustainable green technologies. The green IoT concept has attained research focus in recent years, in this context, as the IoT continues to grow as a disruptive technology, entering all walks of human endeavor. It is expected that at the global level, the number of IoT connected devices will reach a staggering 75 billion by 2025. This emphasizes that energy conservation principles if suitably deployed to IoT devices, communication processes (algorithms, software, and systems), computing techniques employed, and smart strategies for adapting energy efficiency in IoT will enable the new technology to result in lower greenhouse gas emission and sustainable outcomes. One of the key components in green IoT is smart sensors, which reduce the energy consumption in transmitting sensed data to the control and processing nodes, wherein the overall carbon footprint of the IoT deployment is reduced considerably. We in this chapter consider the deployment of such smart sensors in the context of green IoT implementation. It is observed that the implementation of green IoT, to a large extent, depends on the use of smart sensors and devices, greener communication systems and protocols, and green computing. Thus, it is evident that the development of smart sensors is the need of the hour in implementing the smart grid, smart transportation, and smart cities. We consider the research and development that is progressing the world over on smart sensors, greener ICT, energy-efficient cloud computing models, and algorithms. The green IoT encompasses the following elements—smart sensors and communication/computing devices (that consume less energy, and thus contribute a lesser extent to the carbon footprint), energy-efficient algorithms and

doi:10.1088/978-0-7503-2707-7ch15  

protocols, and above all sustainable engineering practices. The 5Rs principle in sustainable engineering, viz. reduce, reuse, recycle, recover, and reserve is practiced in developing smart sensors and devices too, that are widely used in smart grids, smart transportation, and smart cities. As an example, the RFID tags used in sensing devices are made out of recyclable and biodegradable materials such as paper. Recycling of electronic waste (or e-waste) is another alarming issue faced by civic bodies. As far as possible, one should use smart materials that are biodegradable, sustainable and have low energy consumption. In the case of ICT deployment, one should always focus on the development and use of more energy-efficient devices, algorithms, and protocols resulting in a smaller carbon footprint. Distributed computing platforms such as the green cloud can be used to preserve energy consumed in computing. Smart communication protocols and algorithms are also widely employed to ensure that data is transmitted only if it is essential to do so, preventing passing on redundant, unclean data to the sensor network.

## 15.1 Introduction

The evolutions of disruptive technologies like the cloud and grid computing, wireless sensor networks, and the IoT, resulted in researchers focusing on more power-efficient sensor devices and computing systems. Thus, smart sensors and green cloud became a reality. The notion of smart sensors is that they always try to limit the power consumed by the sensor devices while maintaining communication among themselves and the central node. One of the formidable challenges in the deployment of wireless sensor networks is the sustenance of the sensors, which are usually powered by batteries. Optimal communication strategies are adapted by the sensors such that they can last longer in action while being powered by batteries. More efficient algorithms were subsequently invented that would more effectively ensure that the sensors remain active longer when powered by batteries. Figure 15.1 depicts a typical IoT and cloud computing and the long-range wide area network (LoRaWAN) scenario. It is quite common that most of the sensors used in IoT are powered on batteries that last 5 to 10 years. In an IoT deployment, the computation load of processing the aggregated sensorial outputs are done in the green cloud computing environment, which ensures that the power consumed by the devices is minimal, again. It should be emphasized that the sensor nodes are connected to the cloud computing system through the LoRaWAN. Subsequently, the mobile and desktop clients can access the processed information by connecting to the cloud using smartphones and/or desktops. The practice of green IoT results in sensors that survive longer without replacement of their batteries. This is achieved by a two-pronged strategy-by reducing the power consumption of the sensors and enabling energy harvesting.

### 15.1.1 The internet of things: prospects and challenges

The IoT envisages total connectivity of consumer electronics devices, home appliances, smart sensors powered by small batteries. The IoT is a system of intertwined computing and communication devices, mechanical and digital

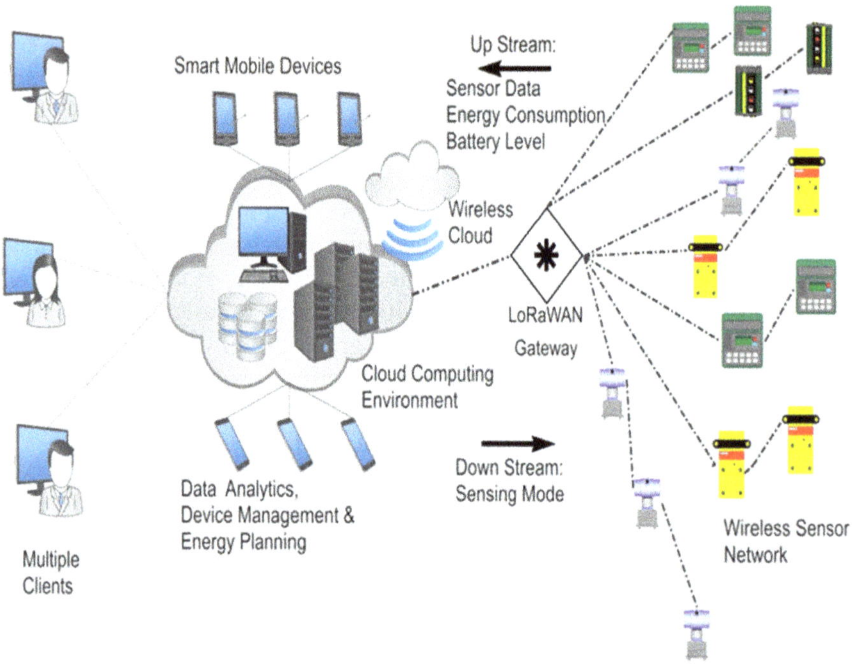


**Figure 15.1.** The IoT, LoRaWAN and cloud environment.

appliances, objects, animals or human beings having unique identifiers (UIDs). All the above entities can transfer data over the Internet without human intervention.

The IoT often refers to sensors, actuators, or any other physical device connected to the Internet. The IEEE 802.11 standard depicts the wireless connectivity technology which is very popular [1]. Applications of IoT include home and building automation, factory automation (which culminated in Industry 4.0), smart cities, smart transportation systems, smart grid and energy management. The Internet Protocol Version 6 (IPv6) over low-power wireless personal area networks (6LoWPAN) adaptation layer and low-power wireless fidelity (Wi-Fi) supported by the IEEE 802.15.4 standard make IoT a reality. The 6LoWPAN adaptation layer provides the specific needs of Internet protocol (IP) for wireless sensors. Several companies are working on reducing the power consumption of Wi-Fi transceiver devices, thus enabling years of battery life [1]. The IEEE 802.15.4 standard is the starting point for several low power wireless connectivity solutions. This includes 6LoWPAN, Zigbee, Thread and many more. The IEEE 802.15.4 wireless communications standard supports communications at a maximum data transfer rate of 250 kbps over distances up to about 10 m. The cost reduction will require highly embedded device solutions in deploying wireless devices, the overall concept of the IEEE 802.15.4 standard has been implemented to permit this concept. The devices we encounter in IoT enabled smart buildings to be grouped into three: AC powered devices such as personal computers (PCs) and home appliances, where the power consumption for wireless communications is not an issue, rechargeable devices

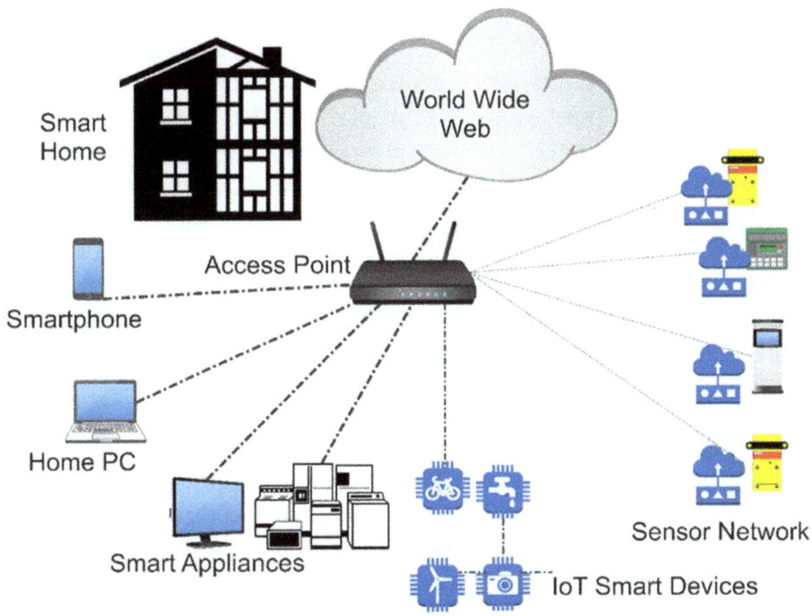


**Figure 15.2.** Connected devices in a smart home.

(smartphones and laptops, that can be powered for a few days from rechargeable batteries) and battery-powered devices including sensors used for detecting humidity, temperature, smoke, and motion. These sensors need to run several years with standard batteries without replacement. As is evident from the illustration, (see figure 15.2), the network of sensors and home appliances (including smartphones, laptops, desktops, and other smart utilities) all connect to the World Wide Web (WWW) through a wired or wireless access point. A typical scenario of connected devices in a smart home is illustrated in figure 15.2. Note that in a typical smart home/building environment, the various sensors and the smart devices/appliances, smartphones, and home PCs are connected to the WWW through a common wired or wireless access point. The cloud computing (CC) hardware and software is in turn connected to the Internet and does most of the high-end signal processing.

### 15.1.2 LoRaWAN® technology

The LoRaWAN® is a kind of low power wide area network (LPWAN) technology. The LoRaWAN is a popular public network protocol developed by the LoRa Alliance® for carrier networks. The LoRa Alliance® include the world's leading technology companies like Cisco, HP, IBM, Foxconn, Semtech and Sagemcom, and the world's leading product companies like Bosch, Diehl, Schneider, and Mueller and several SMEs and start-ups. It is based on spread spectrum modulation techniques, which is derived from the chirp spread spectrum (CSS) technology. LoRaWAN technology has become the de facto technology standard of IoT devices globally. The physical range of LoRaWAN is over 10 km. The key features of

LoRaWAN include long-range (less than 15 km), low power consumption (5–10 year expected battery life), low cost, high security (with embedded end-to-end 128-bit AES encryption of data), and geolocation, which enables indoor/outdoor tracking without global positioning system (GPS). The LoRaWAN technology is ideally suited for a host of smart technologies—smart cities, smart homes/buildings, smart environment, smart lighting, smart metering, smart agriculture, smart industries (Industry 4.0), smart logistics, and smart transportation. It should be noted that LoRaWAN is not the only technology working on the principles of LPWAN. There are several other competing technologies like Symphony Link™ developed by Link Labs which perform better for certain applications, addressing the performance limitations of LoRaWAN.

### 15.1.3 Organization of the chapter

In this chapter, we mainly focus on the smart sensors deployed in the green IoT. We begin with a detailed literature survey on smart sensors and the deployment of green IoT. We also study in-depth various strategies adapted in green IoT usage, especially in smart cities and smart grid deployment. We present a literature survey in section 15.2, which is followed by a detailed study on the principles of smart-low power sensors in section 15.3. We further investigate various algorithms and strategies used for making the green IoT concept practically viable. We compare and analyze the tools and techniques used for green IoT and energy-efficient computing in section 15.4. In section 15.5, we further investigate on how to enhance the performance of smart sensors used in green IoT in the context of smart grids, smart transportation systems and other use cases. We conclude the chapter in section 15.6 by summarizing the ideas developed in previous sections.

## 15.2 literature survey on smart sensors

There is a huge cache of literature available on smart sensors. A basic paper on smart sensor deployment in IoT is by Chun-Ming Huang *et al* [2]. Design and development of smart sensors play a major role in the deployment of IoT. It deserves the paramount attention of researchers who try to develop and design tools and techniques that will make green IoT a reality. One of the often-quoted definitions of the smart sensor is as 'the combination of a sensor with signal conditioning, embedded algorithms, processing circuits and digital interface'. The above-mentioned paper depicts mainly three kinds of smart sensors. They are field programmed gate array (FPGA) based solutions, smart sensor on chip (SSoC) solutions, and smart sensor in package (SSiP) solutions [2]. They are illustrated in figure 15.3 (a–c), respectively. The paper by Dimitrov and Valkovski discusses the use case of optical acoustic sensors for smart city applications [3]. The major merits of optoacoustic sensors are a high level of signal-to-noise ratio (SNR) and wideband coverage. They are also free from electromagnetic interference.

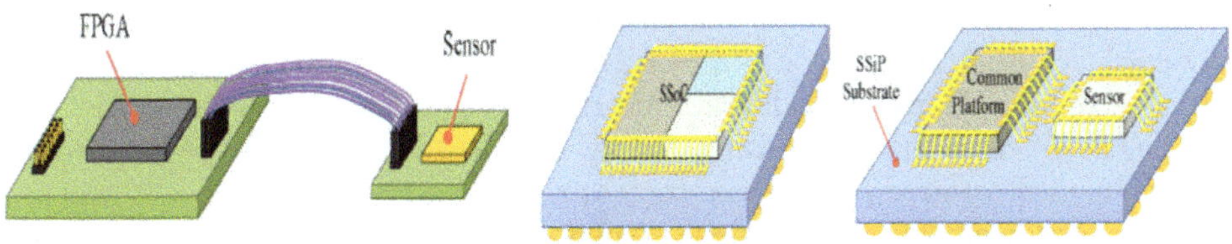


**Figure 15.3.** (a) FPGA based smart sensor, (b) SSoC, (c) SSiP.

Optical sensors are highly reliable and offer excellent performance in use cases such as accelerometers, pressure level meters, and microphones [3]. Using opto-acoustic sensors, one can design and build accelerometers to measure accurately the acceleration of moving objects. Thus, they can be used in smart transportation systems. The textbook authored by Suryadevara and Mukhopadhyay is a comprehensive resource on smart homes; their design, implementation and issues [4]. The book discusses in detail the deployment of smart wireless network in a home environment for monitoring the wellness of elderly people. In another paper authored by Miao *et al*, the working of green cognitive body sensor networks (green CBSNs) their architecture and applications are discussed in detail [5]. The human-centered BSN is found to be the future direction in the deployment of IoT. The green CBSNs gather the human psychological/physiological data from wearable devices, which are usually smart clothing-based devices, with an independently built active green sensor. The wearable device is usually smart clothing or a portable intelligent electronic device. It can gather various psychological and physiological data of human beings by using a multi-modal sensor device. The main component of green CBSN is the body sensor. The body sensor converts the biological, physical, and chemical parameters from the physical world into digital signals [5]. Constituent modules of a sensor node are the sensor, processor, wireless communication, and the power supply modules. The IEEE open access paper authored by Zhu *et al* describes the intricacies of the green IoT for the smart world [6]. In a smart world, various intelligent devices such as smart watches, smartphones, computers and other smart home appliances collaboratively serve human beings. The IoT connects every object in the smart world [6]. This paper discusses the hot green ICT implementations. These include green radio frequency identification (RFID), green CC, green wireless sensor networks, green machine-to-machine communications (M2M) and green data centers (DC). The paper further summarizes the general green ICT principles. The communication protocols available for the IoT include IEEE 802.15.4, ultra-wide bandwidth (UWB), Wi-Fi, Bluetooth, Z-wave, LTE-Advanced, and near-field communication (NFC). We can visualize the following application scenarios for green IoT: smart homes, industrial automation, smart healthcare, smart cities, and smart grid [6]. The recent advancements in micro-miniaturization, RFID and information processing, sensor technology, wireless sensor networks, and smart instrumentation are aiding the easy deployment of green IoT. The paper authored by Habibzadeh *et al* discusses dedicated and undedicated smart city sensing systems [7]. In yet another paper authored by Zeb *et al*, a low-cost, fully printable and flexible, compact, chipless RFID tag is proposed [8]. The tag is being used as a humidity sensor and uses HP photo paper as the substrate. It may be noted that the RFID tag is biodegradable and thus eco-friendly. An overview of the green IoT is given in the paper authored by Albreem *et al* [9]. The paper also deals with the life cycle of green IoT. Due to the tremendous $CO_2$ emissions, as well as the impacts on environment and hygiene, renewable or green technology is attracting the attention of several researchers globally [9]. The evolution of 5G wireless technology in recent years has resulted in 1000 times more data being handled by communication networks. Figure 15.4 depicts the five efficient technologies supported by 5G communication systems. The paper authored by Shaikh *et al* considers several alternative technologies that enable the implementation of green IoT whereby the carbon footprint is reduced [10].

Green IoT can be defined as the 'energy-efficient procedures (hardware or software) adopted by IoT either to facilitate reducing the greenhouse effect of existing applications and services or to reduce the impact of the greenhouse effect of IoT itself' [10]. The major technologies enabling the green IoT are green tags (active and passive RFID tags which are small in size and are biodegradable), green sensing networks, and the green Internet. The paper by Chae-Seok Lee *et al* discusses an energy-efficient RFID protocol. The novel techniques discussed in the paper avoid overheating the environment [11]. The short paper by Kang *et al*, on analysis of types and importance of sensors in smart home services treats issues associated with sensors used in smart homes [12]. Smart homes use sensors to measure physical parameters such as humidity, light intensity, temperature, and movement. The short paper by Gervais-Ducouret treats the details of next-generation smart sensors [13]. The IEEE 1451 standard defines smart sensors as 'sensors with small memory and standardized physical connection to enable the communication with processor and data network'. It may be emphasized that smart sensors containing signal conditioning and embedded functions are more and more acceptable to the user community. They are widely used in consumer electronics, smartphones, and industrial appliances and are gradually replacing the basic sensors [13]. The smart sensors contain sensors/micro electro mechanical systems (MEMS), analog-to-digital converter (ADC), signal conditioning circuits, embedded processors, and a digital interface. Smart sensors are

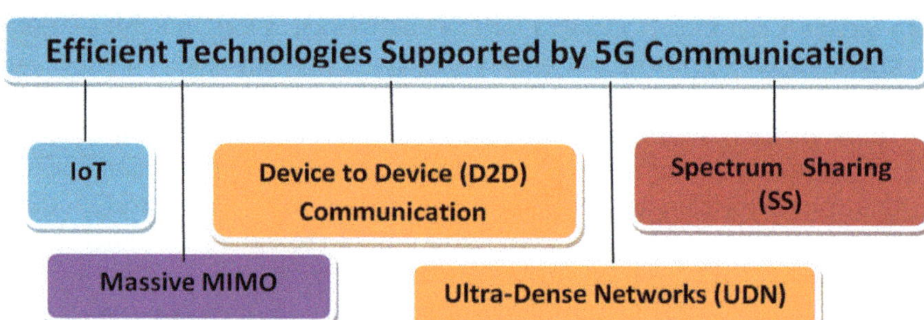


**Figure 15.4.** Five efficient technologies supported by 5G communication systems.

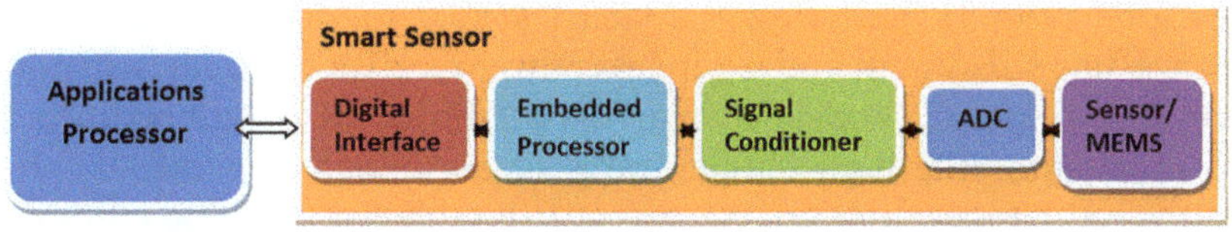


**Figure 15.5.** The structure and components of smart sensors.

highly popular among use cases such as smartphones and tablets. The structure and components of the smart sensor are illustrated in figure 15.5. With smart sensors, only the data that is pre-processed by the signal conditioning circuits and the embedded processor, after removing all redundancies is transmitted; thus, they save both valuable bandwidths of the communication channel and life of the battery. The paper authored by Lung *et al* deals with the management of emergencies in smart cities using smart sensor networks [14].

The authors observe that the steps to prevent disasters include mitigation, prevention, preparedness, response, and recovery [14]. The paper by Tzosu *et al* provides an overview of contemporary smart materials and their applications in sensors and actuators [15]. The details of the development of a new hybrid system in package (SiP) microsensor using high-frequency acoustic wave devices are discussed in the 2016 IEEE Sensors Conference paper authored by Thomas *et al.* The sensor is used for the real-time monitoring of airborne fine particulates ($PM_{2.5}$) [16]. The hybrid particle sensor consists of a solidly mounted resonator (SMR) device based on zinc oxide (ZnO) which is interfaced to an application-specific integrated circuit (ASIC) chip based on CMOS technology [16]. The paper authored by Song *et al* describes sensors used in smart grids. The paper describes the timing, sensing, communication and intelligence specification of sensors for smart grids. It proposes a general model of smart sensors for smart grids meeting these requirements [17]. The authors propose the sensor interface standards used in smart grids and the need for testing interoperability. They suggest a passive interoperability test method for smart sensors used in smart grids to achieve and assure sensor data interoperability. Thus, interoperability issues can be addressed effectively in smart grids applications [17]. The standard definition of smart grids is that 'electrical power grids that apply information, advanced networking, and real-time monitoring and control technologies to lower costs, save energy, and improve security, interoperability, and reliability'. Smart sensors are a crucial entity of smart grids that provide real-time data and status of the grids. Thus, real-time monitoring, control, and protection of the grid are enabled. One of the main challenges faced by the smart grid technology is the real-time exchange of sensor data and the interoperability of the constituent elements [17]. The design, development and deployment of a new fully autonomous self-powered wireless sensor for real-time traffic surveillance in smart cities are discussed in the 2017 IEEE Sensors Conference paper by Balid and Refai [18]. The basis of self-powered wireless sensors is the multi-disciplinary integration of modern ultra-low-power embedded microelectronic devices with smart sensors and wireless sensor networks supported by intelligent algorithms [18]. To enable remote charging of the sensors' primary batteries, the smart sensor module also incorporates an energy-harvesting subsystem and a wireless power receiver enabling self-powering of the sensor unit. A surface acoustic wave (SAW) device-based temperature sensor is deployed to monitor the high temperatures of aero engines wirelessly. It may be noted that wired temperature sensors are not suitable for such use cases due to the difficulty in the smooth transfer of measured signals. Such a SAW-based high-temperature sensor capable of measuring temperatures up to 800 °C with high accuracy and repeatability is discussed in the 2019 *IEEE Sensors Journal* paper authored by Weng *et al.* The substrate used for the SAW temperature sensor is built on 41 degrees Y-cut $LiNbO_3$ along with $SiO_2$ passivation [19]. In the 2019 IEEE Sensors Applications Symposium paper authored by Terkesli *et al*, a smart sensor based on the absorption spectroscopy that meets the demands of both textile dyeing process and industrial specifications is presented. The reported sensor has two path lengths, 1 mm and 20 mm. The electronic circuitry of the smart sensor is integrated into a very compact body. The whole device can communicate wirelessly. The smart sensor is designed, built, deployed and finally tested in a textile dyeing machine [20]. It is to be emphasized that out of the huge energy bill of US $250 million spent on computers every year globally, only a mere 15% (US $ 37.5 billion) is spent on real computing. The rest of the power is just wasted. Thus, it is evident that if one can save the wasted energy on computing and computing hardware, one can prevent several tonnes of carbon emissions into the atmosphere every year. The paper by Abugabah and Abubaker studies the prospects of green computing in reducing the carbon emissions to the atmosphere [21]. Popli *et al* in their survey paper published in *IEEE Access* treat the concept of energy-efficient narrow band IoT (NBIoT) [22]. It is expected that IoT will be able to handle a huge network of billions of connected devices to offer a myriad of intelligent, ubiquitous services to the user community. It will make human life more comfortable but at the cost of huge energy consumption and a resulting larger carbon footprint. Hence there is a real need for tools and techniques that will reduce the energy consumption in computing and communications and consequently the carbon footprint. The idea of green IoT is exactly oriented towards this goal. It may be noted that technologies such as Bluetooth low energy (BLE), ZigBee, Wi-Fi, Z-wave, worldwide interoperability for microwave access (WiMAX), IPv6 over 6LoWPAN etc, offer short-range communications though they consume less power [22]. The IEEE standard protocols *IEEE 802.11ah* and *IEEE 802.11ax* can support several hundreds of devices that consume very low power. Several technologies can offer long-range. They include SigFox, NBIoT, LPWAN, LoRa, long term evolution for machines (LTE-M), and extended coverage global system for mobile IoT (EC-GSM-IoT). All of them can provide long-range. NBIoT is a technology that can support more than 52k connections/channel, both single tone and multi-tone transmissions [22]. Authors Ostal *et al* in their paper, 'Inexact arithmetic circuits for energy efficient IoT sensors data processing' discuss a host of energy-efficient data processing circuits for smart IoT sensors [23]. Inexact or approximate computing offers a very promising solution for data processing which is highly energy-efficient. In inexact computing, the accuracy of computation on the specific application requirements is tuned appropriately to reduce power consumption. Approximate circuits have been designed to 'trade the unnecessary accuracy for a substantial reduction in power, time delay, and hardware resources' [23]. The paper authored by Lin *et al* considers several objectives of the sensor selection problem in the IoT system, including optimization of energy consumption in communications, energy harvesting, energy balancing on all devices, other green concerns, and quality of service (QoS) [24]. The idea behind green IoT is to reduce the energy consumption of IoT devices and to make the environment safe with regards to the reduction of carbon footprint. The IEEE open access paper by Arshad *et al* proposes five principles that can realize green IoT [25]. They are the following: *reduced network size*, *usage of selective sensing*, *usage of hybrid architecture*, *policy towards lower consumption of energy*, and *intelligent trade-offs* [25]. The *IEEE Access* paper by Shirvanimoghaddam *et al* discusses self-powered green IoT using piezoelectric energy harvesting [26]. Many of the IoT devices are installed in hard-to-reach locations, where regular maintenance and replacement of batteries powering the devices are unviable. A feasible solution to this problem is to scavenge and harvest energy from the environment (say, by using solar photovoltaics) to power the devices [26]. Energy harvesting techniques for green IoT may also include energy sources such as thermal, RF signals, and vibrations, in addition to solar energy. The following eight characteristics have been identified for energy sources powering IoT devices: *scalability, maintenance-free, mobility support, long lifetime, flexibility, low cost, sustainability* and *environment friendliness* (*biodegradability*). The direct piezoelectric effect can also be used for energy harvesting in IoT devices [26].

## 15.3 Smart low power sensors—design and applications

Conventional sensors convert physical variables into electrical signals (voltages or currents). smart sensors, on the other hand, do more than that; they have been developed to be able to pre-process the sensed data before it is transmitted. The smart sensors more intelligently connect to the IoT environment. While the basic functionality of the sensors is an essential pre-requisite, smart sensors need to possess the following additional properties to perform as constituent components of IoT:

- they should be inexpensive so that they can be used in large quantities;
- they must be physically small in size to 'disappear discreetly into any environment';
- they must be wireless, as a wired connection is typically impossible in most cases;
- they must be self-validatable and self-identifiable;
- they must consume very little power so that they can survive for several years without a battery change, or manage with energy harvesting;
- they must be rugged and robust, thus limiting or eliminating maintenance;
- they must be self-diagnostic and self-healing;
- they must be self-calibrating or ready to receive calibration commands wirelessly;
- they must have the capability to data pre-processing, to reduce the load on gateways, cloud computing resources and programmable logic controllers (PLCs).

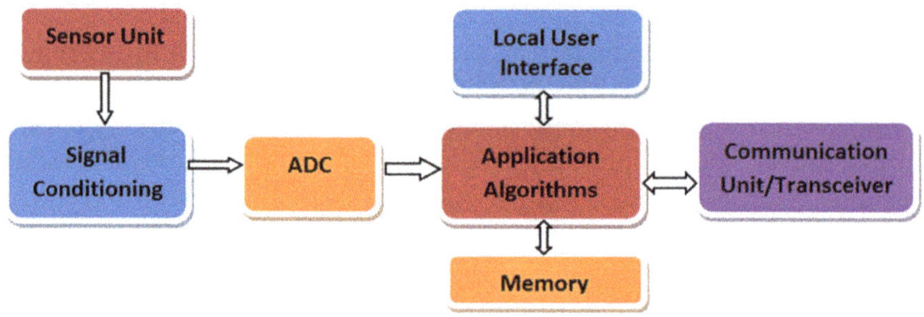


**Figure 15.6.** The building blocks of smart sensors.

Figure 15.6 shows the basic building blocks of a smart sensor. The application programmes/algorithms are computed by an in-built microprocessor/microcontroller unit (MPU/MCU). The smart sensor can do data preprocessing, filtering, compensation, and any other process-specific signal conditioning tasks. Now, let us look into the definition of a smart sensor: '*Smart sensors are built as components of the Internet of Things (IoT) that convert the real-world variable that they're measuring into a digital data stream for transmission to a gateway*'. The intelligence of the microprocessor/MCU can be used to reduce the computational load/overhead on the IoT devices. As an example, the data for calibration of the sensors is automatically set up for any changes in the measurement scenario. The MCU can also detect any drift in performance of the sensors and raise an alarm before any catastrophic failures occur. The smart sensor can work on '*exception mode*' wherein it transmits data only when there is a significant change in the measured physical parameter from the previous values. This will significantly reduce the computational load/overhead on the central processing unit and the power requirements of the smart sensor. This is a significant benefit since the sensor is usually powered by batteries and makes use of energy harvesting techniques from its environment. The smart sensor can also be made having the self-diagnostic capability if there is a redundant sensor module in it. A practical smart sensor was developed by Texas Instruments® which uses their ultra-low-power TI MSP 430MCU®. When installed properly, fault-detectors mitigate operating costs and interruptions in service by giving valuable pointers on a failed node in the network. Moreover, the smart sensing device enhances the safety in operations and diminishes damage to the equipment in its purview by eliminating the necessity of hazardous fault-detection procedures. Fault-detectors are usually located close to the sensing devices and are mostly powered by inexpensive batteries, and often suitable energy harvesting methods are used to enhance the life of the battery module. Thus, the low power operation of fault-detectors is a desirable feature.

### 15.3.1 Fabrication of smart sensors

We will now consider various techniques and tools used for the fabrication of smart sensors. To obtain the full potential of the IoT ecosystem, the smart sensor fabrication techniques should aim at reducing the overall size (form factor), weight,

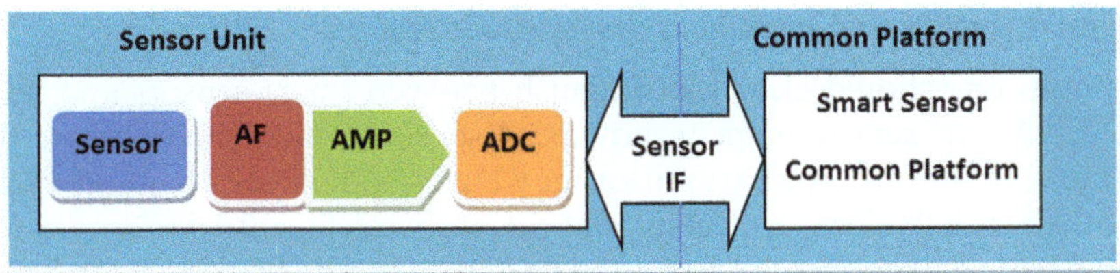


**Figure 15.7.** The architecture of smart sensors.

power consumption and the price of the sensing element and the overall system. The same strategy should apply to the design and fabrication of the packaging adapted, which at present amounts to nearly 80% of the overall price of the device and the form factor. smart sensors are fabricated by integrating MEMS sensor elements and complementary metal oxide semiconductor (CMOS) based integrated circuits (ICs). The CMOS ICs offer the functionality of biasing, filtering, signal magnification, and other signal processing applications. The originally used wafer-level vacuum packing (WLVP) technology, could build only discrete sensors. On the other hand, smart sensors are quite often fabricated by interconnecting discrete MEMS chips to CMOS ICs through the substrate of the package or board using a bonding technique known as multi-chip integration. A better approach could be interconnecting the CMOS IC and the MEMS sensors directly, without using connecting layers in the package or board. This technique is called 'system-on-chip (SoC)'. On comparison to the discrete multi-chip packaging, SoC packaging is predominantly more intricate. However, it brings in a reduction in device parasitics, smaller footprints, larger interconnect densities, and lower package costs. The seminal paper by Huang *et al* gives a detailed discussion on the fabrication of smart sensors as an integrated device [2]. The architecture of smart sensors is illustrated in figure 15.7.

The smart sensor, as shown in figure 15.7, consists of two parts—the sensor unit and the smart sensor common platform. The primary component, namely the sensor unit, comprises the sensor, the analogue front-end processing unit (AF), the amplifier (AMP), and the ADC. The secondary component of the smart sensor is the smart sensor common platform. The sensor interface (IF) interconnects the two parts.

### 15.3.2 Other advantages of having smart sensor intelligence

The smart optoelectronic sensor can detect the structure of objects and any variations in them. This will happen in the sensing element itself, autonomously, and not in any peripheral processing element. This will enhance the throughput of processing and will eventually reduce the processing load/overhead on the central processing element or PLC. Another merit is that the flexibility in manufacturing the sensing unit is improved, which will be a competitive advantage in the present world. Note that intelligent sensors can be programmed remotely with relevant parameters, whenever a variation in the product is needed. For even batch sizes of a single unit, the fabrication, quality inspection, packaging and dispatch can be set at the rates applicable to mass production. Hence each user gets a personalized, customized product and attention. This will be a highly appreciable factor from the viewpoint of the customer.

### 15.3.3 Smart sensors and intelligent sensors—differences

At this point, one should clearly distinguish between smart sensors and intelligent sensors [27]. Smart sensors refer to advanced sensors which can do more than the intended task of sensing. A smart sensor may have sophisticated features, but it may not have the restrictive properties that make it 'truly intelligent.' An *intelligent sensor* has the distinctive capability to detect the varying conditions in its environment and can respond to them appropriately. An intelligent sensor, as implied by its name, is a sensing device that is capable of performing many intelligent functions as part of its designated function or task. It may further be emphasized that an intelligent sensor can do the following tasks in addition to its basic functionality—self-identify, self-adapt, self-test, and self-validate. Intelligent sensors are well aware of the environment in which they are deployed and they can deal with a wide variety of surroundings. Based on the stimulus from the external environment (surroundings), an intelligent sensor is capable of managing its functionality autonomously. This indicates that an intelligent sensor has the architecture to enable advanced self-learning, adaptation to changes in stimulus, and signal processing. All these functionalities are incorporated into one integrated circuit. Quite often, an intelligent sensor needs dedicated hardware known as signal conditioning/processing circuits to supervise and control itself and other appliances in its purview. Authors Taymanov and Sapozhnikova clearly distinguish between smart and intelligent sensors in a book chapter [28]. Both smart and intelligent sensors are capable of functioning for several years, powered by batteries, making use of suitable energy harvesting circuits. The energy harvesting unit consists of an antenna unit and the energy conversion unit. They sometimes use solar photovoltaics (solar energy conversion) or RF energy to DC converters.

### 15.3.4 Energy harvesting techniques in smart sensors

It may be noted that smart/intelligent sensors always have the provision to operate from energy sources that harvest energy from their environments. In this subsection, we will look into various methods of energy harvesting usually used by smart/intelligent sensors. In the *IEEE Microwave Magazine* paper, authors Kimionis *et al*, discuss one of such techniques, in the case of zero-power wireless sensors for smart objects [29]. Wireless sensors and IoT devices depend on batteries for powering the sensors. The batteries often have a limited lifetime. It is a real burden on the users to keep the batteries manually replaced or recharged to keep the sensing devices up and running. There are two distinctive classes of zero-power wireless sensors—all passive structures and low-complexity electronics. In all passive structures, the sensors vary their electrical characteristics according to the sensed variable. In the case of low complexity electronics, the sensing module harvests energy from its environments [29]. An illustration of a zero-power wireless sensor node is shown in figure 15.8.

The end device modules in a smart sensor are illustrated in figure 15.9. Note that it consists of an MCU, sensors, RF transceiver unit and the powering unit which contains an energy harvesting circuit. The energy harvesting unit powers the sensors,

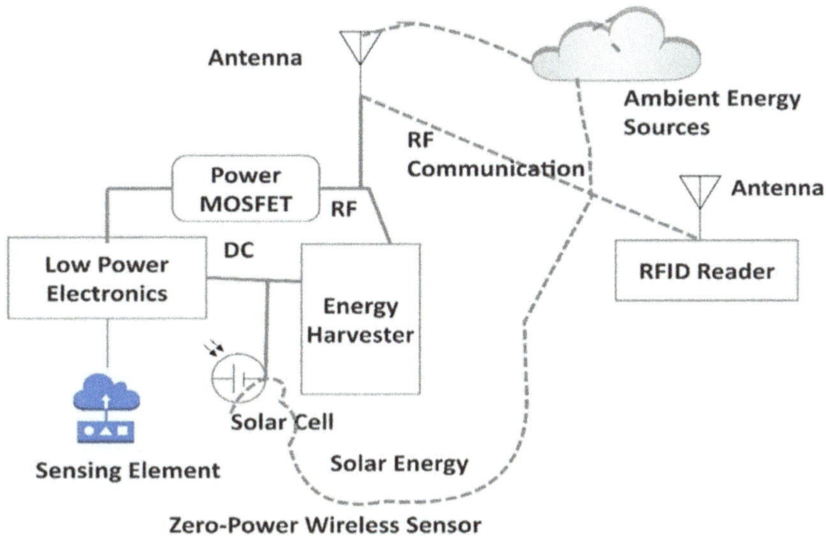


**Figure 15.8.** A zero-power wireless sensor node.

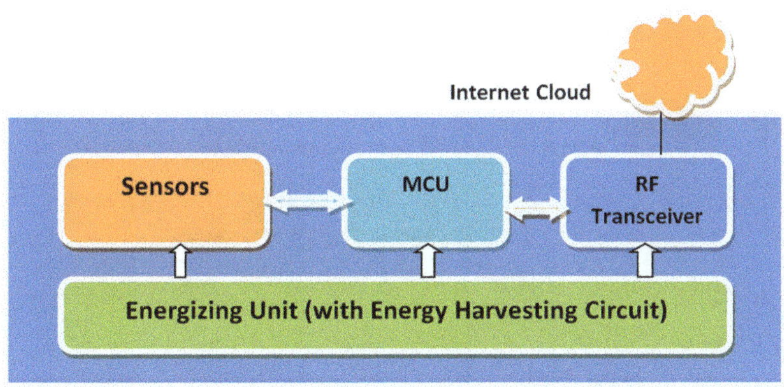


**Figure 15.9.** Modules in a smart/intelligent sensor unit.

the MCU and the RF transceiver unit. The RF transceiver is in turn connected to the Internet cloud.

### 15.3.5 Internet cloud

Figure 15.10 illustrates how the wireless sensors and connected to the LoRaWAN or other LPWAN gateway, and subsequently to the CC environment. The *end device (ED)* can be a wireless sensor. It is shown that there are several such devices for the wireless sensor network which is connected to the CC environment through the LoRa gateways [30]. The authors Misra *et al* describe an IoT based wireless energy harvesting unit, in reference [31]. The RF energy harvester uses several energy-efficient voltage doubler stages in an RF to DC converter. Although there are

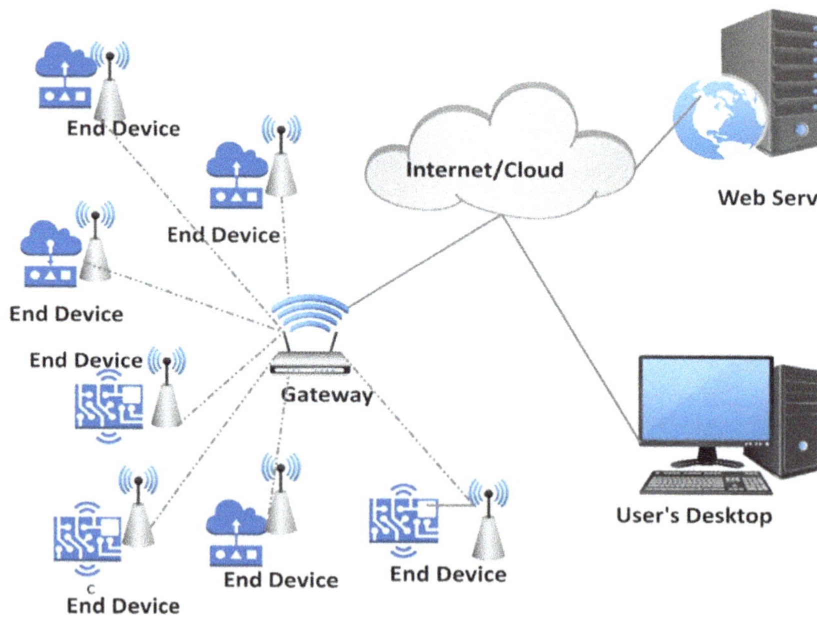


**Figure 15.10.** Wireless sensors connected to a gateway and the cloud.

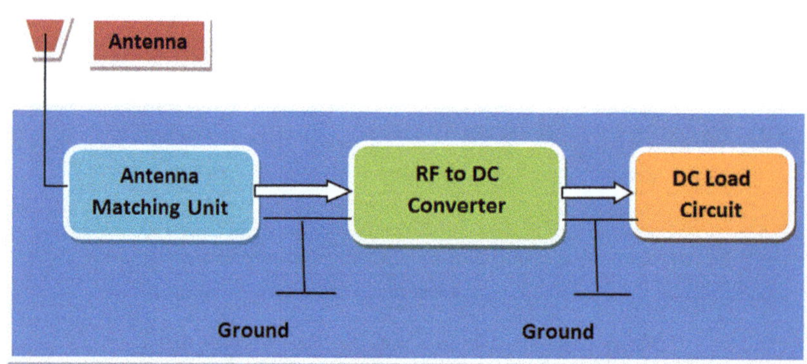


**Figure 15.11.** The basic building blocks an RF energy harvesting unit.

several energy harvesting techniques such as light (photovoltaics), vibration (kinetic), motion (piezoelectric), wireless (RF) and thermal, the *wireless energy harvesting (WEH)* is the most prominent energy harvesting method. The reasons for the wide popularity of WEH are its simplicity in structure and implementation and wide accessibility. The WEH is done making use of the ambient RF energy in the environment. The basic building blocks of a typical RF energy harvesting unit are illustrated in figure 15.11. The working is quite simple. The RF energy collected by the antenna is fed to the antenna matching unit (AMU). The AMU is functionally an impedance matching circuit that matches the impedance of the antenna to that of the RF-to-DC converter circuit. This enables maximum power transfer. The AMU feeds the RF to DC converter circuit, which transforms the RF energy to the appropriate DC voltage. The output of the RF to DC converter power the DC load. The major concerns of the RF energy harvesting unit are energy conversion efficiency, size and cost. The RF to DC converter circuit includes a voltage multiplier circuit using HSMS2850 zero-bias Schottky diodes (Villard voltage multiplier circuit) [31].

The output voltage obtained at the seventh stage of the Villard voltage multiplier circuit is 2.273 V [31]. The circuit diagram of a single stage of the Villard voltage multiplier is shown in figure 15.12. An important characteristic of the Villard voltage multiplier circuit is that as the number of diode-capacitor stages increases, the DC output voltage also increases. However, the ripple voltage generated also increases. Also, after the fifth stage, the enhancement in the DC output voltage is negligible [31]. Thus, in a practical circuit, one will employ up to five diode-capacitor stages, which is considered to be an optimal solution.

## 15.4 The green IoT—prospects and challenges for smart sensors

The energy consumption of the devices and circuits in an IoT environment is a highly challenging research domain. In a review paper, Albreem *et al* delve deep into the green IoT [9]. Miao *et al*, in their *IEEE Sensors Journal* paper describe the architecture, energy harvesting and smart clothing-based applications of green Cognitive Body Sensor Networks [5]. The authors predict that the human-centric Body Sensor Network will evolve as one of the most promising future directions in the technology of IoT, due to the recent innovations in embedded microelectronics, wireless communication technology, and mobile internet technology. Combining the above with green sensors results in the development of green CBSN. The paper also summarizes some of the open issues and challenges in the security of IoT, optimization of health big data algorithms, energy-saving and energy harvesting. Arshad *et al*, in their survey paper in *IEEE Open Access* investigate the scope of green IoT practices for 2020 and beyond [25]. They have identified the following

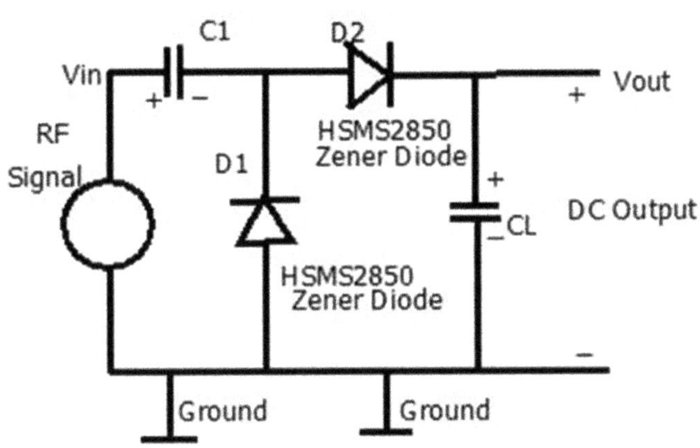


**Figure 15.12.** The circuit of Villard voltage multiplier.

application domains for green IoT: power distribution, transportation, mining and garments industries, food supply chain, smart cities, smart homes and buildings. Some of the major issues identified by the authors on IoT deployment are data secrecy and privacy. One of the most common cases is that of RFID tags, which can be used to follow a person without his/her consent and knowledge, which will arise some amount of distrust. The massive use of active RFID tags consumes large amounts of energy. Also, the massive volume of sensor data emanating from the active RFID tags are to be processed by huge data centers that also consume lots of energy [25]. Green IoT and green computing try to address these issues. Green IoT mainly focuses on efficient information and communication technology (ICT) practices. Green IoT tries to reduce the emission of $CO_2$ and other greenhouse gases into the atmosphere. A concept of green IoT is shown in figure 15.13. For implementing green IoT practices one has to adopt several strategies and techniques. These include the use of green RFID tags, green data centers, green sensor networks, and green CC. Energy harvesting from environments is quite important when one deals with green IoT practices using wireless sensor nodes. green data centers and green cloud are yet another strategy towards making the IoT green. Data centers play a pivotal role in making IoT green. Intelligently selecting servers and geographically distributing the computing load will also help to achieve the goal. Controlling the behavior of sensors, including making them send the data

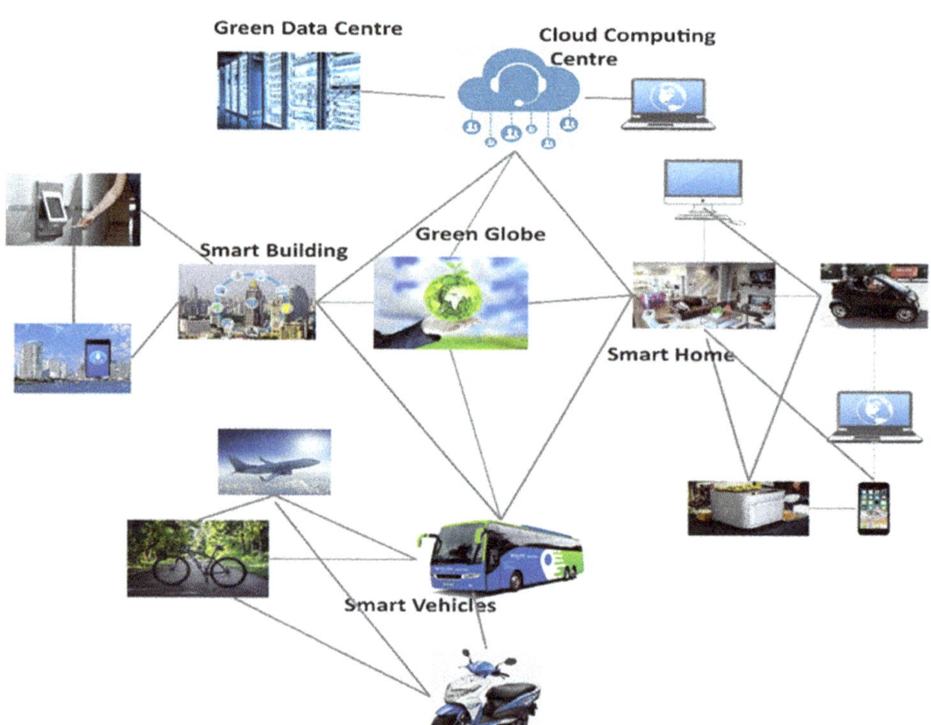


**Figure 15.13.** The concept of green IoT.

selectively, sleep scheduling, compressed sensing will make the sensing units green. Controlling the devices and appliances in smart homes by an intelligent operating policy also contribute quite a lot to the objective of green IoT. Table 15.1 summarizes the strategies to be adopted to achieve green IoT.

**Table 15.1.** Strategies for green IoT.

| Technology focus | Software/ hardware based | Energy saving method | Compromises made |
|---|---|---|---|
| Data center (DC) | Software | Context-aware, demand-based allocation of servers | Needs more resources for QoS |
| Data center (DC) | Software | Workload distribution among geographically dispersed data centers. | Higher complexity |
| Cloud computing | Software | Reduction in data path | Drop-in QoS |
| Cloud computing | Software | Dynamic packet downloading using access points to shrink latency and traffic | Compromise in QoS |
| Processor | Hardware | Task scheduling and optimization using multiple processor cores | High cost and complexity |
| Sensors | Software | Selective sensing, sleep scheduling, and compressed sensing | Privacy and energy overheads for context-aware sensing, extra resources for QoS |
| Sensors | Hardware | Reduced processing at the sensor end | Communication delays in extreme cases |
| Sensors | Hardware | Categorization of objects in layers | Communication delays between sensor nodes |
| Sensors | Hardware | Segmenting the networks and assigning specific tasks to each | Transmission latency |
| Sensors | Hardware | Dividing workload among different types of nodes (sensor and relay) | Communication delays |
| Sensors | Software | Combining media access control (MAC) with wireless smart sensors to reduce communication between nodes | Communication delays |
| RFID tags | Hardware | Use of passive sensors | Communication delays |
| ICs | Hardware | Mitigating network traffic using SoC | Higher cost |
| Smartphones | Hardware | Recycling of used parts to upcycle them | Chance of wastage of recycled parts |
| Smartphones | Software | Prediction of energy consumption for different applications | Data Privacy and Security |

It is reasonable to classify the approaches towards green IoT as hardware-based, software-based, awareness-based and policy-based [25]. Though some of the suggestions are productive in the sense of reducing the carbon footprint, the overall performance of the system is affected. Hence, a judicious selection of the techniques is often warranted. In their review paper, 'Green Internet of Things: a study of technologies challenges and applications' authors Nazish and Banday, discuss the various challenges and applications of green IoT [32]. The authors speculate that the extensive use of IoT devices may increase the emission of greenhouse gases and aggravate global warming, which can lead to climate change and other undesirable situations. Therefore, sustainable, eco-friendly, low-energy consuming solutions are to be invented. Low-energy variants of conventional IoT solutions are available and the use cases of such green IoT hardware and software are to be promoted. The enabling technologies for green IoT are green sensor networks, green data centers, green CC systems, green RFID tags, and green machine-to-machine (M2M) communications [32]. green sensors and localization of computing load by deploying intelligent/smart sensors is one approach towards achieving green IoT. The authors identify the following challenges in green IoT practices.

Data security and privacy is one such major concern. Scalability of the devices used another major challenge. Handling of failures and reliability of the hardware and software is yet another challenge. Maintaining the trust of the user community is another bigger challenge the developers face. The applications of green IoT are many. They have been widely used in smart homes/buildings, smart grids, smart waste management, smart irrigation and agriculture, smart healthcare, and smart transportation [32]. smart cities, smart grids, and smart transportation have evolved as major subject disciples in recent years. The four different modes of green sensors are illustrated in figure 15.14.

A recent book edited by Thomar and Kaur, Green *and Smart Technologies for Smart Cities*, treats the concept of smart cities and their various constituent technologies in sixteen chapters [33]. The paper authored by Wang *et al*, 'From IoT to 5G I-IoT: the next generation IoT-based intelligent algorithms and 5G technologies', discusses the implementation of the next-generation IoT, named as intelligent IoT (I-IoT) using the emerging fifth-generation communication technology, 5G [34].

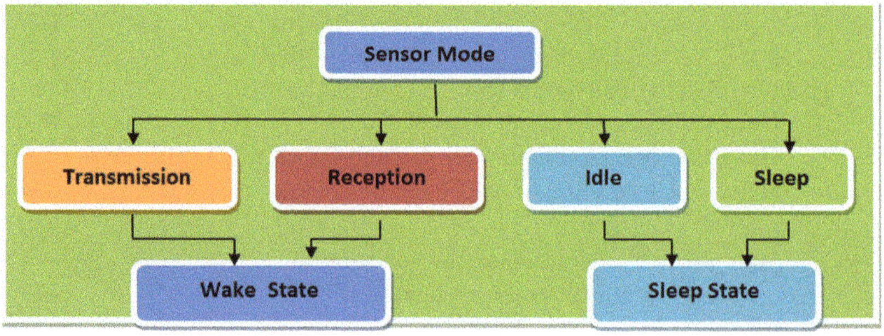


**Figure 15.14.** The four different modes of a green sensor.

### 15.4.1 Smart sensors for smart grids—problems, prospects and challenges

In the seminal paper authored by Burunkaya and Pars, 'Smart meter design and implementation using ZigBee based wireless sensor network in smart grid', the design and implementation of *wireless sensor home area network (WSHAN)* with smart meters using ZigBee protocol are presented [35]. By definition, smart grids are 'electrical power grids that incorporate information processing, advanced networking and real-time control and monitoring technologies to bring down operational costs, conserve energy, and enhance security, interoperability, and reliability'. For this purpose, they make use of smart sensors which present data and status of the power grids in real-time, for concurrent monitoring, surveillance, control and protection. One of the major operational challenges of smart grids is the exchange of sensor data and interoperability. There are specific requirements for the smart sensors in sensing, timing, communications and intelligence [17]. Wireless sensor networks play a crucial role in the deployment of smart grids [35]. Smart energy metering is one of the constituents of smart grids. The National Institute of Standards and Technology (NIST), USA proposed the following general expectations and requirements from smart grids: highly reliable power source, improved quality, increased capacity and efficiency, enabling self-healing and self-monitoring, enabling of automated maintenance, diminished emissions of $CO_2$ and other greenhouse gases, increased security, provision for consumer selection, and ushering new applications, products and systems. The sensors used in smart grids should meet the following requirements: highly accurate timing and timing synchronization to Universal Time Coordinated (UTC), fast data processing capacity and accessibility of intelligent algorithms for generating synchronized frequency, phasor, and rate of change of frequency (ROCOF) estimation from the measured voltage and current signals along with time synchronizing signals, highly accurate and sensitive measurements of voltage and current magnitudes and phase angles. They should also enable ultra-fast secure and reliable network communications and standards-based data transmission, standardized interfaces and test methods to help achieve smart sensor interoperability and plug-and-play (PnP) capability, a wide range of sensors with high bandwidth/dynamic range such as measuring voltage from 600 V to 69 kV of medium voltage, current from amperes to kilo-amperes, frequencies from 50 Hz to 5 MHz, multiple sensing capabilities of electrical and physical parameters including voltage, current, power flow, temperature, weather and climate, intelligent capabilities for sensors such as self-description, self-identification, self-localization, self-testing, self-awareness, self-compensation, self-diagnostics, and self-calibration [17]. Authors Yassine *et al* in their IEEE open access paper, 'Smart meters big data: game theoretic model for fair data sharing in deregulated smart grids' rightly point out that smart metering based on analysis of big data emanating from smart sensors deployed in smart grids can provide valuable insights into consumer power usage patterns [36]. It is observed that the power grid should be incorporated with the sophisticated solutions of power utility companies (generators and distributors), to secure greater reliability, availability and speed in response to outages and fault occurrences, as well as fault-tolerance and resilience [37]. The IEEE Standards

Association's documents [38, 39] highlight the technology forecast for smart grids and the associated communications technology.

### 15.4.2 Smart cities—concept, technologies and challenges

The concept of smart cities is fairly old. Smart cities incorporate recent innovations in smart grids, sustainable engineering, and wireless sensor network-enabled I-IoT to build state-of-the-art environments for humans. The paper authored by Shamsir *et al*, 'Applications of sensing technology for smart cities', discusses various aspects of using smart sensors as an enabling technology for smart cities [40]. The very rapid urbanization of human society and the aftermath of the industrial revolution have resulted in huge consumption of natural resources like water, air, soil, and fossil fuels and natural gases leading to several environmental issues. The concept of smart cities tries to address the above-mentioned problems by making use of the advances in smart sensors and 5G wireless communication technology, thus making human life more sustainable and eco-friendlier. Smart sensing has a vital role in this perspective right from the inception stage to the design and deployment of infrastructure to automatic monitoring and control of the entire structure [40]. The following constituents are present in the smart city paradigm-smart infrastructure, smart surveillance systems, smart transportation systems, smart healthcare, smart resource management, and smart services (including smart e-governance). The smart cities are characterized by their intelligent and sustainable infrastructure by incorporating advanced, green technologies. The smart homes and buildings in a smart city provide a more comfortable, energy-efficient, economically viable, and environment-friendly environment to the residents [40]. The recent innovations in the development of smart sensors have played an increasingly prominent role in the quality of our everyday life. The smart sensors can collect data, detect critical conditions in the environments, along with the smart and intelligent signal conditioners and processors that are integrated into them. smart sensors can process the detected data and take necessary actions autonomously, thus eliminating the need for human intervention, thereby reducing the price tag and energy utilization [40]. The major challenges are the availability of inexpensive sensors, which are biodegradable and hence environment-friendly, and resilient, needing very little electrical power for their operation.

### 15.4.3 Smart transportation systems—problems and prospects

Smart transportation systems are an integral part of smart cities. It may be seen that principles of big data analytics are widely applied to several application scenarios like the efficient planning, designing and deployment of smart control systems, smart transportation systems, smart healthcare, smart cities, smart communities, and smart grids. However, the analysis of big data for smart control systems offers many challenges and issues, if one resorts to using conventional engineering tools and techniques. These challenges include the problems associated with the fast processing of big data in real-time and efficient decision making and management, based on the results. Authors Jan *et al* in their paper in *IEEE Wireless Communications*, discuss a model for analyzing transportation data with *Hadoop* and *Spark* to handle real-time data processing originating from transportation systems [41]. The system is grouped into four layers: data acquisition and collection, network, data processing, and application. Smart transportation systems are a major constituent of the smart city. It requires interconnected, intelligent instrumentation for enhancing the interactions of citizens and boosting safety on roads. Smart vehicles and nearby roadside units (RSUs) are outfitted with hardware modules (sensors and data processing units) for detecting, processing, analyzing, and communicating real-time data and statistics making a *Vehicular Adhoc NETwork (VANET)*. The huge chunk of data communicated between vehicle-to-vehicle (V2V) and vehicle-to-infrastructure (V2I) necessitates efficient data collection, cleansing, processing, and transmission of aggregated information to specific receiving nodes [41]. The following services have been identified as part of smart transportation-monitoring of traffic congestion, smart parking, and smart traffic signaling/smart lighting. Most commonly, the necessary traffic-related data is collected using RFID tags mounted on the vehicles. Also, GPS is employed to track and locate the vehicles on the move. The survey paper by Hettikankanama and Vasanthapriyan, 'Integrating smart transportation system for a proposed smart city: a mapping study', reviews a large number of research papers on smart transportation systems [42].

## 15.5 Enhancing smart sensors for a smart world

The prospects of smart sensors are enhanced to a larger extent by making them *intelligent*. As discussed in detail in subsection 15.3.3, *intelligent sensors* are much more than being merely *smart*. They have special purpose signal cleansing and preprocessing capabilities as well as the agility and capability to respond to the changes in the environment in a truly intelligent way. Modern artificial intelligence (AI) algorithms such as *concurrent neural networks (CNNs)*, *deep learning* and *machine learning* are implemented in intelligent sensors. They reduce the computational overhead on the central controlling node drastically. Also, the communication latency is reduced to a large extent as it communicates only data that is non-redundant and useful for further processing. Intelligent sensors, more or less, make use of principles *compressive sensing* as well. A compressive sensor senses the data in a non-linear fashion, reserving a lesser number of bits to represent sparse data. Intelligent sensors find greater acceptability in the deployment of modern industrial automation, which is often called 'Industry 4.0'. A fairly good discussion of smart/intelligent sensors is given by authors Taymanov and Sapozhnikova in reference [28]. *Zero-power sensing* is another hot area of research and development activity by which energy harvesting from the environment is done to reduce the dependence on the energy from batteries. Smart and intelligent sensors will invariably have provision for energy harvesting from the environment. In their *IEEE Microwave Magazine* paper, authors Kimionis *et al*, deep delve into the functioning of such zero-power active sensors [29]. Another issue to be addressed is the biodegradability of the sensors, after they are decommissioned or dismantled after years of operation. This aspect is very important in the context of sustainable development.

## 15.6 Summary and conclusions

We have considered some of the most promising developments in the field of green and smart sensors and their use cases in green IoT applications. We have done a detailed review of current literature on the IoT, smart sensing and intelligent sensing. The concept of green IoT is discussed and technical details of smart and intelligent sensors are given. Applications of green IoT like smart homes and buildings, smart healthcare, smart grids, smart transportation, smart waste management and smart cities are discussed. Rapid urbanization and the fourth industrial revolution, *Industry 4.0*, have resulted in a greater strain on the consumption of natural resources and pollution of the environment, globally. An RF energy harvester unit, based on the Villard voltage multiplier, using 5–6 stages of diode-capacitor networks was discussed in detail. It is generally observed that green IoT, deployed with the help of smart and intelligent sensors, low-energy communication devices and protocols, energy harvesting from environments, smart *in situ* processing, and advanced green data centers and green CC will be the technology of the future. The availability of low-power inexpensive MCUs sharing the sensor data processing overload from the data centers and CC systems offer a viable alternative and decentralized data processing will continue to evolve in the coming decade. Energy harvesting from the environment will make more smart sensors thrive on inexpensive long-lasting batteries that can be left unattended with practically zero maintenance for several years in the field. We have also considered the recycling of materials used in green IoT devices so that the technology becomes eco-friendly and sustainable. Smart printable RFID tags made of biodegradable paper are a prominent solution in this regard. The whole idea behind green IoT is to reduce energy consumption in the devices, so that the emission of greenhouse gases is reduced, thus paving the way to the diminished carbon footprint. Meeting the United Nations' Sustainable Development Goals (SDGs) is effectively undertaken by implementing green IoT worldwide. As smart and intelligent sensing continues to evolve and several disruptive technologies are being developed, one major challenge to be faced by the user community is the data overflow due to sensors and abundant use of devices connected to the IoT. However, the more recent developments in data sciences and data analytics are offering feasible solutions to the deluge of data. A new branch of science, viz. big data analytics has started offering viable solutions to the plethora of data emanating from the smart/intelligent sensors, globally. It is observed that one of the major use cases of big data analytics is indeed processing of data streams from I-IoT. It is predicted that the use of the green IoT devices and smart monitoring devices will give a new fillip in the design and deployment of smart cities and smart grids and several other use cases.

## References

[1] Tozlu S, Senel M, Mao W and Keshavarzian A 2012 Wi-fi enabled sensors for internet of things: a practical approach *IEEE Commun. Mag.* **50** 134–43

[2] Huang C-M *et al* 2019 A smart sensor development platform and its system demonstration *Proc. 2019 IEEE Symp. Circ. Syst. (ISCAS)* 1–4

[3] Dimitrov K and Valkovski T 2019 Using optical acoustic sensors for smart city applications *Proc. 2019 Int. Conf. Creat. Busi. Smart Sustain. Growth (CREBUS) (March)* 1–4
[4] Suryadevara N K and Mukhopadhyay S C (ed) 2015 *Smart Homes: Design, Implementation and Issues* (Berlin: Springer)
[5] Miao Y, Wu G, Liu C, Hossain S M and Muhammad G 2019 Green cognitive body sensor network: architecture, energy harvesting and smart clothing based applications *IEEE Sens. J.* **19** 8371–8
[6] Zhu C, Leung V C M, Shu L and Ngai E C-H 2015 Green Internet of Things for smart world *IEEE Access* **3** 2151–62
[7] Habibzadeh H, Qin Z, Soyata T and Kantarci B 2017 Large scale distributed dedicated and non-dedicated smart city sensing systems *IEEE Sens. J.* 7649–58**17**
[8] Zeb S, Habib A, Amin Y, Tenhunen H and Loo J 2018 Green electronic based chipless humidity sensor for IoT applications *Proc. 2018 IEEE green Technol. Conf. (green Tech)* 172–5
[9] Albreem M A M, El-Saleh A A, Isa M, Salah W, Jusoh M and Azizanand A. Ali M M 2017 Green internet of things (IoT): an overview *Proc. 4th IEEE Int. Conf. Smart Instr., Measure. App. (ICSIMA)* 1–6
[10] Shaikh F K, Zeadally S and Exposito E 2017 Enabling technologies for green internet of things *IEEE Syst. J.* **11** 983–94
[11] Lee C-S, Kim D-H and Kim J-D 2014 An energy efficient active RFID protocol to avoid overhearing problem *IEEE Sens. J.* **14** 15–24
[12] Kang B, Kim S, Choi M-i, Cho K, Jang S and Park S 2016 Analysis of types and importance of sensors in smart home services *Proc. 2016 IEEE 18th Int. Conf. High-Perf. Comp. Commun.; IEEE 14th Int. Conf. Smart City; IEEE 2nd Int. Conf. Data Sci. Syst.* 1388–9
[13] Gervais-Ducouret S 2011 Next smart sensors generation *Proc. 2011 IEEE Sensor App. Symp.* 1–4
[14] Lung C, Buchman A and Sabou S 2018 smart city emergency situations management system based on sensors network *Proc. 2018 IEEE 24th Int. Symp. Design Technol. Electron. Pack (SIITME)* 288–91
[15] Tzou H S, Lee H J and Arnold S M 2004 Smart materials, precision sensors/actuators, smart structures, and structronic systems *Mech. Adv. Mater. Struct.* 11 367–93
[16] Sanju Thomas M, Cole F H, Villa-Lopez J W, Gardner J, Peters and Jan Theunis A 2016 Low-cost acoustic microsensor based system in package for air quality monitoring *Proc. 2016 IEEE Sensor Conf.* 1–3
[17] Song E Y, Patrick G J F and Lee K B 2017 Smart sensors and standard-based interoperability in smart grids *IEEE Sens. J.* **17** 7723–30
[18] Balid W and Refai H H 2017 On the development of self-powered IoT sensor for real-time traffic monitoring in smart cities *Proc. 2017 IEEE Sensor Conf.* 1–3
[19] Weng H, Duan F L, Zhang Y and Hu M 2019 Hu, high temperature SAW sensors on $LiNbO_3$ substrate with $SiO_2$ passivation layer *IEEE Sens. J.* **19** 11814–8
[20] Terkesli I, Demircan C, Demirci I, Yavuz S and Taygun M 2019 Industrial spectrophotometric smart sensor towards real-time optimization of textile dyeing process *Proc. 2019 IEEE Sensor App. Symp.* 1–5
[21] Abugabah A and Abubaker A 2018 Green computing: awareness and practices *Proc. 2018 4th Int. Conf. Comput. Technol. App.* 6–10

[22] Popli S, Jha R K and Jain S 2019 A survey on energy efficient narrowband internet of things (NBIoT): architecture, application and challenges *IEEE Access* **7** 16739–76

[23] Ostal M, Ibrahim A, Chible H and Valle M 2018 Inexact arithmetic circuits for energy efficient IoT sensors data processing *Proc. 2018 IEEE Int. Symp. Circ. Syst. (ISCAS)* 1–4

[24] Lin C-C, Deng D-J and Lu L-Y 2017 Many-objective sensor selection in IoT systems *IEEE Wirel. Commun.* **24** 40–7

[25] Arshad R, Zahoor S, Shah M A, Wahid A and Yu H 2017 Green IoT: an investigation on energy saving practices for 2020 and beyond *IEEE Access* **5** 15667–81

[26] Shirvanimoghaddam M, Shirvanimoghaddam K, Abolhasani M M, Farhangi M, Barsari V Z, Liu H, Dohler M and Naebe M 2019 Towards a green and self-powered Internet of Things using piezoelectric energy harvesting *IEEE Access* **7** 94533–56

[27] Yurish S Y March 2010 Sensors: smart vs intelligent, *Sensors Transducers J.* **114** I–VI

[28] Taymanov R and Sapozhnikova K 2018 What makes sensor devices and microsystems 'intelligent' or 'smart'? *Smart Sensors: MEMS Intelligent Sensor Development Microsystems for Industrial Applications* ed S Nihtianov and A Luque 2nd edn (Cambridge: Woodhead Publishing) 1–22

[29] John Kimionis W, Su J, Hester J, Bito X, He T-H, Lin and Tentzeris M M 2018 Zero-power sensors for smart objects *IEEE Microwave Mag.* 32–47

[30] Fialho V and Fortes F 2019 Low power IoT network sensors optimization for smart cities applications *Proc. 2019 Int. Conf. Smart Energ. Syst. Technol. (SEST) (September 2019)* 1–6

[31] Misra D, Das G and Das D 2018 An IoT based wireless energy harvesting using efficient voltage doubler stages in a RF to DC converter *Proc. 2018 4th Int. Conf. Comput. Commun. Auto. (ICCCA)* 1–5

[32] Mir Nazish and Banday M T 2018 Green internet of things: a study of technologies challenges and applications *Proc. 2018 Int. Conf. Auto. Comput. Eng. (ICACE–)* 210

[33] Pradeep T and Kaur G (ed) 2020 *Green and Smart Technologies for Smart Cities* (Boca Raton, FL: CRC Press)

[34] Wang D, Chen D, Song B, Guizani N, Yu X and Du X October 2018 From IoT to 5G I-IoT: the next generation IoT-based intelligent algorithms and 5G technologies *IEEE Commun. Mag.* 114–20

[35] Burunkaya M and Pars T 2017 A smart meter design and implementation using ZigBee based wireless sensor network in smart grid *Proc. 2017 4th Int. Conf. Electric. Electron. Eng.* 158–62

[36] Yassine A, Shirehjini A A N and Shirmohammad S 2015 Smart meters big data: game theoretic model for fair data sharing in deregulated smart grids *IEEE Access* **3** 2743–54

[37] Burian R, Gontijo M and Alvarez H 2019 Robustness and reliability in smart grid solutions *Proc. 2019 the 7th Int. Conf. Smart Energ. Grid Eng.* 59–63

[38] The IEEE smart grid research group 2013 IEEE Grid Vis. 2050 Roadmap (IEEE PES, IEEE Standards Association) 1–15

[39] Bush S F, Goel S and Simard G 2013 The IEEE smart grid research group *IEEE Vis. smart grid Commun. 2030 Beyond Roadmap (IEEE Communications Society IEEE Standards Association)* 1–19

[40] Shamsir S, Mahbub I, Islam S K and Rahman A 2017 Applications of sensing technology for smart cities *Proc. 2017 IEEE 60th Int. Midwest Symp. Circ. Syst. (MWSCAS)* 1150–3

[41] Jan B, Farman H, Khan M, Talha M and Din I U August 2019 Designing a smart transportation system: an internet of things and big data approach *IEEE Wireless Commun. Mag.* **26** 73–9

[42] Hettikankanama H K S K and Vasanthapriyan S Integrating smart transportation system for a proposed smart city: a mapping study *Proc. 2019 IEEE Int. Res. Conf. Smart Comput. Syst. Eng. (SCSE)* 196–203

# Chapter 16

## Basic modeling of 'cognitive sensors' based on literary study of 'thick description' of human behavior

**Valiur Rahaman and Sanjiv Sharma**

Sensors are the spirits of modern and postmodern technologies. Where there is no sensor, there is no technology. The emergence of the automatic machine is caused by thinking about sensors or controls. From machine control to human control, sensors developed automation of human and non-human society. This chapter delineates the findings of research review in the history and philosophy of technology conceptualizing a novel idea of how humanities experts can help develop sensors to measure constituents of human behavior such as beliefs, values and attitude. Literary theory is one the integral branches of literature under humanities. It is observed that applications of literary theory and cultural studies may be utilized to pave the paths of people pursuing science and technology for its advancement. Intelligent and human sensors have revolutionized the recent development. However, social psychology of a person or social human behavior remained undiscussed in terms of how sensors can detect the problems in human social behavior. This is simply because machines embedded with human/intelligent sensors are not taught about or do not have embedded constituents of human social behavior. Hence, this chapter delineates the application of Gilbert Ryle's study of human social behavior and Clifford Geertz's study of semiotics of human social behavior known through a model they call 'thick description'.

The application of 'thick description' is observed as a relevant literary theoretical model for establishing a basic model for defining a 'cognitive sensor' to be used for human security in more effective ways without stemming any loss to human life. In doing so, a brief history of literature related to the development of cognitive sensors for their uses, their success, and failure, as well as their ecological effects raising questions for human security at large would be an integral part of our discussion. The discussion would raise issues related to human social behavior reading by the

doi:10.1088/978-0-7503-2707-7ch16  

sensors and how the implementation of theories of Gilbert Ryle and Clifford Geertz help develop cognitive sensors to control the prevalence of human crimes, and affirm human security through digital litigation.

## 16.1 Introduction: humanities-inspired technology

Ada Lovelace, a well-known collaborative mathematician, computer scientist and a daughter of a leading Romantic poet Lord Byron, worked with Charles Babbage, known as the father of computers on his Analytical Machine. She wrote about what she called 'poetical science' emphasizing collaboration of science and the poetics (humanities). Bringsjord and Ferrucci [1] related Lovelace's idea of this collaboration with AI-creativity; Kernan [2] exemplifies how technology 'actually affects individual and social life.' For Feenberg [3], literary writers and critics have contributed to the development of science and technology: Lear; Madden [4, 5] from Aristotle's *Problemata Mechanica*; Moorman [6]; Miller [7] on Descartes's Cartesian oval; to Thomas Carlyle's Carlyle Circle; Shakespeare, Johnson, and Carsten [8–10] on Jacques Derrida [8, 9, 11, 12] and his deconstructive mathematics; yet in the history of technology or philosophy of technology we seldom come across these facts. This is a problem for which this chapter establishes humanities inspired technology for society (HITS) as a requisite for advance sensor technology. In the chapter, we explain the findings on how literary arts do have reflections of technology and medical science and how literary studies or study in literature can help us think better about the world of technology.

Tracing the contribution of literary authors, critics, philosophers of language, traditional knowledge and social scientists in the history of science and technology is one of the requirements of research in HITS. Computational studies have, now, started intersecting many research areas of humanities and social sciences and attempted to replicate human behavior to machine behavior. This attempt of mechanical replication has warmed the research communities to welcome experts in humanities and social science. We live in the age of IBM Watson led by David Ferrucci and his team. IBM Watson, the machine, is a living example of the result drawn from working in human behavior, language, world knowledge system, and sentiment reactions. Other such examples are arguments related to Putnam [13], the ratiocination of Sherlock Holmes directs readers to real-life detective deduction resulting in 'the mind of BRUTUS, a storytelling machine'. In a nutshell, it's a replication of the most intelligent mind of the universe if such a creature exists. It is one of the best examples of exposition of power of the sensor based in IoT and AI to date.

The strangeness of the machine is not because of only technological knowledge but also the knowledge in humanities and social sciences. For example, the foundation of natural language processing is the knowledge of human language, its parent and proto-language [14] like *Devanagari* and its phonetics. The knowledge of aesthetics plays a crucial role in sentiment embedding process to a machine or sentiment analysis or behavior analysis making an expert understand the *rasas* or human emotions. The present chapter fills the gap with exposition of the knowledge system of how literary/cultural theory, which is an integral part of literature studies (one of the disciplines in humanities) helps develop sensors to study human behavior.

## 16.2 The function of literary theory for medical sciences and technology

Pandit and Rahaman [15] describe 'advanced humanities', of which medical humanities remains its integral branch of knowledge, offering researchers and scholars an opportunity to witness humanities beyond its traditional premises and circumferences of studies. Literature, being a sub-branch of humanities, has gone far across its functions and become the source of knowledge system for the world of natural sciences and technology, nursing and medical sciences. Charles W Eliot [16] compiled viewpoints of literary artists about their works to be published, ranging from William Caxton to H A Taine, and it is observed how they are concerned about every aspect of past, present and future of human lives, be it science, logic, government, policies, law, philosophy, psychology, habits of human mind, astrology, anthropology, and what not. Perhaps it was the reason why Jacques Derrida defined literature as 'everything'. As observed above, Aristotelian theory of rhetoric is still a foundation for all the models of communication in electronics; Descartes, the philosopher and literary artist developed a principle called 'Cartesian oval' which is an integral part of geometry; Thomas Carlyle, a literary artist, developed what is known in mathematics as 'Carlyle mircle'; Bacon's 'Preface to *the Novum Organum*' still works as an epistemic shadow of what we know as 'critical pedagogy' or 'democratic education' and 'clinical psychology'.

### 16.2.1 Health science in literature

The development of 'cognitive sensors' demands not only the knowledge of technology but also the knowledge of health security through calculating the side-effects of the sensors and envisioning the prevention system. Heikenfeld *et al* [17] studies identification of the epidermis, affirming its need while developing and projecting or implementing a sensor for two reasons: epidermis as information barrier and security reasons in human life. Heikenfeld *et al* [17] state, for example, 'Epithelia like the epidermis are common in other areas and organs of the body where a barrier function is required': in wearable sensors. Literary knowledge also works here. There is plethora of literary arts that are the clusters of medical conditions represented by literary characters: Shakespeare's *Hamlet, Othello, King Lear*, and *Macbeth*; Dostoevsky's *Crime and Punishment*, James Joyce's *Ulysses*, J M Coetzee's *Disgrace* all represent socially disturbed and criminal mindset behaviors. They are the examples of how to study brain and behavior.

### 16.2.2 Cognitive poetics of health: health imageries

Authors are writing about the findings and the procedures of study during the COVID19 pandemic lockdown. What common things we observed in all these creative specimens are human fear of contagion conditions, escapism from a fatal life-in-death, riot conditions and foretelling of the literary artists. Hence, when we apply Valiur Rahaman's [18] 'cognitive poetics of health' (conceptualized and coined by Valiur Rahaman during COVID19) as an approach to literature, we have to demonstrate these observations in comparison to the literary arts/works like *The Decameron* by Giovanni Boccaccio (1313–1375) and *The Canterbury Tales* by Geoffrey Chaucer (1343–1400). Portraying the fear of contagion increased vices such as self-centrism, avarice, greed, and corruption, which paradoxically led to infection and thus to both moral and physical death; *A Journal of the Plague Year* by Daniel Defoe (1659–1731), which is a long, detailed narrative of events, anecdotes, and statistics regarding the Great Plague of London of 1665, *The Plague* by Christina Rossetti, *The Plague* by Camus and *The Stand* (1978) by Stephen King are all post-apocalyptic literary works based on epidemics or a pandemic of a weaponized strain of influenza that kills a large population across the world. The survivors, with societal behaviors and positive attitude, united in groups, establish a new system in society.

These literary arts require experts in literary theory who do not apply any theorists' principles to study arts but develop or model a new theory, a new approach, a new principle to study deeply any arts for providing solutions to the problems raised by technology and its users in terms of any issues which may badly affect human and post-human society. This human problem-solving function is the function of literary theorists in the twenty-first century.

### 16.2.3 Identifying neurocognitive conditions: analysis of Keats' *Ode to a Nightingale*

Hippocrates (c. 460–c. 370 BC) the philosopher, historian and literary prose writer wrote *Of the Epidemics* (400 BC), symptomatizing tuberculosis (TB) as the symptoms of what he called 'phthisis', which literally means things that cause decay or 'wasting away', known as an epidemic related to lung diseases in his time. The symptomatology of phthisis as he described the epidemic behavior and attacking time in words as 'early in the beginning of spring, and through the summer, and towards winter, many of those who had been long gradually declining, took to bed with symptoms of phthisis … Many, and in fact, most of them died…Phthisis makes its attacks chiefly between the age of eighteen and thirty-five. Pains about the head and neck, and heaviness of the same along with pain, occur either without fevers or in fevers'. If we observe verse-lines of Keats' poems *Ode to a Nightingale* (1819): 'My heart aches, and a drowsy numbness pains/My sense, as though of hemlock I had drunk,/Or emptied some dull opiate to the drains/One minute past, and Lethe-wards had sunk' exemplifies the medical condition of the poet who was suffering from TB, in the beginning of 19th century, even when the word tuberculosis was not coined nor any bacterial diagnosis could be possible. This symptomatology of TB in a poetic composition of Keats was exposed sixty years before a German physician and Noble laureate Dr Robert Koch announced the discovery of *Mycobacterium tuberculosis*, the bacteria that causes TB in March 1882. This is an example of how diligent reading of literary arts helps exemplify the cognitive neurological factors of human behavior. Joseph Collins, the neurologist, read literature for illustrating neurological conditions of the patients. He was one of the first reviewers of James Joyce's *Ulysses* (1922) and wrote *The Doctor Looks at Literature*, a finest example of reading literature in terms of medical psychoneurology. Table 16.1 represents how words used in a literary art reflect actual symptoms of TB or phthisis or Hippocrates' epidemic symptoms resulting in cognitive neurological conditions and factors.

Ramachandran [19] studies 'anosognosia' (1998) as one of the neurological phantasmatic effects in human life and explains some of its symptoms in dialogic or narrative forms to depict how the subject does not feel or realize or actualize his/her sickness in spite of being sick like paralytic in real-life as explained in the words of Ramachandran [19] 'the patients' indifference to their predicament not only is an impediment to rehabilitation of the weak arm or leg, but often leads them to unrealistic future goals.' Here Keats' words reflect what may be called reverse anosognosia due to his perception of paralytic limbs whereas he is not a paralytic person. Keats' words 'drowsy numbness pains' and 'hemlock I had drunk/or emptied some dull opiate to the drains' are fine examples of 'reverse anosognosia'. And, the poet wants to escape from the conditions in which he really is and he has to be, yet he ignores the condition with strong desire and longs for a peace which is not possible in his life at all as he accepts. Sufferers with COVID19 have a similar issue. For people suffering from first aid symptoms of CORONA the virus that causes COVID-19 is transmitted through droplets generated when a CORONA infected person coughs, sneezes, or exhales. The number of affected patients in India, recorded on 18 May 2020 is 90 927, confirmed died 2872 people, and in the World 4.69 million, confirmed died 313K. Even after knowing this disastrous pandemic, people try to escape from diagnostic centers, and checking stations. This anti-quarantine escapism that leads to mortality is a result of the denial Ramachandran [19] called 'indifference to predicament'.

Rahaman [18] after reading literature in terms of cognitive sciences or neuro-pandemic terms, medical conditions, societal unrest, public unawareness, and results of anti-quarantine escapism breaking lockdown laws made under COVID19 pandemic, of which we the human and post-human beings are suffering, can at least lead us to its prevention and control. The entire history of literary arts related to epidemics and pandemics addresses how social irresponsibility of a person or a few persons bring disaster and decay and at the same time a few socially inspired or humanity inspired persons bring out new societies where the care of the self becomes social care to prevent and control a pandemic.

The values of literary theory applied to literary arts increases when it promotes technology. Sam Halliday [20] studies Hawthorne, Melville, Twain, James, Whitman, Margaret Fuller, Henry Adams, Mary Shelley, Hardy and Kipling as representative of electricity's significance. Campolo [21] says that Heidegger thinks that technology objectifies the world in terms of its potential usefulness to the human subject.

**Table 16.1.** Neurological factors of human behavior exemplified through reading poetry.

| The poetic words used in Keats' *Ode to a Nightingale* | Hippocrates' epidemic symptoms | Cognitive behavior | Basic neuro-cognitive factors |
|---|---|---|---|
| Heart aches | Cardialgia, convulsions | | The amygdala is the cause of the formation of fear and **anxiety**-related memory resulting hyperactive in **anxiety** disorders. ippocampus, thalamus, and hypothalamus. Plexuses torso |
| Drowsy numbness pains | Pains about the head and neck | Sensation malfunctions | |
| Hemlock I had drunk or emptied some dull opiate to the drains | Nausea, 'convulsions occurring in persons attacked with frenzy' | Anosognosia | Anosognosia |
| Cannot see what flowers are at my feet… soft incense hangs upon the boughs | 'paralysis, mania, and loss of sight' | | |
| Weariness | Heaviness of the whole head | Melancholia | |
| Fever, fret, sad | Occur either without fevers or in fevers | | |
| Last gray hairs | These cases end in paralysis, mania, and loss of sight | Ageing | |
| Grows pale | Convulsions occurring in persons attacked with frenzy, and having vomiting of verdigris-green bile | | |
| Nightingale 'requiem become a sod' | The sweet call of death, | Escapism from Reality | State of Anxiety |
| 'Dies', 'easeful death' | Many, and, in fact, most of them died | Mortality disorders | |

### 16.3 Sensors are the spirits of modern and postmodern technologies

Sensors receive information and act/respond according to the given commands by that information. Before defining modern sensors, the first author of the chapter clarifies why humanities researchers in collaboration with researchers in science and technology write about 'cognitive sensors', though the subject belongs to the realm of science and technology. The first author of the chapter, as Principal Investigator of the AICTE-NPIU CRS Project being carried out at MITS Gwalior India, while completing the project hypothesized that technology is made for serving human and post-human society while Fellows [22] observed that technology is developed for its own show of attainments, caring little about the impact on both human and post-human society. Technology is no longer dependent on determinism but now has to reflect what Donna Haraway [23] called 'a historical system depending on structured relations among people'. In his study, the author also observed keenly its reason and that is development of technology with zero consultation with humanities knowledge. For Rahaman [24] this was the motivation behind the emergence of the concept of HITS, and humanities inspired computational studies. The casual flow chart explains HITS.

Figure 16.1 represents the objective of the chapter, which is that sciences and their application in a technological guise are born out of the study of nature and humans and are ready to pay back to their sources of birth in their best futuristic trends with security and care. In the case where they fail to assure security and care of the human, society and nature, there is no value of any technological advances. This perhaps inspired the philosophers of science and technology to study psychology, linguistics and aesthetics, and they recently started working in neurological philosophical enquiries. This transdisciplinary approach to science and technology has brought revolutionary changes in perspectives to observe the knowledge system deconstructing isolationist concept of two cultures, i.e. humanities and sciences; sciences and technology etc. Sensors have added ease to the hi-tech daily life for safety and controlled society.

#### 16.3.1 Security, technology and literary arts

Technology fails in two major conditions: first, when it starts harming natural aesthetics while helping human beings and causes imbalances made between the existences of man and nature; second, when it harms human beings and all their cognates directly. Aldous Huxley [25, 26] foresees a futuristic technologically advanced world, whose citizens are environmentally engineered into an intelligence-based social hierarchy and where reproductive technology develops a

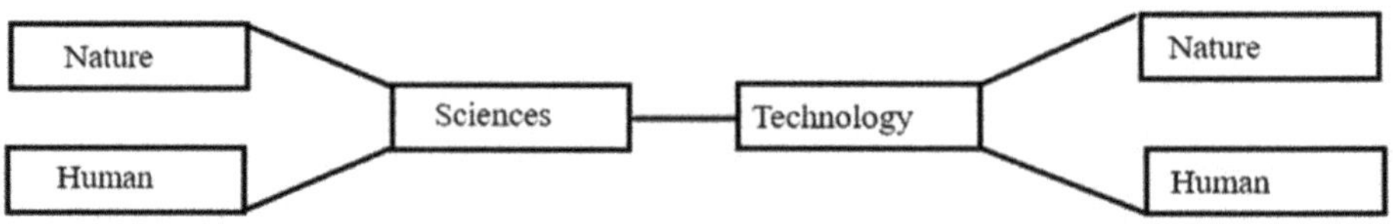


**Figure 16.1.** Casual flow chart of etiology of science and technology and result.

dystopian society. Huxley [25] reflects apocalyptic conditions under terrestrial and extraterrestrial threats. Technology wins for two reasons: first, when it saves natural heritage, nature-inspired human life and humans from future mishappenings due to, for example, BTEX (a highly volatile compound predominantly that exists in an environment due to petroleum products causing dizziness, nausea, headaches), detecting harmful ethylbenzene, benzene, toluene, xylenes vapors using electrochemical gas sensors. The utilization of sensors revolutionized industrial automatization which alarms humans about the dangerous impacts of the results, described by Mir and Khachane [28], of products such as helium balloons (full of tiny helium atoms in spite of oxygen and nitrogen), dispensing gases, welding gases, LPG and medical oxygen-emission, also harmful though they are used usually for good reasons and occasions. Hence such environment detectors are used for screening the toxic gases in hazardous environments. Zhang *et al* [29] describe human sensors that detect human bodies in a given space; $CO_2$ sensors are often used for detecting trapped human/dead bodies in rubble. Hence, sensors in technology are just like the spirit/soul in a human body which is nothing without a soul. Human health monitoring sensors are a unique development in technology. Narayan and Verma [30] describe graphene-based sensors, for example, for pharmaceutical drugs, electroencephalograph signal. Szunerits [31] discusses real-time measurement of the human body such as body temperature, heart rate, respiration rate, and blood glucose and pressure.

### 16.3.2 Thick description and cognitive sensor

Thick description is basically used in anthropology, ethnography, folk and culture studies. Geertz who was much interested in the study of philosophy of human nature followed the general philosophy of Weber and observed that human beings are creatures [32] 'suspended in webs of meaning that they themselves have spun'. The observation of the 'webs of meaning' in the beliefs, habits of mind, attitude, and behavior needs interpretative art not scientific law to understand. Geertz [32, 33] takes culture as webs to be interpreted in search of meaning. The method of such art of interpretation is 'thick description'. Hence, the 'thick description' is a methodological tool to utilize for retrieving a knowledge system of 'the webs of meaning' reflected through habits of mind or cultures of human society. The authors used thick description as a method to interpret human behavior and attempt to model a thick descriptor in the form of cognitive sensor which may detect the reasons of behavior interpreting a person's beliefs, faiths, cultural background, which are the formative factors of the human values that affect human attitude leading to human behavior.

Figure 16.2 represents how human life is governed by the formative factors of human behavior: first, 'belief, faith, tradition, culture', second 'values' and the third 'attitude' leading to human behavior. It is represented how tradition and belief are webbed together formulating the human values which haunt human attitude till the attitude converts into behavior. Hence, to comprehend human behavior and to interpret it, a sensor must come across the represented behavior of an individual.

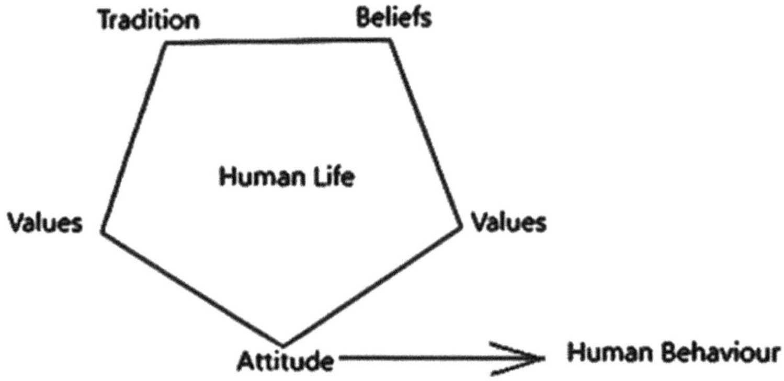


**Figure 16.2.** The formation of human behavior.

Here the question is how an anthropologist or ethnographer or literary reader or an artist or student of Nature itself helps develop any model to develop technology. In responding to the question, the authors applied a discourse-investigation approach to philosophy pf technology and arts. It is observed by literature review that even being anti-scientific in approaches to life and against the ages of technology and science, works of writers and artists yield inspiration and designs to new technology. Science and magical realists' arts enforce the need for new technology. For examples, the term 'consensual hallucination' originates in William Gibson's *Neuromancer*. Gibson's book had a direct influence on science-fiction films such as *The Matrix*. Ernst Kapp [34] in his magnum opus popularized the theory of 'organ projection' to various areas of the scientific world. The technology is advanced only through pursuing the scientific replication of the natural world. Ryle [35] denies the notion that every natural phenomenon is reflected in technology but establishes the concept that a few of natural phenomena may have resemblance to technological replications. Ryle [35] asserts that nature inspires 'mechanical laws', and that nature is either one big machine, or else a 'conglomeration of machines'. He identifies man-made machines such as clocks. Ryle [35] thinks that some machines always have a ghostly existence of natural systems, for example solar systems. These go on by themselves and repeat indefinitely the same series of movements.

Ryle's arguments for a machine's relationship exists in terms of imitation. A human being is a 'non-replicable universal identity'. Human identity cannot be replicated but a certain resemblance with mechanical or technological apparatus is hopefully identifiable. Man is a complex phenomenon, everyman is a complex phenomenon, everyman's complexity is due to cultural background. The cultural background is understandable through human semiotics of behavior and gestures. And for comprehension of text one has to read, so human gesture must be read through signs in behavior.

#### *16.3.2.1 Cognition of human behavior patterns*

In the work by Ortner [33], Geertz defines the human condition in terms of the constant symanticization to human life governed by 'culture.' Geertz [32] established culture in terms of formulation of culture which needs not social laws to define it but needs the behavior of a person. Comprehension as known is a cognitive aptitude of an individual. Cognitive sensors may be used for such study of humans as ethnography studies human behavior. We model a cognitive sensor and its possible uses based on the ethnographic theory of interpretation of culture propounded by Clifford Geertz [32] who defines ethnography as an interpretative subject not to understand it. Every culture is brought about by people's behavior, which is not the result of a day or two but of years. Dealing with human behavior is trying to sense reasons for the behavior. This sensing we call cognitive behavior reading (CBR) or sensing. Human behavior is the enigmatic result of mental exercises for something. Geertz, in order to understand human behavior, has developed the theory called 'thick description'. It's a way to understand the constituents of human gestures. What causes a human gesture, thought or unthought? What are the gestural signals of good or evil sentiments or emotions of a person? Behavior is a social action, it is an articulation of cultural forms. The cognitive sensor we use in these symbols, these articulations. Cognitive sensors do not identify the person through gesture only but why he is in that gesture. Thick description observes keenly these constituents of gesture as symbols.

Figure 16.3 is a model that represents sensoring areas in the constituents of human behavior. Goethe [16] wrote in 'Introduction to Propylaeon' that the human figure immeasurable is not comprehensible only through observation superficially but also deeply with ifs and buts.

*16.3.2.2 Modern sensors to date*

Sensors are studies in terms of their transfer function, span, correctness, calculative, calibrations, hysteresis, saturation, repeatability, dead band, resolution, special properties, impedance, excitation, dynamic characteristics, environmental factors, reliability, application characteristics, uncertainty and accuracy. In these terms there

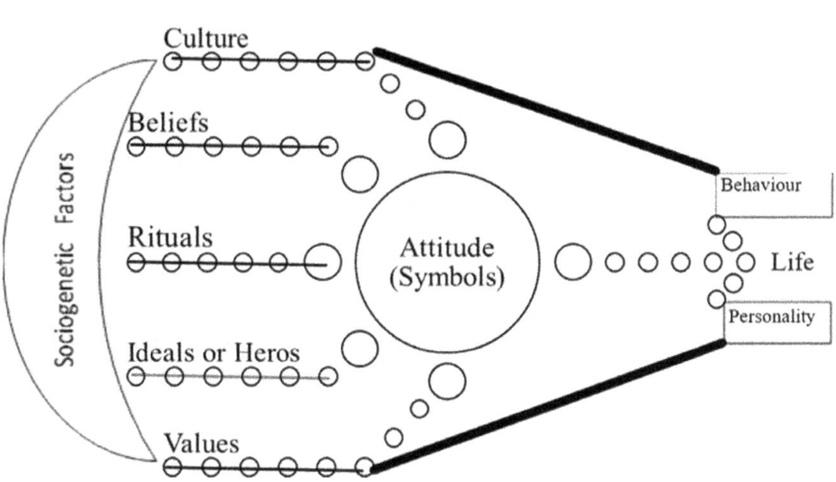


**Figure 16.3.** Represetation of the constituents of human gesture and behavior as symbols.

is nothing related to human behavior but physical behavior of matter, i.e. sensor. Hruska [36] studies development of modern sensors such as innovations of the sensor area for measuring the composition and concentration of the gas mixture, modern temperature measurement sensors and MEMS. Physical principles of sensing, optical components of sensors, temperature sensors, optical sensors, electric and magnetic sensors, mechanical sensors, acoustic sensors, chemical and biological sensors, radiation sensors, smart sensors and actuators are the sensor studies to date. The most novel and developing sensors are the sensors for cognition of human nature, and must not be confused with human sensors or intelligent sensors, but are different from them.

Before the outset of the concept 'cognitive sensors' we explain the basic terms known as sensor, detector, actuator and transducers as they are mistakenly used synonymously by beginners. Sensors are the machine of data acquisitors and reflectors of actual reaction set for the acquisition. Fraden [37] gave a differential explanation of the terms. The terms sensor and detector are synonymously used, but while detector stresses on qualitative features, sensor is known for its quantitative nature of measurement. The term sensor is distinguished from transducer [37] in terms of actions they perform. Fraden [37] defines a sensor as 'a device that receives a stimulus and responds with an electrical signal.' A sensor, hence, does not function by itself; it is always part of a larger system that may incorporate many other detectors, signal conditioners, processors, memory devices, data recorders, and actuators. According to De Marez and De Moor [38], the wireless sensor network enabled smart environment is one of the most popular, dynamic, and promising applications in the IoT era. Intelligent sensors, and human sensors are now developed based on AI architectural frameworks and their uses in many other fields including litigation and cyber-crimes.

## 16.4 Basic modeling of cognitive sensors

Based on the above arguments, this chapter proposes the concept of literary knowledge engineering under the umbrella term of HITS and present basic architectural frameworks of cognitive sensors. Hence it delineates that the unmathematical, approgrammatic, and non-algebraic concept of cognitive sensors would be known for three promising qualities:

a. quality of service (QoS): possibility of the mutual interactions between applications and network services;
b. quality of information (QoI): the interactions between applications and network services for receiving and filtering and transmitting information accurately;
c. quality of experience (QoE) [39]: letting both the users and mechanical activators in it experience the input and output of the interactions between applications and network services.

Figure 16.4 represents a key model of basic needs and function of cognitive sensors. K1, 2, 3 help ovalizing factors to embed a knowledge system in the sensor

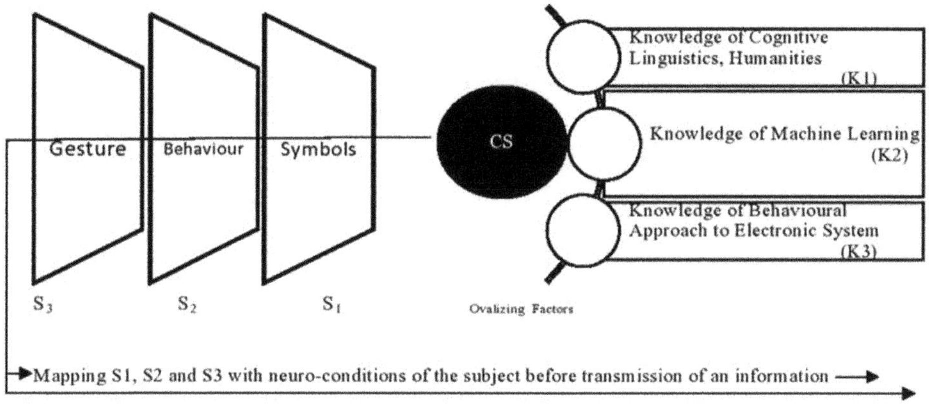


**Figure 16.4.** Framework for analyzing triplite communications of cognitive sensors due to knowledge and ovalizing factors.

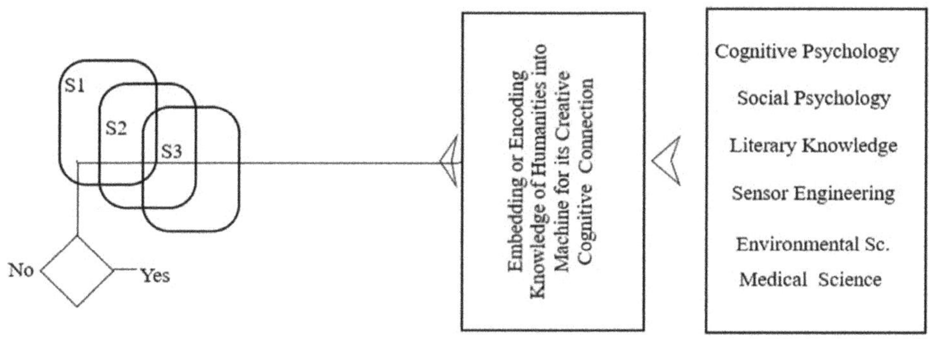


**Figure 16.5.** Framework of knowledge engineering needed for cognitive sensors.

components. Ovalizing factors are the directing processes through which knowledge is engineered through behavior modeling of electronics for the preparation of cognitive sensors which primarily receives information about symbols, behaviors and gestures of a subject mapping symbols with behavior and gesture (S1, S2, S3) and their neuro-conditions or brain-functioning in the subject.

S1 (sensing symbols of the subject); S2 (sensing behavior mapped with symbols of the subject) and; S3 (sensing gesture mapped with both behavior and symbols) are the target study of the cognitive sensors. The computational thick description for engineering means to receive comparative information of etiologies of S1, S2, S3.

The next figure represents the possibility of humanities knowledge through machine learning for behavioral modeling of electronic systems to be used as components of cognitive sensors.

In figure 16.5, S1, S2, S3 are the target study of the cognitive sensor applying humanities knowledge and other knowledge systems as mentioned in figure 16.5. Transmission of the constituents of human behavior system (S1–S3) to the sensor system help make the machine more creative and cognitive. The more information of the constituents is in a sensor system, the more human the sensor would be. The modeling requires further study of micro-modalities, modulation, factors of information barriers, challenges and prospectus. The authors literature reviews explore descriptive research in the field of 'cognitive sensors' development from the standpoints of neuro-critical or cognitive literary studies.

## 16.5 Conclusion

The reading of literary theory of thick description, thus, helps us to propose a novel model of cognitive sensors for human security and other utilities in the future. Experimentation based on this model is yet to come. This model is proposed for industry-based initiatives where competent knowledge engineers objectively understand the components of making the sensors for understanding human gesture and compare them with behavioral patterns resulting from human gestures.

## Acknowledgments

Authors acknowledge the assistance from NPIU (A Unit of MHRD) for funding the project titled 'Cognitive Literary Studies for the Progress of Computational Cognitive Modelling: A Humanities Inspired Approach to Technological Advances' under TEQIP-III AICTE-CRS. We also acknowledge Barkha Singh for her consistent remarks related to legal issues and cognitive sensors. The chapter represents one of the rubrics of the research findings.

## References

[1] Bringsjord S and Ferrucci D A 2000 *Artificial Intelligence and Literary Creativity Inside the Mind of BRUTUS* (Mahwah: Lawrence Erlbaum Associates)

[2] Landow G P 1993 *Hypertext: Contemporary Critical Theory and Technology* (London: The Johns Hopkins Press Ltd)

[3] Tyler V 2006 *Democratizing Technology: Andrew Feenberg's Critical Theory of Technology* (Albany, NY: State University of New York)

[4] Lear J 1982 Aristotle's philosophy of mathematics *Philos. Rev.* **91** 161–92

[5] Madden E H 2015 Aristotle's treatment of probability and signs *Philos. Sci.* **24** 167–72

[6] Moorman R H 1943 The influence of mathematics on the philosophy of Spinoza *Natl. Math. Mag.* **17** 296–307

[7] Miller L G 1957 Descartes, mathematics, and God *Philos. Rev.* **66** 451–65

[8] Shakespeare S 2013 The word became machine: Derrida's technology of incarnation *Derrida Today* **6** 36–57

[9] Johnson C 2005 Derrida: the machine and the animal *Paragraph* **28** 102–20

[10] Carsten Strathausen 2009 The philosopher's body: Derrida and teletechnology *New Centen. Rev.* **9** 139–64

[11] Campolo L D 1985 Derrida and Heidegger : the critique of technology and the call to care *J. Am. Acad. Relig.* **53** 431–48

[12] Glendinning S and Eaglestone R 2008 *Derrida's Legacies: Literature and Philosophy*. (New York: Routledge)

[13] Putnam H 1960 Minds and machines Dimensions of Minds ed S Hook (New York: New York University Press) pp 138–16

[14] de Saussure F 2011 *Course in General Linguistics* ed M Perry and H Saussy (New York: Cambridge University Press)

[15] Pandit R and Rahaman V 2019 Critical pedagogy in digital era: understanding the importance of arts and humanities for sustainable IT development *Proceedings of International Conference on Digital Pedagogies (ICDP) 2019* (Delhi: SSRN)

[16] Eliot C W 1910 *Prefaces and Prologues to Famous Books* vol 39 (New York: PF Collier and Sons)

[17] Heikenfeld J, Jajack A and Rogers J *et al* 2018 Wearable sensors: modalities, challenges, and prospects *Lab Chip* **18** 217–48

[18] Rahaman V 2020 Epi/pandemic in literature: a study in medical humanities for COVID 19 prevention plenary speaker *National Webinar on Literature & Epidemics (May 2020)* (Gujarat, India: MK Bhavnagar University) pp 16–7

[19] Blakeslee S and Ramchandran V S 1998 *Phantoms in the Brain* (New York: William Morrow and Co.)

[20] Halliday S 2008 *Science and Technology in the Age of Hawthorne, Melville, Twain, and James: Thinking and Writing Electricity* (New York: Palgrave)

[21] Campolo L D 1985 Derrida and Heidegger: the critique of technology and the call to care *J. Am. Acad. Relig.* **LIII** 431–48

[22] Fellows R 1998 Philosophy and technology (Cambridge: Cambridge University Press) pp 13–112

[23] Haraway D J 2003 *Manifestly Haraway* (Minneapolis, MN: University of Minnesota Press)

[24] Rahaman V and Sharma S 2020 Reading an extremist mind through literary language: approaching cognitive literary hermeneutics to R N Tagore's play *The Post Office* for neuro-computational predictions *Cognitive Informatics, Computer Modelling, and Cognitive Science* (New York: Academic) pp 197–210

[25] Bloom H 2004 *Bloom's Guide to Aldous Huxley's 'Brave New World'* (New York: Chelsea House)

[26] Aldous Huxley 2000 *Brave New World Revisited* (New York: Rosetta Books)

[27] Sekhar P K and Subramaniyam K 2014 Detection of harmful benzene, toluene, ethylbenzene, xylenes (BTEX) vapors using electrochemical gas sensors *ECS Electrochem. Lett.* **3** 3–6

[28] Mir A and Khachane A 2018 Sensing harmful gases in industries using IOT and WSN *Proc. 2018 4th Int. Conf. Comput. Commun. Control Autom. ICCUBEA 2018* pp 1–3

[29] Zhang D, Sessa S, Kasai R, Cosentino S, Giacomo C, Mochida Y, Yamada H, Guarnieri M and Takanishi A 2018 Evaluation of a sensor system for detecting humans trapped under rubble: a pilot study *Sensors* **18** 852

[30] Huang H, Su S, Wu N, Wan H, Wan S, Bi H and Sun L 2019 Graphene-based sensors for human health monitoring *Front. Chem.* **7** 1–26

[31] Szunerits S and Boukherroub R 2018 Graphene-based biosensors *Interface Focus* **8** 20160132

[32] Geertz C 1973 The Interpretation of Cultures (New York: Basic Books) pp 3–412

[33] Ortner S B 2017 Obituaries *Am. Anthropol.* **109** 786–9

[34] Kapp E 2018 *Elements of a Philosophy of Technology: On the Evolutionary History of Culture* (Minneapolis, MN: University of Minnesota Press) p 121

[35] Gilbert R 2009 *Concept of Mind* (Oxon: Routledge)

[36] Hruska F 2015 *Trends in the Sensor Development BT-Intelligent Systems in Cybernetics and Automation Theory* ed R Silhavy *et al* (Cham: Springer)

[37] Fraden J 2016 *Handbook of Modern Sensors*
[38] De Marez L and De Moor K 1970 The challenge of user-and QoE-centric research and product development in today's ICT-environment *Obs.* **1** 1–22
[39] Harman G 1990 The intrinsic quality of experience *Philos. Perspect* **4** 31

www.ingramcontent.com/pod-product-compliance
Ingram Content Group UK Ltd.
Pitfield, Milton Keynes, MK11 3LW, UK
UKHW051905100926
8556IPUK00005B/772

9 780750 327053